Routledge Companion to Real Estate Development

Real estate development shapes the way people live and work, playing a crucial role in determining our built environment. Around the world, real estate development reflects both universal human needs and region-specific requirements, and with the rise of globalization there is an increasing need to better understand the full complexity of global real estate development. This Companion provides comprehensive coverage of the major contemporary themes and issues in the field of real estate development research. Topics covered include:

- social and spatial impact
- markets and economics
- organization and management
- finance and investment
- environment and sustainability
- design
- land use policy and governance.

A team of international experts across the fields of real estate, planning, geography, economics and architecture reflect the increasingly interdisciplinary nature of real estate studies, providing the book with a depth and breadth of original research. Following on from the success of the textbook *International Approaches to Real Estate Development*, the *Routledge Companion to Real Estate Development* provides the up-to-date research needed for a full and sophisticated understanding of the subject. It will be an invaluable resource to students, researchers and professionals wishing to study real estate development on an international scale.

Graham Squires is Associate Professor and Head of Property Group at Massey University, New Zealand.

Erwin Heurkens is Assistant Professor of Urban Development Management at Delft University of Technology, the Netherlands.

Richard Peiser is the Michael D. Spear Professor of Real Estate Development at the Harvard Graduate School of Design, USA.

Routledge Companion to Real Estate Development

Edited by Graham Squires, Erwin Heurkens and Richard Peiser

LONDON AND NEW YORK

First published 2018 by Routledge
2 Park Square, Milton Park, Abingdon, Oxon OX14 4RN

and by Routledge
605 Third Avenue, New York, NY 10017

First issued in paperback 2021

Routledge is an imprint of the Taylor & Francis Group, an informa business

Publisher's Note
The publisher has gone to great lengths to ensure the quality of this reprint but points out that some imperfections in the original copies may be apparent.

British Library Cataloguing-in-Publication Data
A catalogue record for this book is available from the British Library

Library of Congress Cataloging-in-Publication Data
Names: Squires, Graham, editor. | Heurkens, E. (Erwin W. T. M.), editor. | Peiser, Richard B., editor.
Title: Routledge companion to real estate development / [edited by] Graham Squires, Erwin Heurkens, Richard Peiser.
Description: Abingdon, Oxon ; New York, NY : Routledge, 2018. | Includes bibliographical references and index.
Identifiers: LCCN 2017005843 | ISBN 978-1-138-91434-6 (hardback : alk. paper) | ISBN 978-1-315-69088-9 (ebook : alk. paper)
Subjects: LCSH: Real estate development.
Classification: LCC HD1390 .R68 2018 | DDC 333.73/15—dc23
LC record available at https://lccn.loc.gov/2017005843

ISBN 13: 978-1-03-209658-2 (pbk)
ISBN 13: 978-1-138-91434-6 (hbk)

Typeset in Bembo by
FiSH Books Ltd, Enfield

Contents

Contents

Contents

Figures

Tables

Contributors

Nadia Alaily-Mattar is a research associate at the Technical University of Munich (TUM). She is an architect, and graduate of the American University of Beirut. She holds a PhD from University College London and a Master's from the London School of Economics.

Frank Apeseche is a professor and former co-faculty director of the MDes program in real estate at the Harvard Graduate School of Design. His courses include Building and Leading Real Estate Enterprises and Entrepreneurship, Advanced Real Estate Finance, Environment Economics and Enterprise, and Independent Studies. Frank is also Chairman of Amstar World, a countercyclical real estate investment manager with operations on three continents. Previously he was Chief Executive Officer of the Berkshire Group, the operating parent of several established public and private real estate subsidiaries. During his 26 years at Berkshire, Frank led four companies to their initial public offerings, successfully launched a number of private firms and sponsored 14 institutional funds with $7 billion of capitalization. Frank received his BA from Cornell University and his MBA from the University of Michigan.

Sven Bienert (PhD, MRICS, REV) is Head of the IRE|BS Competence Center of Sustainable Real Estate at the University of Regensburg. He studied real estate economics and business administration with a specialization in finance at the Leuphana University Lüneburg. He wrote his PhD thesis "Impact of Basel II on real estate project financing" at the Albert Ludwig University of Freiburg. The thesis was awarded a distinction. Professor Bienert worked for many years in top positions for leading real estate consulting companies. Parallel to this he was active as a managing director for Probus Real Estate GmbH headquartered in Vienna until 2013. He is editor and author of real estate specialist books and has won numerous research awards. Furthermore, he works as an independent consultant and is initiator of numerous projects in cooperation with different valuation associations such as RICS, TEGoVA, FIABCI, gif, ÖVI, ARE, a member of various sustainability committees of DGNB, gif, ZIA, DVFA, ICG, and Sustainability Fellow of the ULI. In June 2015 he was appointed to the supervisory board of the ZIMA Holding AG. Since November 2015 he has been a board member of the Corporate Governance Initiative of the German Real Estate Industry (ICG).

Stephan Bone-Winkel is CEO of BEOS AG, a development and asset management company he co-founded in 1997. BEOS manages more than 2.1 million square meters of commercial real estate, with a team of 125 people and offices in Berlin, Hamburg, Cologne, Frankfurt, Stuttgart, and Munich. In 2010, 2012, and 2015, BEOS raised three light industrial real estate funds for German institutional investors with a total investment volume of over €2 billion. Stephan graduated from the University of Cologne with a degree in business administration.

From 1990 to 1993, he held a research assistant position at the European Business School (EBS), where he also earned his PhD. Until 1996, he acted as managing director of the EBS real estate academy in Berlin. He then worked for the Deutsche Bank AG as a property developer until the establishment of BEOS. Since 2006 Stephan has been an honorary professor of Real Estate Development at the University of Regensburg (IREBS). He has authored numerous publications on real estate development and asset management, and he received the Leadership Award from ULI Germany in 2014 and was appointed "Head of the year" by Immobilien Manager in 2015.

Peter Hendee Brown (AIA, AICP, PhD) is an architect, planner, and independent development consultant based in Minneapolis. He has worked in architecture, government administration, real estate development, and as an owner's representative for public, private, and non-profit clients. Peter's professional and academic work focuses on large, complex, public–private urban redevelopment projects. Peter teaches private sector development at the Humphrey School of Public Affairs at the University of Minnesota and is the author of *How Real Estate Developers Think: Design, Profits, and Community* and *America's Waterfront Revival: Port Authorities and Urban Redevelopment*.

Edwin Buitelaar (PhD) is research programme leader in urban development at the PBL Netherlands Environmental Assessment Agency and research fellow at the Amsterdam School of Real Estate. He publishes regularly in international journals on issues of land, housing, real estate, property rights and planning law.

Suzanne Lanyi Charles is an assistant professor at Cornell University in the Department of City and Regional Planning and the Baker Program in Real Estate. Her teaching and research examine physical, social, and economic changes in inner-ring suburban neighborhoods. Her research has received research grants from the US Department of Housing and Urban Development, the Real Estate Academic Initiative at Harvard University, and Harvard University's Joint Center for Housing Studies. She holds undergraduate and graduate degrees in architecture, and she received her doctorate from Harvard University.

Sofia Dermisi is the Runstad Endowed Professor in Real Estate and Chair of the Interdisciplinary Group for the MSRE at the University of Washington. She holds a degree in Planning and Regional Development Engineering from the University of Thessaly, Greece and Master's and doctoral degrees from the Harvard Design School. Her research agenda has focused on office markets of major downtown locations, with three areas of emphasis: sustainability, disasters (terrorism, life–safety issues), and market analysis. She has earned many research awards, including the prestigious William Kinnard Award from the American Real Estate Society, six best manuscript awards, and a best practice award from BOMA/International. Dr Dermisi received external grants from NSF, RERI, IDOT, and BOMA/Chicago. She is the Vice Program Chair of the American Real Estate Society, editorial board member of the *Journal of Sustainable Real Estate*, and editor of the Data/Methods/Technology section of the *Journal of Real Estate Literature*.

Johannes Dreher is a researcher at TUM. He is a geography graduate of the Goethe University of Frankfurt. He has worked as an analyst for CBRE Germany.

Alex Duval is the founder of Duval Design and Duval Development, companies that provide architectural design and real estate development solutions for public, private, and nonprofit clients. He has over 18 years of experience. Prior to founding the companies, Duval was a director with Portman Financial, the private investment office of the Portman family with assets under management of over $1.5 billion; a director with Portman Holdings, a real estate development company that has developed over 50 million square feet of space; and a project manager with the architecture and engineering firm John Portman & Associates, a design firm responsible for iconic projects around the world. The Urban Land Institute, Princeton Architectural Press and others have published his original research and writing on urbanism. He has exhibited work at the Venice Biennale of Architecture and has been active as an invited speaker at Harvard, Yale, Columbia, Georgia Tech, and other institutions of higher education.

Erwin Grommen (MSc, MA) works at the Central Government Real Estate Agency in the Netherlands. He holds a Master's in Public Administration from the Radboud University Nijmegen and also holds a Master's in Geography and Spatial Planning from the University of Luxembourg.

David Hamilton is a partner at Geobarns, LLC and frequent lecturer at the Harvard Graduate School of Design. With Richard Peiser he is co-author of *Professional Real Estate Development: the ULI Guide to the Business* (3rd edn).

Valerie (Lynn) Hammett lectures at Clemson University and has had a diversified career in the field of real estate. Her 30 years of experience have encompassed various aspects of real estate including title examination and commercial valuations. Her practical experience includes valuations for an array of property types including multi-family housing. She has been consulted as an expert witness to address ownership inconsistencies. Her research focuses on policy influences in real estate development activities, particularly the effect of disaster policy on real estate recovery and risk. She is also interested in behavioral real estate and the triple bottom line of sustainability within the real estate community.

Erwin Heurkens (PhD, MSc) is Assistant Professor of Urban Development Management at Delft University of Technology, the Netherlands. He is co-editor (with Graham Squires) of *International Approaches to Real Estate Development* and author of *Private Sector-Led Urban Development Projects*. He has been rewarded UKNA research secondments at the University of Southern California (USA) and University of Hong Kong, and has been a visiting scholar at the University of Birmingham (UK) and University of Technology Sydney (AUS). His main research interest involves urban development and regeneration, real estate development, public–private partnerships, international comparative studies, sustainable urban redevelopment, and urban planning. His research and publications have been focusing on understanding how private and public sector actors co-deliver sustainable urban areas and real estate projects in various institutional contexts worldwide.

Paul H. K. Ho (PhD) is Principal Lecturer and former Head (2007–2012) of the Division of Building Science and Technology, City University of Hong Kong. Before joining the university, he worked as a project manager with leading property developers for many years. His extensive experience covers the whole property development process. Dr Ho is an active researcher and his research interests include real estate development, construction management and economics, building information modeling, and sustainability. He has served as a member of scientific committees and on the editorial boards of various international conferences and academic journals. He has also enthusiastically contributed to relevant professional institutions and government committees.

Colin Jones is an economist and has been Professor of Estate Management at Heriot-Watt University, Edinburgh, since 1998. He formerly worked at the universities of Manchester, Glasgow and the West of Scotland. His research interests span commercial, industrial and housing market economics, and urban/planning policy. Colin was winner of the Royal Town Planning Institute's 2013 award for excellence in academic spatial planning research. He has published over 60 papers in academic journals and edited/written seven books, including *The Right to Buy, Housing Markets and Planning Policy, Challenges of the Housing Economy: An International Perspective*, and *Office Markets and Public Policy*.

Sophia M. Kongela (PhD) is currently a lecturer and Head of the Department of Land Management and Valuation at Ardhi University in Tanzania.

Erwin van der Krabben is Professor of Planning and Property Development at Radboud University, the Netherlands and Professor of Real Estate at the University of Ulster, Northern Ireland. His research and publications focus on land policies, municipal finance, the interaction between planning and land and real estate markets, and property rights, while his work covers topics in these fields both in Europe and in Asia.

Moses M. Kusiluka (PhD) is the Commissioner for Lands for Tanzania. Prior to this current position, Dr Kusiluka was a lecturer and Head of the Department of Real Estate Finance and Investment at Ardhi University in Tanzania.

Chris B. Leinberger is a land use strategist, teacher, developer, researcher and author, balancing business realities with social and environmental concerns. He is the Charles Bendit Distinguished Scholar and Research Professor, George Washington University School of Business; Chair of the Center for Real Estate & Urban Analysis at GWSB; Nonresident Senior Fellow at the Brookings Institution; founding partner of Arcadia Land Company, a New Urbanism development firm; founder of LOCUS; Responsible Real Estate Developers and Investors; and former owner of RCLCo, an international real estate advisory firm. He has written *The Option of Urbanism: Investing in a New American Dream* and *Strategic Planning for Real Estate Companies* and has contributed chapters to 15 books.

Emil Malizia (PhD, FAICP) is Professor of City and Regional Planning at the University of North Carolina-Chapel Hill. His expertise spans the related areas of regional economic development, real estate development, and urban redevelopment. His current research and practice focuses on vibrant centers of employment in the USA with emphasis on real property performance, economic development, and public–private partnerships. He is author or co-author of five books and over 170 scholarly articles, monographs, and other publications. He received his baccalaureate from Rutgers University and his Master's and doctoral degrees from Cornell University.

John McDonald earned the PhD in economics from Yale University in 1971 with a specialization in urban economics, and joined the Department of Economics at the University of Illinois in Chicago in that same year. He left UIC in 2009 to become the Gerald W. Fogelson Chair in Real Estate at Roosevelt University. He is the author of 11 books and was elected a fellow of the Regional Science Association International in 2008. He served as North American editor of *Urban Studies* and editor of the *Journal of Real Estate Literature*. He was awarded the David Ricardo Medal by the American Real Estate Society in 2013, and retired in that same year.

Michael Nadler holds the Chair of Real Estate Development at TU Dortmund, which is one of the leading technical universities in Germany, with 35,000 students. His research interests are the integration of development projects in urban planning, innovative financial instruments for urban and real estate investments, appraisal methods, and decision-support systems for (new/existing) neighborhood development projects. In addition to his university position, he has had a strong professional background in the real estate and policy sector for almost 25 years. He holds a postdoctoral lecture qualification from TU Kaiserslautern, a PhD from Heinrich-Heine-University Düsseldorf and a Diploma from the University of Cologne.

Kathy Pain is the Professor of the Real Estate Development and Director of Research for Real Estate and Planning at the Henley Business School, University of Reading. Her research specializes in real estate development opportunities associated with economic globalization and the challenges for governance. Her engagement includes government bodies in the UK, Europe, North America, the United Arab Emirates, and the Asia Pacific region. She has served on the Royal Institution of Chartered Surveyors Vision for Cities Task Group and is a member of the US–China Centre for City Competitiveness and the UK All Party Parliamentary Group for Smart Cities.

Richard Peiser is the Michael D. Spear Professor of Real Estate Development at the Harvard Graduate School of Design. He is co-author of *Professional Real Estate Development: The ULI Guide to the Business* (3rd edn) one of the Urban Land Institute's all-time best-selling books. Before coming to Harvard, he taught at USC where he founded the Lusk Center and the MRED Program. At Harvard, he founded the Real Estate Academic Initiative and the Advanced Management Development Program (AMDP). Over the years he has developed houses, apartments, industrial, and land in Texas, California, and China. He received his PhD from the University of Cambridge.

Raul Pérez is Associate Professor at the University of Alicante, Spain. His main background is as a building engineer and economist so as his main research interest revolves around housing economy from the economic and technical perspective. In addition, issues related to teaching–learning methodologies and also energy of the housing markets are areas in which his research is being developed. He has participated as an author in the writing of several books and also in national and European research projects.

Richard Reed (PhD) is the Chair of Property and Real Estate at Deakin University (Melbourne). He has conducted research into many aspects of sustainability in the built environment, has wide academic experience and has published extensively internationally. In addition he has industry experience in the public and private sectors and has presented at numerous national and international property conferences as a keynote speaker. He has authored many property-related books including *Property Development* and *A Greener House*. He has assisted industry and professional bodies to better understand how sustainability can be embedded into the built environment.

Hilde Remøy (PhD) is Associate Professor of Real Estate Management at the Department of Management in the Built Environment, Faculty of Architecture and the Built Environment, Delft University of Technology. She joined the TU Delft as a PhD student in 2005, and was appointed staff member in 2009. Hilde Remøy undertakes scientific and practice research, focusing on real estate market mismatches, sustainable building adaptations, and adaptive reuse of (heritage) buildings. She frequently publishes research articles, reports, and books. She is a member of the editorial board of *Corporate Real Estate Management* and *Zeitschrift für Immobilienökonomie*, and board member of the European Real Estate Society (ERES).

Karim Rochdi heads the corporate solutions division of BEOS AG, a value-investment-focused real estate developer and asset manager. Before joining the Berlin-based BEOS AG, he served as a senior researcher at the International Real Estate Business School of the University of Regensburg. He is the co-author of several articles investigating the role played by real estate in international stock markets. He graduated from the University of Stuttgart, with a degree in industrial engineering and had study visits at the University of California, Santa Barbara, as well as the California State University, East Bay as a Fulbright Scholar. He earned his PhD from the University of Regensburg, concentrating on real estate investments.

Rogerio Santovito (MBA, MEng) is Senior Researcher on Urban Planning Engineering at the Universidade de São Paulo, Brazil. Since 2013, he has also worked at the Competence Center of Sustainable Real Estate, IRE|BS Institute at the University of Regensburg.

Karl-Werner Schulte (PhD, HonRICS) is a real estate professor and Academic Director for the Center for African Real Estate Research at IRE|BS International Real Estate Business School at Regensburg University, Germany.

Petros Sivitanides is an associate professor and Director of the Real Estate Department at Neapolis University in Paphos, Cyprus. He holds a doctoral degree in Real Estate Economics from Massachusetts Institute of Technology. He has extensive research experience with some of the leading consulting firms in the field of real estate in the USA and Europe and has published several articles in international journals in the field of real estate. His research and publications have focused on the pricing of real estate, the intertemporal behavior of the office market, capitalization rates, real estate portfolio structuring, and real estate risk.

Graham Squires (BA Econ, MA, PhD) is Associate Professor and Head of Property Group at Massey University, New Zealand. He is the author of *Urban and Environmental Economics*, co-author of *Building Procurement*, and co-editor of *International Approaches to Real Estate Development*. He has been a visiting scholar at the University of California, Berkeley, sponsored by the Fulbright US–UK Visiting Scholarship Program. He is also responsible for leading the property research group that engages with government policy makers, high-profile property practitioners, and accredited professional bodies (Royal Institution for Chartered Surveyors, Property Institute of New Zealand, New Zealand Valuers Registration Board). His main interests include: property economics and finance; property policy; and affordable housing.

Paloma Taltavull de La Paz is Professor at the University of Alicante, Spain, and Visiting Scholar at the University of California, Berkeley, Goldman School of Public Policy (2011–2012) and Georgia State University (2012, 2015). Her main research interest is in housing economics including macro and microeconomics perspectives of housing prices, demand and supply determinants. Real estate economics, education matters, and also energy efficiency in housing markets are further areas in which her research is developing. She is the author of more than 50 academic articles and several books and a participant in various pan-European and international research projects.

Bo-sin Tang is Professor at the Department of Urban Planning and Design at the University of Hong Kong, and was previously Professor and Associate Head of the Department of Building and Real Estate at the Hong Kong Polytechnic University. He received his PhD in Urban and Regional Planning from the London School of Economics and received education in economics, business administration and real estate in Hong Kong and the USA. His research interests cover land use planning, property development, and institutional analysis.

Alain Thierstein is Professor of Spatial Development at TUM. He holds a PhD in economic sciences from St Gallen University. He has held positions at St Gallen and ETH Zurich. He is also a partner of EBP Schweiz AG, Zurich.

Raymond G. Torto (PhD, CRE) is currently a lecturer on market research and urban economics at Harvard University's Graduate School of Design. Previously he was a founding principal of Torto Wheaton Research (TWR), an internationally recognized commercial real estate research company. TWR was acquired by CBRE and is now branded as CBRE Econometric Advisors. Torto served as CBRE's Global Chief Economist from 2008 to 2014. He is a co-recipient with Bill Wheaton of the 2007 James A. Graaskamp Award for Real Estate Research Excellence. Torto holds a PhD in Economics from Boston College.

Bing Wang (PhD) is Associate Professor in Practice of Real Estate and the Built Environment at the Harvard University Graduate School of Design (GSD). Wang is the faculty director for the Master's degree program in Real Estate and the Built Environment at the Harvard GSD. She also directs the School's activities in advanced education for senior real estate executives and the joint real estate management program between the Harvard Business School (HBS) and the GSD. Wang's teaching, research, publications, and practice as a real estate investor, architectural designer, and urbanist articulate a broad dimension of the interconnectedness between design and real estate, and between form and finance. She received her BArch from Tsinghua University and her MArch in Urban Design and Doctor of Design degrees from Harvard University.

Fabian Wenner is researcher and PhD candidate at TUM. He holds degrees in planning from TU Dortmund and the London School of Economics.

Sara Wilkinson (BSc, MA, MPhil, PhD, FRICS, AAPI) is a Chartered Building Surveyor, a Fellow of the Royal Institution of Chartered Surveyors (RICS), and a member of the Australian Property Institute (API). She has worked in UK and Australian universities. Her PhD examined building adaptation, whilst her MPhil explored the conceptual understanding of green buildings. Her research focus is on sustainability, adaptation in the built environment, retrofit of green roofs, and conceptual understanding of sustainability. Sara sits on professional committees for RICS to inform her research and ensure direct benefit to industry. Her research is published in academic and professional journals and Best Practice Guidance Notes to practitioners.

Elaine Worzala (CRE, FRICS) is the Director of the Corky McMillin Center for Real Estate and a finance professor at San Diego State University. She has been a professor and directed centers at the College of Charleston, Clemson University, and Colorado State University. She was an academic program director at Johns Hopkins University and the University of San Diego. All three of her degrees are from the University of Wisconsin-Madison. She is active in the leadership of the American Real Estate Society and the International Real Estate Society. Dr Worzala has received numerous awards in research, teaching, and service.

Acknowledgments

Graham Squires: I would like to thank Erwin and Richard who have been dependable and wise co-editors during the compiling of the book. The team at Routledge have been supportive as ever, with special thanks to Ed Needle. Most of all, thanks go to the contributors; the companion would not have materialized without your work. Thank you!

Erwin Heurkens: I would like to thank greatly the various authors for their valuable original contributions to this Companion. Thanks also to co-editors and companions Graham and Richard for the pleasant collaboration on this international undertaking, and Routledge for giving the opportunity for publishing a Companion on the topic of real estate development. And last, but not least, I would like to thank my students for their valuable research on and shared interest and enthusiasm for contemporary urban real estate development across the globe.

Richard Peiser: I would like to thank my many friends who have contributed chapters to this book. I am especially grateful to my colleagues at Harvard – Bing Wang, Ray Torto, and Frank Apeseche – not only for their participation in the book but also for their many contributions to my own thinking on the grand subject of real estate development. A special thank you to my co-author and colleague David Hamilton who has been my partner in crime on the past two editions of our ULI book as well as the chapter in this Companion. I would also like to acknowledge the support of several long-time friends who have influenced me over the years and who represent the best of the real estate industry – Linda Law, Sam Plimpton, Wayne Barwise, Franz Colloredo-Mansfeld, John Bucksbaum, Joe O'Connor, Stan Ross and Gerald Hines. Lastly, I would like to thank my students who have shared their excitement and unique perspectives about real estate development from every corner of the world.

1

Real estate development

An overview

Graham Squires, Erwin Heurkens, and Richard Peiser

Introduction to the Companion

Real estate development is a significant factor in shaping the built environment. It shapes the way people live and work, and by doing so determines and enables human activity and its evolution. Around the globe, real estate development forms a contemporary reflection of societal needs and the market demand in particular geographies. Nonetheless, due to globalization we increasingly witness similar needs and demands for space despite institutional differences existing between countries and cities. Within each context real estate development is concerned with innovations, such as introducing new forms of development finance, setting up new types of development partnerships, or incorporating sustainable solutions within development processes. As such, the practice and science of real estate development are subject to constant change, reflected in a diversity of topics that are addressed in both academic publications and professional practice. As an applied and practice-oriented science, real estate development studies provide answers to current and future societal questions and market needs, and by nature can be seen as a multidisciplinary endeavor. These intriguing notions ask for a better understanding of the real estate development phenomenon in all its complexity.

Nature of the book

The *Routledge Companion to Real Estate Development* introduces the prospect of a new possibility for the study of real estate development on an international scale to enhance knowledge, understanding, and curiosity. This Companion provides in-depth and selective content on the theme of real estate development, aimed at doing justice to its diversity. Its intention is to enhance the depth of learning, and encourage a stronger professionalization of practice and policy in the built and natural environment. The Companion builds on work from specific and comparative analysis in *International Approaches to Real Estate Development* (Squires and Heurkens, 2015) by gathering key themes on real estate development, in order to generate a global in-depth compendium on the subject. This Companion also forms a science-oriented addition to the *Professional Real Estate Development: The ULI Guide to the Business* (Peiser and Hamilton, 2012) and textbook-oriented *Real Estate Development:*

Principles and Process (Miles et al., 2007). It provides a global coverage of different key current themes in real estate development with each of the original chapters functioning as more accessible versions of journal articles, for a broad audience. Moreover, there was ample space to explore the topic of real estate development in a multifaceted way, since without development, there is no new real estate or redevelopment. Development is a core part of all aspects of real estate. Therefore this Companion should be a welcome addition to books dedicated to real estate such as the *Routledge Companion to Real Estate Investment* (MacGregor et al., 2017).

Purpose of the book

This book is intended to be an advanced companion in the area of real estate development, but accessible to readers with little or no prior knowledge of key real estate issues in an international context. It should be a useful addition to the libraries of students, professionals and academics/researchers in real estate development and affiliated academic fields and professional domains such as urban planning, architecture and design, real estate valuation, investment, geography, economics, finance, and environmental ecology. For the majority of undergraduate and postgraduate programs this Companion offers opportunities for students and academics to learn about and discuss major contemporary themes and issues in the field of real estate development with global coverage. In addition to university courses, practitioners should find this Companion of interest as it provides an overview of key current issues in real estate development practices all over the globe with best practices and indicative case studies being incorporated in most chapters. Professionally this text can be used for understanding real estate development and is a source of knowledge for essential tasks carried out by mid-career policy-makers and practitioners.

Themes of the Companion

This book is divided into the seven main themes related to real estate development that we think are essential in understanding the complexity and diversity of real estate development. Within each of these parts selected authors write about specific real estate development subjects in different global locations, and at national, regional, urban-metropolitan-city, and local development scales. In this way the Companion provides an overview of original contributions from authors worldwide that covers the broad field and varying features of real estate development, at the same time allowing room for more context-specific information.

Social and spatial impact of real estate development

The socio-spatial impacts of real estate development are a useful way to understand the various externality effects of development activity on a neighborhood and city scale and on real estate markets. Decisions about prioritizing the (re)development of real estate within inner cities for instance triggers spatial investments and redefines how people use space. In this light it is, for instance, also important to understand the impact of real estate flagship projects in terms of the social, spatial and economic added value of such developments. The opposite is also true, as for example infrastructure provision and the people moving to live within cities direct real estate development and investment. This reciprocal relationship between social and spatial impacts and real estate development is the subject of various chapters in this section.

Chapter 2 by Bo-sin Tang focuses on the relationship between metro railway and real estate development in Hong Kong in leveraging property values. Using a framework of new institutional economics it explains why the "rail-plus property" (R+P) development model is more than simply using property incomes to subsidize infrastructure development. The chapter discusses the institutional arrangements necessary to make such combined development models successful and elaborates on spatial-environmental impacts in Hong Kong and the transferability of the R+P model to other countries.

Chapter 3 by Suzanne Charles elaborates on the phenomenon of residential real estate development called "teardowns" consisting of replacing older, single-family housing with larger new construction variants with the potential to transform the physical, economic, and social character of American suburbs. The chapter reveals property and neighborhood characteristics associated with teardown redevelopment, examines the spatial clustering of teardowns, and describes how the physical form of teardowns varies across different types of suburban neighborhoods. Suzanne Charles finds that teardowns with the lowest ratio of new house floor area to original area are smallest in very affluent suburbs, with much larger ratios in middle-income neighborhoods. The chapter concludes with a discussion of the potential for suburban gentrification brought by the phenomenon.

Chapter 4 by Nadia Alaily-Mattar, Johannes Dreher, Fabian Wenner, Alain Thierstein investigates the relationship between public real estate development projects and urban transformation and in particular flagship architectural projects. It argues for a multi-scalar approach to understand the range of impacts of such projects as a way to legitimize large amounts of public investment for these kinds of development. In doing so, the chapter will assist policymakers to comprehend the wider benefits for competitive positions of cities and allocate scarce public financial resources more effectively.

Markets and economics of real estate development

This section of the book focuses on market forces and wider economic distribution of resources that may not necessarily follow market logic in real estate development. Connections between real estate development and (amongst other areas) employment, land, overbuilding, and orientation are made. Special attention is given to the relation between real estate development and the materials, labor, land, political, financial, investment and occupier market. Moreover, attention is paid to the effects of the latest global financial crisis, or Great Recession, that have profoundly influenced real estate development and investment. This section demonstrates the interrelationship between market drivers and economic benefits of real estate development by exploring various geographical contexts.

Chapter 5 by Raymond Torto elaborates on the investment and development behavior following the Great Recession. It investigates the operation of the global commercial real estate investment management business by comparing the commercial real estate market in the pre-2007 and end 2015 periods. Subsequently, insight is given into how changed circumstances influence the investment management of commercial real estate assets. The chapter concludes with a discussion on the prospect of an emerging commercial real estate bubble and the consequences for the business of real estate development.

Chapter 6 by Moses Kusiluka, Sophia Kongela, Karl-Werner Schulte focuses on land policy changes and property market evolution in some African countries. More particularly it elaborates on the process of institutional reforms in property rights and land administration in Tanzania and Kenya. With market reforms and transparency improving, Kusiluka, Kongela, and Schulte argue that regional and foreign investment in real estate is increasing in many African countries providing opportunities for economic diversification of African real estate markets. With annual growth rates in GDP in the 5–7 percent range, Africa will continue attracting more investors from around the world.

Chapter 7 by Paloma Taltavull de La Paz, Raul Pérez dives deeper into the relationship between residential prices and housing development, focusing on the role of construction costs. By making use of a specific database of real estate development projects, and applying an average cost model and new supply model, construction-cost–house-price relationships are identified. The findings reveal the existence of scale economies in the construction costs, and that extra construction costs in housing amenities increase housing prices.

Chapter 8 by Petros Sivitanides contains a retail market analysis for development projects which cope with e-commerce, multi-channel marketing, and other new frontiers in retailing. It provides real estate developers with a four stage model for more accurately projecting retail sales in new shopping centers. Sivitanides discusses how e-commerce's increasing share in total retail sales and the resultant competition with bricks-and-mortar retailing changes the role of physical stores so as to complement the shopping experience. Making use of real estate market analysis stages including defining the project's trade area, analyzing competing retail developments, shifts in consumer patterns, and volume of online sales, he demonstrates how to make more accurate projections of a project's true sales potential and supportable square footage by line of trade in a way that takes into account all crucial new-frontier aspects.

Organization and management of real estate development

This section puts forward individual and institutional aspects of organization and management in real estate development. Particular interest is devoted to trends in public–private partnerships engaged with real estate development, development management skills, and strategy-making. Special attention is given to the ability of a working partnership as a multidisciplinary approach to govern real estate development in the changing institutional, market and political environments. In addition, this section contains contributions that look deeper into the necessary project leader and manager competencies to run a real estate business and operate successfully within real estate development. This section contains contributions that are original and foremost have a strong forward-looking perspective towards organizational and managerial aspects of real estate development.

Chapter 9 by Erwin Heurkens introduces the reader to the increasingly global phenomenon of private sector-led urban development, by sketching the changing roles of and relationships between the state and the market in real estate development. Conceptual characteristics and typologies are supplemented with empirical findings of private sector-led urban development practices in different institutional settings. Insight is given into the variety of private sector actors that can lead urban real estate development projects, ranging from real estate developers, investors, communities to corporations. The conceptualizations and findings offer the opportunity to understand the importance and implications of a more prominent role of private actors in sustainably developing real estate and urban areas in cities across the globe.

Chapter 10 by Frank Apeseche zooms in on how to build and lead high performance real estate companies, a subject not widely understood and described in the literature. Based on professional experience and conceptual insights, this chapter forms a welcome contribution in identifying necessary managerial skills and organizational capacities to operate successfully in real estate business. The chapter explores the critical ingredients for building and leading real estate companies, such as acquiring personal managerial skills, building competitive strategies to access capital, hiring and developing talent, and creating a sustainable, high performing organizational culture. He discusses the stages of evolution that entrepreneurs must go through in order to compete successfully and the patterns of communication flow that mark the difference between low-performance and high-performance companies. He explains how to develop a high-performance culture that maximizes the strategic competitiveness of the overall company rather than individual departments.

Chapter 11 by Paul Ho introduces strategic management systems for real estate development that enable development organizations to respond to internal and external strategic issues during development processes. These management systems include proactive, reactive, planned, and ad hoc variants that can be adopted by different types of developer. Based on a survey in Hong Kong, this study indicates that understanding emerging strategic issues and anticipating threats and opportunities are critical for the success of real estate developers.

Chapter 12 by Michael Nadler deals with the topic of organizing public–private partnerships for real estate development through urban finance innovation. Despite increasing interdependencies between public and private organizations in achieving sustainable urban development, misunderstandings, delays, controversies, and failures seem to be part of everyday practice. Given this, the chapter presents an alternative PPP approach by applying finance innovations to support integrated urban real estate development projects that are financially feasible and can render public favored externalities.

Chapter 13 by Edwin Buitelaar, Erwin Grommen and Erwin van der Krabben analyzes the institutionalization of organic urban development strategy in the Netherlands by building on the concept of the self-organizing city, which recognizes the importance of new types of actors such as local communities and entrepreneurs in (re)developing real estate and urban areas. By using theories from institutional change the chapter provides insights into understanding new planning and development practices and aims to identify to what extent they exceed the threshold of institutionalized behavior in one city.

Finance and investment for real estate development

This section explores the development finance and investment aspects of real estate development. It demonstrates selected contemporary instruments and mechanisms on offer that are appearing in various practices. The section starts by retracing some models of finance for real estate development that have been used over the past decades. Then, special attention is given to new and promising financial instruments and applications for real estate development. Examples include new financing mechanisms for affordable housing provision and value investment approaches supported by concrete practical applications in a variety of contexts.

Chapter 14 by Colin Jones offers an historical, evolutionary, and cyclical perspective on models of development finance from the 1950s to the present day. It introduces the reader to the traditional model of incremental debt funding and repayment, and its dependency on market conditions and property cycles. In addition, alternatives are discussed such as forward funding/equity sharing partnerships, non-recourse or limited recourse loans, and shares, bonds, and commercial paper issuance. Jones elaborates on the relationship between finance and development by making an argument for the trend towards large scale development projects as a result of the broad availability of global development finance.

Chapter 15 by Christopher Leinberger explores the future of finance and investment of real estate development and introduces a significant structural trend towards walkable mixed-use urban real estate. This chapter challenges the mismatch between real estate as a long-term asset class being evaluated and underwritten using short-term finance techniques. Leinberger proposes a structural shift in development financing by means of educating the investment community, upgrading development/asset management skill sets, place-based inventory trends and performance metrics, amongst others, to "re-tool" finance and investment. Walkable communities with mixed-use developments require more equity, especially patient equity, than most developers think. New underwriting standards as well as fundamentally different business strategies that emphasize place-management are needed to deal with the increasingly popular mixed-use forms of development.

Chapter 16 by Graham Squires introduces mechanisms for financing affordable housing development. It provides lessons from a San Francisco City and Bay Area case study. Direct, fiscal, and monetary types of mechanisms are put forward in the shaping of affordable housing development markets covering a multitude of finance vehicles and instruments. The chapter concludes with a combination of "weighted-blended" plus "tiered level" finance as an appropriate mechanism frame for developing affordable housing.

Chapter 17 by Stephan Bone-Winkel and Karim Rochdi presents a value investment approach to real estate development, in an era of increased importance of redeveloping and adapting existing real estate to changing (flexible) uses and tenants' needs. By presenting a case study from Berlin, and using a securities analysis model, Bone-Winkel and Rochdi compare the value investment approach to the core investment approach of institutional investors and conceive it as a contrarian strategy for developers, independent of location and cycle. They emphasize the life cycles of all real estate, including core office properties and discuss the opportunities of upgrading business parks and industrial buildings for small- and medium-sized businesses that dominate the market. They illustrate their approach with a case study of a multi-tenant and multi-use business campus in Berlin, which they turned around into a highly successful investment through branding, revitalization, landscaping, and re-tenanting.

Environment and sustainability in real estate development

The collection of chapters on sustainable real estate development covers contemporary debates about environmental, economic, and social problems and solutions. This section elaborates on current sustainability and resiliency issues within different real estate sectors and different built environment scales. Most notably since the millennium, research and practice in real estate development and adjacent fields show an increased concern with the environment and realizing property in a sustainable manner. It has become a central and crucial aspect of real estate development. This chapter contains original contributions about the transformation and redevelopment of offices, the importance of rating systems, drivers and opportunities for sustainability, and resiliency measures within new building and community developments.

Chapter 18 by Hilde Remøy and Sara Wilkinson discusses the importance of sustainable real estate transformation through conversions of existing real estate stock. Within the context of predicted climate change and areas undergoing change causing obsolete land uses and vacant buildings, reuse is an inherently sustainable option as it reduces the amount of waste and raw material usage. The chapter investigates the nature of office conversions in Sydney, as well as drivers and barriers for useful conversion, resulting in key lessons applicable in other market and urban contexts.

Chapter 19 by Sofia Dermisi focuses on the sustainability of office buildings. As holistic building sustainability assessment systems are becoming the norm, three primary certification systems have emerged in the United States, Europe, and Canada for rating the sustainability performance of office buildings. However, quantitative data on the financial cost–benefits ratio is scarce. In this chapter, Dermisi investigates the main financial drivers, construction cost premiums, and payback periods for investments in sustainable office buildings based on academic and professional research. The chapter provides insight into the reasons developers, tenants, and owners pursue sustainable development strategies, the financial incentives and grants, regulatory issues, assessment of true market benefits, and developer objectives. The lessons are beneficial for real estate developer and investor decision-making. However, they are influenced by construction costs, certification rating systems, and market rents and valuations that differ depending on jurisdiction, location (urban versus suburban), and building type.

Chapter 20 by Elaine Worzala and V. Lynn Hammett explores how public policy can influence redevelopment during the post-disaster recovery period. It looks at the case of the 2004 hurricane season in Florida and resulting establishment of the Low Income Housing Tax Credit projects. More specifically, it investigates the construction locations of the LIHTC projects in terms of their resilience to storm surges from category 3 or 5 hurricanes. The findings reveal that proactive site planning from both public and private stakeholders is necessary to diminish vulnerability to flooding and real estate development and investment risks.

Chapter 21 by Sven Bienert and Rogerio Santovito investigates the drivers and opportunities of sustainability in real estate development. It looks at the increasing pressure from society and tighter regulatory framework which have placed the real estate industry's contribution to sustainable development at the forefront of discussions, especially with regard to reducing GHG emissions. A number of instruments and solutions are introduced that could assist the real estate development sector to deal with climate change events and simultaneously ensure long-term business success.

Chapter 22 by Richard Reed examines the concept of sustainability and why it has now been fully embedded into the built environment with reference to new and existing buildings, rather than viewing sustainable attributes as "optional" or a sustainable building as "different." Reed acknowledges that although most stakeholders in the real estate market are typically profit-seeking with the underlying aim of "maximizing shareholder returns," sustainability can now contribute both directly and indirectly to bottom line profit rather than cost.

Design in real estate development

Design has always been at the core of real estate development. The quality of design determines the extent to which a project successfully serves or fails to serve its target market. Design concerns the visualization or conceptualization of ideas for and by different stakeholders involved in real estate. Therefore, good design is often subjective, although it is often argued that good design sells and that it represents a certain (economic) value. Within real estate development, various actors have different views on what good or valuable design entails, and the creative process and methods to achieve it are often ill-understood. As such, this section conceptualizes what differentiates superior design from pedestrian design and how it contributes to value creation in real estate development.

Chapter 23 by Peter Brown discusses actor perceptions of good design from a professional viewpoint. Good design increases the economic value of a real estate development project and its intrinsic value to its community, but good design is in the eye of the beholder and often conflicting interests from various actors exist in any development project. Brown considers good design from three perspectives: high vs functional design; the monetary vs intrinsic value of property; and what users of real estate really want. An example of achieving good design is illustrated with a high-rise condominium project in Chicago.

Chapter 24 by Alex Duval elaborates on the principles of good design for real estate development that would help avoid creating "generic cities". The chapter discusses the causes for this global phenomenon and the pivotal role that developers and designers play in being responsive to societal needs and local circumstances in order to be innovative in designing buildings. Duval presents 27 principles for generating unique designs for differentiated real estate development. Examples of superior design by prominent developers are then presented for office, residential, industrial, retail, and lodging in five US cities.

Chapter 25 by Bing Wang presents a framework for value creation in design and real estate. Design and real estate are inherently intertwined, but this linkage is not easy to dissect or analyze without encountering the obvious danger of oversimplification. Wang analyzes value-adding strategies through selected case studies and dissects conceptual overlaps and differentiations between design and real estate. The chapter examines divergent perspectives in evaluating design and value creation in real estate development beginning with aspects of value creation, and then elaborating on four different methodologies for measuring real estate value with respect to design. The chapter concludes with a framework for guiding value creation through design thinking: the emotional narrative, critical thinking and cultural production, key components of city making, urban scale and context, social change and equality, and environmental ecology.

Land use policy and governance of real estate development

The final section of the book deals with the relationship between land use policy and governance and real estate development. A continuing neoliberal orientation of planning and land use policies across the globe eventually affects how real estate development is shaped, regulated or stimulated in specific places. Policy directives on property will be put forward, whilst providing the general shift in governance structures such as the continuing devolution to lower levels, as exemplified by a move towards localism in many countries. This section covers land use policy and governance aspects in relation to real estate development as diverse as governing vibrant city centers, challenges for effective future governance, local politics and land use controls, and trends in land use and government policies.

Chapter 26 by Emil Malizia offers insight into public–private development perspectives as a governance vehicle to foster urban revitalization in US city centers. It elaborates on project-level and strategic-level approaches that could assist in attracting market investment in weak and strong city centers. Malizia argues that export/traded sector companies should be given priority over the development of public amenities within policies, to increase the vibrancy of inner cities, as these constitute the most viable part of the metro economy's base. The chapter features different financial mechanisms that cities can use to subsidize selected types of development and concludes with a case illustration showing how different public options for improving feasibility affect the rates of return.

Chapter 27 by Kathy Pain explores the challenge for effective governance of real estate development in a future urban world. As the twenty-first century is a time of dynamic global change and cities are the places where this change is most fundamental, it is important to understand the part that real estate development plays in supporting their functioning as centers of productivity and innovation. The chapter concludes with pressing challenges for sustainable real estate development and offers a governance agenda for the coordination of policies, regulatory mechanisms and decision-making to address urbanization and real estate development issues.

Chapter 28 by John McDonald elaborates on the relationship between local politics, land use controls, and real estate development. The chapter provides an overview of the real estate development process from setting community goals to seeing actual development on the ground, and examines the political, legal, and economic context in which development actors must operate. Finally, cases in New York, Chicago, and Philadelphia illustrate how a real estate developer, a mayor, and a city planner can play a leading decisive role in real estate development.

Chapter 29 by David Hamilton and Richard Peiser deals with trends in land use and government policies affecting real estate development in the United States. Hamilton and Peiser provide examples about the increased regulation and policies at regional and state levels that affect real estate development projects, particularly if these have significant economic and fiscal impacts. While regulation remains primarily a local government prerogative, major gateway cities are creating regional air quality and water quality districts to oversee negative externalities that affect the environment. In addition, the chapter presents an overview of a number of recent changes affecting development finance, both on the private and public side. Examples such as EB-5, crowdfunding, private equity funds, and REITs present a broad array of equity sources. The chapter concludes with examples of mechanisms for public–private development such as community benefits agreements and tax increment financing that offer the reader insight into new policy trends related to real estate development finance.

References

MacGregor, B., Schulz, R., Newell, G., and Green, R. K. (2017) *Routledge Companion to Real Estate Investment*, London: Routledge.

Miles, M. E., Berens, G. L., Eppli, M. J., and Weiss, M. A. (2007) *Real Estate Development: Principles and Process*, 4th edn, Washington, DC: Urban Land Institute.

Peiser, R. and Hamilton, D. (2012) *Professional Real Estate Development: The ULI Guide to the Business*, 3rd edn, Washington, DC: Urban Land Institute.

Squires, G. and Heurkens, E. (2015) *International Approaches to Real Estate Development*, London: Routledge.

Part I

Social and spatial impact of real estate development

2

Leveraging property values for metro railway development in Hong Kong

Experiences and lessons

Bo-sin Tang

Abstract

Joint development of real property and the metro railway by the Hong Kong MTR Corporation (MTRC) has contributed enormously to urban growth and spatial transformation of Hong Kong over the past four decades. This "rail plus property" (R+P) development model has also made the MTRC one of the most profitable and successful railway operators by world standards. Based on the theoretical framework of new institutional economics, this chapter explains why this model is more than simply using property incomes to subsidize metro railway development. It elucidates how the R+P model embodies an institutional arrangement that can effectively coordinate action of the government, railway operator, developers and other market players in transforming the urban built environment, and properly aligns their different interests and capabilities with appropriate incentives in accomplishing a desirable use of urban space. Successful implementation of the model requires supportive government land use and transport strategy, complementary project planning and development process, and competent and responsible organizations that strive towards making the best use of urban space and producing high-quality urban infrastructure and land development in fulfilling the private needs and the public interest. This chapter will examine the experiences of Hong Kong's metro railway on urban spatial development and the property market, identify the critical success factors and highlight some potential problems of this model.

Introduction

Compact urban form and transit-oriented development help reduce the negative environmental impacts of urbanization and make cities more sustainable. As a high-capacity passenger carrier, metro railway is considered not only as one of the most efficient modes of public transport in terms of energy and urban land consumption, but also as an "integration leader" in possibly creating an integrated, user-friendly and bustling urban form (UITP, 2003). While joint development of land uses and the metro railway generates obvious benefits, a lot of attention is put on the project design of such development (Bernick and Cervero, 1997; Bertolini and Spit, 1998; Cervero, 1998; Zhang, 2007; Cervero and Murakami, 2008) rather than on the institutional mechanism that makes it happen (Tang et al., 2004; UITP, 2009). One of the main obstacles in developing a metro railway system is its high construction and operating costs. How to finance its construction and sustain its operational viability after completion present major challenges to urban leaders and managers. This chapter elucidates the experience of the Hong Kong metro system (or the MTR for short) in addressing these issues and achieving a reasonably successful integration between land use and railway development. The emphasis is about how the Hong Kong MTR model embodies an appropriate implementation structure in organizing a sustainable use of urban space, rather than as a simple funding model for the construction of the metro railway.

MTR: background and performance

The Hong Kong metro railway system is operated by the MTR Corporation (MTRC). Established in 1975 by the Hong Kong government, the MTRC has carried out the missions of constructing and managing, on prudent commercial principles, a mass transit railway service, which now constitutes an integral part of the public transport system in Hong Kong. The Hong Kong government was the sole owner of the MTRC until October 2000, since when approximately 23 percent of its shares have been privatized and traded in the Hong Kong Stock Exchange. As the majority shareholder, the Hong Kong government has committed to maintaining not less than 50 percent shareholding in the company for at least 20 years from the date of the initial public offering in 2000. In December 2007, the railway network of another public railway organization, the Kowloon–Canton Railway Corporation (KCRC), was formally merged with the MTRC, making it a "railway monopoly" in the territory (Yeung, 2008).

The KCRC was a statutory corporation transformed in 1982 from the former Railway Department of the Hong Kong government, which had operated Hong Kong's first railway connecting mainland China since 1910 (Yeung, 2008: 71–77). At the time of its merging with the MTRC, it operated the cross-boundary railway link with the mainland, and provided local railway services connecting new towns and urban districts in Hong Kong and a light rail system in the North Western New Territories of the city. Combining these networks after the merger and in the year of 2014, the MTRC operated a total railway route length of 220.9 km comprising 87 stations on ten lines and 68 light rail stops (MTRC, 2014). It now has a total annual patronage of over 1,900 million passengers using its domestic networks and its market share in franchised public transport exceeded 48 percent. Average weekday patronage of the domestic railway services (excluding the light rail) reached about 4.5 million passengers in 2014 (MTRC, 2014).

This chapter focuses its analysis only on the mass transit railway system because the KCRC railway development was operated under different principles. Before the merger, the MTR system had a total route length of about 91 km, connecting the airport and the densely populated corridors in the main urban area (Figure 2.1). In 2006, it carried over 2.5 million passengers during an average weekday, accounting for about 25 percent (second to public bus services) of the total market share of franchised public transport services in Hong Kong (MTRC, 2006). Apart from railway operations, the MTRC has engaged actively in real estate development. In joint ventures with private developers, the MTRC before the merger developed over 65,000 housing units in high-density residential estates and over 1.4 million

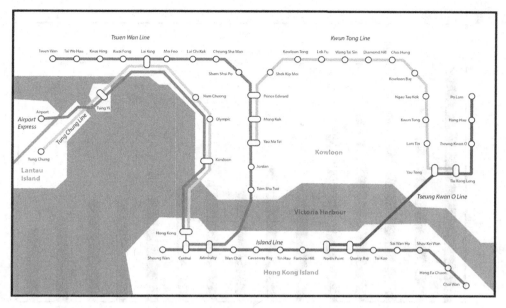

Figure 2.1   MTR system map (before the merger)

Source: MTRC

sq.m. of upmarket commercial and office projects at 25 MTR stations in Hong Kong (2008 data). In 2014, the MTRC held an investment property portfolio of about 212,500 sq.m. of retail space and 41,000 sq.m. of offices. It also provided property management services to over 763,000 sq.m. of commercial space and more than 91,400 residential units (MTRC, 2014). Property development has contributed to its railway business in two ways. First, it provides a major income source to finance the construction of railway projects. Second, property development projects built on or near railway stations assemble a critical mass of railway riders that increases railway patronage.

The Hong Kong MTR network is one of the most successful built-and-operated infrastructure projects by world standards. It provides safe, reliable, efficient, and reasonably affordable transport services to the Hong Kong community. Unlike railway operators in many other cities, the MTRC is a profitable company that demands no operational subsidies from the government towards its daily operations (Figure 2.2). Furthermore, it contributed over HK$103 billion (US$ 13 billion) of financial gains to the Hong Kong government in terms of land revenue, cash dividend and proceeds from public listing, in addition to over HK$73 billion (US$9.4 billion) worth (as of 31 October 2008) of company assets (Chow, 2008). A more recent estimate was that, as at the end of 2012, the net financial gains to the government exceeded HK$226 billion (US$29 billion) including land sale income, annual dividends, IPO receipts and the asset value of government shareholding. This amount did not take into account the increase in property tax revenue (called "rates" in local terms) to the government from private property owners who had received property value enhancement due to improved connectivity to the MTR railway. These achievements are attributed to the business model of the MTRC, which is often described as the "Rail plus Property (R+P) model" invented by the

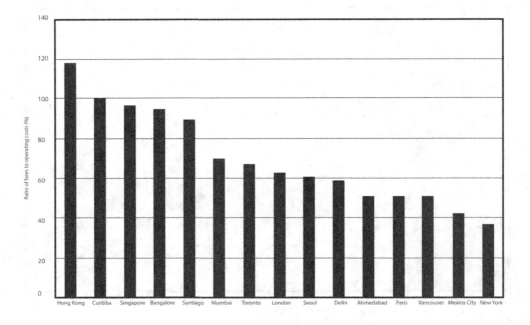

Figure 2.2 Ratio of fares to operating costs of public transit

Source: World Bank (2013: 30)

first chairman of the MTRC, Norman Thompson (MTRC, 2012). As the name suggests, the model entails a combination of both railway and property development. Indeed, property development and investment have made significant financial contributions to MTRC's profits (Figure 2.3). The MTRC is therefore not a railway operator only; the company has become a prominent player in the local property market and has acknowledged itself as "one of Hong Kong's biggest developers" (MTRC, 2012: 99).

The MTR network has exerted considerable impact on the Hong Kong property market. Real estate development on top of or near MTR stations is highly popular with developers, buyers and tenants. Housing estates that are better integrated with the MTR access facilities tend to command a price premium over that of comparable housing projects located slightly further away. Their property prices were found to be more resilient to property downturns and able to escalate more during property booms (Table 2.1). The presence of the MTR network has strongly encouraged private land development and building activities. Completion of new private floor space, especially from non-domestic buildings and along the Island Line, tended to cluster on land within the 400 m (walking distance) radius around an MTR Station (Table 2.2). The urban spatial structure of Hong Kong has evolved with the growth of the railway system, which is now earmarked by the government as the backbone of the local transport infrastructure. According to the recent estimates of the government planning department, about 75% of office/commercial floor space, 46% of industrial/storage floor space, and 43% of housing property is now located within 500 m of a railway station in Hong Kong (Ling, 2014).

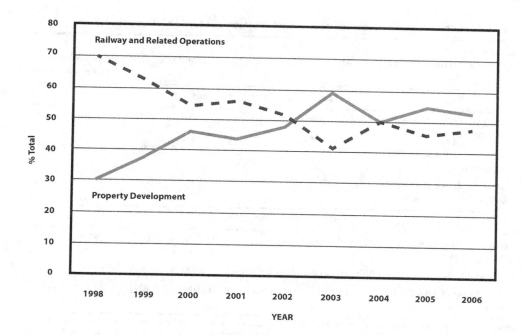

Figure 2.3 Contribution of railway and property to MTRC's operating profits

Source: MTRC annual report, various issues

Table 2.1 Property price comparison of comprehensive housing estates

Housing estates		Differences in housing prices – Estate A vs Estate B (HK$/m² and %)			
Estate A: Better integration with MTR access	Estate B: Longer distance to MTR in same neighborhood	Full period: Jan 1993– Sep 2012	Boom: Jan 1993– Sep 1997	Recession: Oct 1997– Aug 2003	Boom: Sep 2003– Sep 2012
Sceneway Garden	Laguna City	3126* (+ 6.9%)	3761* (+ 6.9%)	2077* (+ 5.5%)	3459* (+ 7.6%)
Telford Gardens	Amoy Gardens	3719* (+ 9.9%)	3942* (+9.5%)	701* (+ 2.2%)	5590* (+14.3%)
Tai Koo Shing	Kornhill	6266* (+ 11.5%)	117 (+ 2.1%)	2760* (+ 6.4%)	11700* (+ 20.0%)

Notes: *Significance level at .000.
Property price comparison is based on average monthly property transaction prices of the housing units in these housing estates during the study period. Data are extracted from property transaction records supplied by EPRC.

Source: Author's analysis

Table 2.2 Completion of private floor space near MTR stations, May 1981–December 2014

Type of buildings/ completion of usable floor area (UFA)	MTR Island Line	MTR Tsuen Wan and Kwun Tong Lines
Residential or composite buildings		
Domestic UFA within 600 m of MTR station (in '000 m²)	5250.76 (100%)	3558.53 (100%)
Domestic UFA within 400 m of MTR station (in '000 m²)	4172.87 (79%)	1806.73 (51%)
Domestic UFA between 400 m and 600 m of MTR station (in '000 m²)	1077.89 (21%)	1751.81 (49%)
Commercial and industrial buildings		
Non-domestic UFA within 600 m of MTR Station (in '000 m²)	5691.42 (100%)	6954.58 (100%)
Non-domestic UFA within 400 m of MTR Station (in 000 m²)	4684.17 (82%)	4338.25 (62%)
Non-domestic UFA between 400 m and 600 m of MTR Station (in '000 m²)	1007.25 (18%)	2616.33 (38%)

Notes: The analysis is based on data from the Monthly Digests of the Buildings Department from May 1981 to December 2014 about completion of new buildings for which occupation permits have been issued. Locations of the new buildings are identified using GeoInfoMap of the Lands Department and the statutory planning portal of the Town Planning Board. MTR Island Line covers all the stations from Chai Wan Station to Sheung Wan Station. MTR Kwun Tong Line covers all the stations up to Kwun Tong Station.

Source: Author's analysis

Institutional arrangement: theoretical explanations

Land use planning has long recognized a synergistic relationship between transport and land development. Metro railway can substantially improve urban transport accessibility and hence the value of the land serviced by the railway system. The rise in land values and property prices can help finance railway construction and sustain its operational viability. But this synergy does not come about naturally. It requires a conscious effort of all stakeholders in organizing the use and transformation of urban space. The R+P model of the MTRC is more than just a juxta-position of railway and property development. It is equally not represented only by the landmark property development projects on top of the MTR stations; they are just the physical outcomes of the model. Nor is it simply the use of real estate incomes to subsidize and finance railway development. Instead, the R+P model represents a unique approach in handling the relationship between land use and railway infrastructure (Tang et al., 2004). It embodies an institutional framework that can effectively coordinate the action of numerous players in transforming the urban built environment and accomplishing a desirable and efficient outcome.

Under the perspective of new institutional economics (NIE), institutions refer to the "rules of the game" that cover the formal rules, informal norms, and their enforcement characteristics (North, 1981; 1990; 1997). Institutions are constructed by human society to govern the social relationship and structure the pattern of social interaction in daily life. The institutional arrangement provides the systems of incentives and constraints that influence individual and organizational behaviour. Depending on circumstances, such behaviour may take the form of "cooperative exchanges" that enhance benefits of all the contracting parties and improve overall economic efficiency, or, conversely, of "exploitative appropriation" of the others resulting in welfare reduction (Hirshleifer, 2001). In the long run, collective outcomes of these strategic interactions can either reinforce or transform the prevailing institutions (Aoki 2000; Campbell, 2004).

Transformation of the urban built environment, as in the case of railway and property development, involves numerous exchanges of resources, information, assets of economic values, promises, and actions between many parties in the process. Whether benefits can be generated and captured through a sensible use and development of the land resources depends on the underlying institutions, especially in connection with the property rights system. Urban land resources, like many common-pool resources, are vulnerable to competitive, free-riding opportunistic actions that can quickly deplete their economic values. Privatization is a possible means to resolve this problem. Given a clear delineation of property rights in a land asset, an individual owner will have an incentive to protect it, enhance its value through deliberate improvements and capture its benefit through voluntary transactions with others. However, the NIE perspective recognizes that market transactions and their enforcement are not cost-free. High transaction costs are "frictions" that can stifle beneficial exchanges and adversely affect the efficient use and allocation of resources. The theoretical implication, therefore, is about how to devise an appropriate institutional arrangement that works towards reducing the transaction costs.

To integrate railway and property development in an urban context embodies extremely high transaction costs. Railway and property involve different business knowledge and expertise. Furthermore, the task of combining railway and above-station property development includes numerous interdependent decisions, assets, and resources, which may not be easily divisible. Imperfect knowledge about the conditions of the development sites will increase the monitoring and search costs. Many stakeholders, such as the government, railway operator, land developers, contractors, and the general public are participating in the development scene. Opportunism, cheating, and non-compliance of these parties tend to increase the negotiation and enforcement costs of the process. This is especially likely to happen when real estate objectives and transport considerations become incompatible and when these two aspects are pursued by separate organizations. The business objective of the other party is "external" to its own. It may not be possible for them to resolve the conflicts through private negotiations in order to allow the synergy of property and transport to take full effect.

There are many possible institutional forms of governing the production of the urban built environment (see, e.g. Alexander, 2001a; 2001b). Below are two possible institutional models for organizing and coordinating the transformation of urban space for a metro railway and its adjoining land development (Figure 2.4).

Model A assigns separate roles and functions to the railway company and land developers respectively. The government provides land or development rights, through competitive sale or direct allocation, to these companies separately. The railway company is primarily responsible for constructing and operating the railway infrastructure while the land developers implement their own real estate projects within the railway catchment areas. Under this model, public-sector decision making, statutory town plans, and government policies and regulations are the principal coordinating mechanisms in bringing together all the key players in developing the sites. The degree of railway and property integration at the project level relies mainly on the quality of interactions between these market players, their interpretations of the government regulatory framework and their compliance with the conditions imposed upon them.

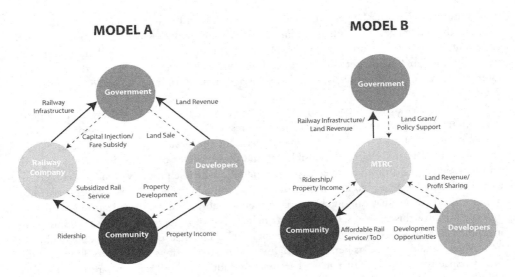

Figure 2.4 Institutional models for integrating railway and property development
Source: Author

Model B presents the R+P business model implemented by the MTRC. This model puts the railway company at the centre in planning, coordinating and managing the use and development of space around the station sites. While the railway company is assigned with the missions of constructing and operating the railway infrastructure, it also receives a government grant of exclusive development rights for the land above and around railway stations. The MTRC has to pay a land premium, at full market value, to the government on the basis of a greenfield site without a railway. It builds the railway and develops the property in partnership with qualified land developers, who are required to shoulder the land premium and share property development profits with the railway company. Project planning and development particulars are determined by this consortium in consultation with the government and other public authorities. The railway company supervises implementation of both railway and property development projects and provides the platform for resolving all possible conflicts of the relevant parties in connection with site development at the project level.

Both models involve leveraging property development incomes to support railway construction. In Model A, this is undertaken through the government by channeling land sale revenues to the railway company in terms of capital injection and/or providing a fare subsidy to support daily operations of the railway after completion. In Model B, however, the subsidy is more subtle and is primarily in the form of an exclusive land grant to the MTRC at the stage of project commissioning. Under a project agreement, the company is required to pay land revenues to the government and also fulfill its contractual duty of building the railway. The MTRC has to make the best use of its land resources in order to generate and capture the additional values derived from integrating railway and property development together.

Both Models A and B can achieve some form of integration between railway and property development. Nonetheless, judging from the theoretical NIE perspective and the empirical evidence, Model B is definitely more capable of generating better integration and more positive development outcomes, primarily because it embodies comparatively lower transaction costs than Model A. The merit of the R+P model is not only that it provides an incentive and constraint structure for the MTRC to maximize the returns from its land resources by means of comprehensive planning, good design and appropriate project timing, but it also provides a proper alignment of the interests, resources, expertise, and decision-making capabilities of all the key parties including the government, property developers and the MTRC.

First, it is appropriate for the government to operate at a strategic level by creating a favorable regulatory environment and setting supportive strategic land use and transport policies that take into account the public interest of the territory. The government gives autonomy to the MTRC in its corporate management such that business efficiency can prevail in its daily operations. In addition to the exclusive land grant to the MTRC, the government also needs to implement a complementary urban growth management policy that encourages higher development density around railway stations, and a public transport policy that restricts competition from other transport modes. Second, the property developers, in pursuing their private interests, operate at the project level, and they are required to build the development projects taking into account site-specific requirements and the deals agreed with the MTRC. Private developers have to compete through tender by offering the best bid to the MTRC in order to get the property development contracts. Finally, the MTRC operates at the intermediary level responsible for monitoring and coordinating the implementation of joint development projects, converting strategic objectives into site-specific requirements, transforming policies into deals and balancing possible conflicts between the public and private interests. In the process, the company performance is also disciplined by the financial market, which exerts an impact on its credit ratings, costs of borrowing, and financial returns to investors. This creates additional incentives as well as constraints to the MTRC towards maximizing the values of all development projects and "internalizing" all possible external benefits generated from railway and property development.

Case study

This section provides a case study of Tung Chung development to illustrate the positive impacts of the Hong Kong R+P Model on spatial planning and the urban built environment. Tung Chung new town was developed as part of the Airport Core Programme in the late 1980s. In 1989, the then colonial government decided to build a new international airport at Chek Lap Kok, located more than 30 km to the west of the central business district, and planned to complete it before the return to China of political sovereignty of Hong Kong in July 1997. Located about 2 km to the south-east of the new airport site, Tung Chung new town was concurrently planned and developed, not only as a new housing area, but also as an airport back-up community. Massive infrastructure investment has been made to build up this new town and develop the adjoining land during the past two and a half decades. Tung Chung is now connected to the main urban areas of Hong Kong by MTR line (Tung Chung Line) and major highways, such as the North Lantau Expressway, the Lantau Fixed Crossing and the Tsing Ma Bridge.

Tung Chung Station is planned to be the town center of Tung Chung new town, and is also the terminal station of the Tung Chung MTR Line. According to the government's initial conceptual master plan, commercial uses were proposed to be developed on and around the station site (Figure 2.5). Strategic transport links such as the North Lantau Expressway and the Airport Express Link (a high-speed train also operated by the MTRC) would run along the southern edge of the major housing areas, pass the commercial zone and connect to the new airport. Several housing zones were proposed to the north-east and south-west of Tung Chung Station. Low- to medium-density residential development was proposed on the waterfront, and several large high-rise high-density housing zones were located on the inland sites. Land parcels for open space, government and community uses were identified on the master plan.

Think about this conceptual master plan for Tung Chung New Town as a typical output of Model A from our theoretical framework in Figure 2.4. Compare it with the development plan of the Tung Chung town center that was subsequently prepared and actually built by the MTRC under the R+P Model (Figure 2.6). Endowed with the exclusive grant of development rights of land on and around the Tung Chung Station, the MTRC did not implement the conceptual master plan of the government. It sought to revise the urban design and spatial planning of the station area with a view to enhancing marketability of its property for buyers and investors, and creating an iconic landmark for visitors and tourists at this strategic location. What the MTRC did was to arrange the array of high-rise residential towers in a curvilinear pattern along the southern edge to take full advantage of the spectacular sea views to the north and mountain views to the south. This creates a visually stunning development identity that strengthens the landmark of this Hong Kong's gateway position (Figure 2.7). The low-rise housing complexes were set back and located on the seaward side to avoid obstruction of sea views to the high-rise blocks. Landscaped public open space was built into green corridors and integrated into the development connecting to Tung Chung Station. The commercial complex is strategically designed to bridge across the North Lantau Expressway and the Airport Express Link to provide a welcoming first impression to Hong Kong's inbound visitors.

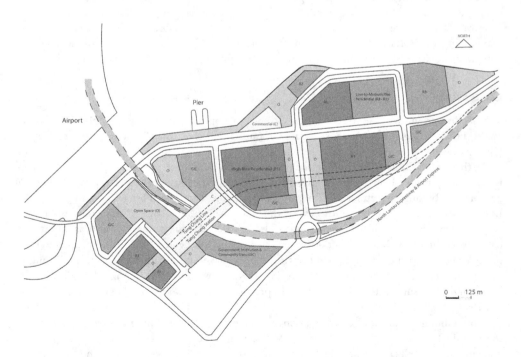

Figure 2.5 Comparison of conceptual master plans for Tung Chung New Town Development: Government's initial master plan

Source: Author, based on a government plan

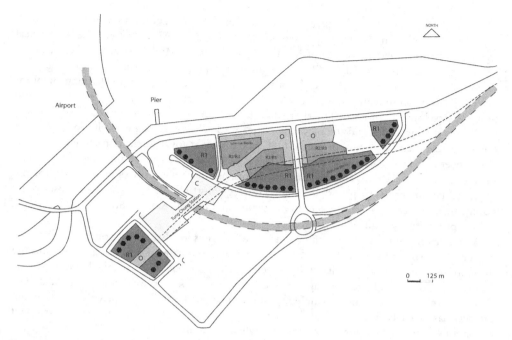

Figure 2.6 Comparison of conceptual master plans for Tung Chung New Town
Development: MTRC's revised master plan

Source: Author

Figure 2.7 Tung Chung development based on MTRC's master plan

Source: Author

Why can Model A (i.e. the government-led approach) not come up with such a design scheme? Can this scheme be implemented through Model A? Following the theoretical arguments above, a major strength of Model B (i.e. the R+P Model) is that it does not only provide the incentives for the MTRC to maximize property returns by means of good planning and attractive design, but it also provides the appropriate institutional arrangement for the company to ensure implementation of the most desirable way to integrate railway and property development projects. The alternative government-led approach, more often than not, lacks both the incentives and the meticulous means to achieve the same result. To implement the alternative design scheme of the MTRC, for instance, government departments need to work together to stipulate and prescribe all the detailed requirements in their regulatory tools, such as zoning, land leases, and building covenants, enforce these provisions and monitor the implementation process by private developers during construction. The transaction costs to the government, under the existing incentive structure and bureaucratic coordinating mechanism, will be prohibitively high.

All regulatory institutions and enforcement mechanisms involve transaction costs of varying degrees. Government planning rules are more effective in terms of regulating broad land use disposition, development intensity, and some elements of the built form such as building height, number of storeys, and site coverage. It is notoriously weak in the aspects of urban design, project innovation and scheme implementation. The marketability of private development projects has never been a principal concern of development regulators. Indeed, government regulations tend to be more cost-effective when they aim at "satisfying", i.e. setting the minimum acceptable standards and requirements, rather than "maximizing", i.e. prescribing all the details and leaving minimal flexibility. Relying on the government to achieve the latter objective works against the incentive structure faced by most civil servants, and its enforcement will encounter enormous difficulties. This case study illustrates how the application of the R+P model at the right place can better realize the full potential of land resources, and deliver a development masterpiece that receives wide public applause. If the initial planning proposals for Tung Chung town center were to be implemented through separate land allocation by the government to property developers and the railway company, it would have ended up as another commonly found high-density development cluster that might be incongruent with its strategic gateway location.

Concluding remarks

Sustainable urbanization requires an efficient and intensive use of land space to accommodate the needs of a fast-growing urban population. Many developing countries are plagued by problems such as traffic congestion, environmental pollution, and urban sprawl. Metro railway provides a high-capacity, energy-efficient, and land-saving mode of urban transport and can therefore bring enormous benefits to rapidly growing cities. The MTR of Hong Kong is internationally famous because not only does it carry the above benefits, but it is also a profitable business venture that relieves the government of operational subsidy. This chapter explains the underlying institutional arrangement that has contributed to these positive outcomes. The essence of the R+P model does not lie in the simple use of land and property revenues to subsidize railway development. Nor is it replicable simply by the allocation of property development sites to a railway operator. The core argument is that the model embodies an institutional and regulatory framework that comprises a supportive government land use and transport strategy, a complementary project planning and development process, and competent and responsible organizations that strive towards making the best use of urban space and producing high-quality urban infrastructure and land development in meeting the private needs and the public interest (Tang et al., 2004).

No model is perfect. While the R+P model has been remarkably successful over the past three decades, it also encountered some recent difficulties. The success of the model relies heavily on a growing urban economy, a healthy real estate market and the availability of government land allocation to the MTRC. These provide the essential setting for the MTRC to capture adequate property income and fare revenues to support railway development. But, as the Hong Kong government is currently facing enormous challenges in finding new land for future development, the opportunity to expand the R+P model into other new districts may be hampered. Furthermore, due to market openness and investment capital inflows, Hong Kong housing prices have climbed to record highs in recent years. Unaffordability of new housing units for many local families and increasing social polarization of the community imply that many households are increasingly unwilling to live away from the main urban areas and employment centers, leading to an adverse impact on MTR ridership. Rising demand for subsidized and affordable housing units also leads to growing public criticisms of the government policy of granting prime development sites to the MTRC for private property development.

Public perception towards the use of the metro railway and the form of high-rise, high-density living is another critical success factor. This is probably a major hurdle in applying the R+P model in many western cities in which low-rise development and auto-dependency are widely accepted by the community. In recent years, rising environmental awareness and community participation in Hong Kong have heightened public concerns about the undesirable impacts of high-rise development at strategic locations such as the waterfront and above railway stations. These development projects are described by some environmentalists as "walled buildings," which are said to be out of proportion to their neighboring buildings, visually intrusive, blocking the sun and wind to the neighborhood, and eliminating street-level activities (Figure 2.8). The public demand for reducing development density may adversely affect the property returns under the R+P model and thus retard further expansion of the metro system.

Another closely related issue is the privatization of urban space. The R+P model requires government allocation and assignment of exclusive development rights of land to the MTRC. This approach allows the company to privatize the land and exploit fully the use of space in partnership with private property developers. This can impose a strong redistributive impact on the urban form by relocating economic opportunities across territory and concentrating development impetus along railway corridors. Concentration of geographical advantages, unless mitigated by counteracting spatial strategy, may lead to increasing territorial inequality, fragmentation of urban space, "splintering urbanism," and social exclusion (Harvey, 1973; Graham and Marvin, 2001). All these contravene the goal of building harmony in a city that is essentially a place for community sharing, interaction and exchange, especially in public spaces. Privatization of public open space land is a worrying phenomenon in Hong Kong because it tends to be disintegrating and discriminating (Tang and Wong, 2008). Like sustainability, public space or open space is a multifaceted concept that serves multiple functions in an urban society. As urban sustainability is built upon an interdependency of human activities and a common destiny of the community, the R+P model may not be truly integrative if the development projects are socially exclusive rather than inclusive, or spatially segregating rather than assimilative. Privatization of urban space, while contributing to urban efficiency, cannot be a total solution. Pushing it to an extreme would endanger the public interest in an urban society unless it is concurrently accompanied by the development of inclusive public space.

Figure 2.8 High-rise, high-density property development which may have a "walled" effect
Source: Author

Acknowledgments

The previous version of this chapter was presented at the Building and Real Estate Workshop of the Hong Kong Polytechnic University in March 2010. Recent research work is supported by the General Research Fund of the Research Grants Council of the Hong Kong Special Administrative Region (Project Number: 717113).

References

Alexander, E. R. (2001a). Governance and transaction costs in planning systems: a conceptual framework for institutional analysis of land-use planning and development control – the case of Israel. *Environment and Planning B: Planning and Design*, 28(5), 755–776.

Alexander, E. R. (2001b). A transaction-cost theory of land use planning and development control: towards the institutional analysis of public planning. *Town Planning Review*, 72(1), 45–75.

Aoki, M. (2000). Institutional evolution as punctuated equilibria, in: C. Menard (ed.) *Institutions, contracts and organizations: perspectives from new institutional economics*. Cheltenham: Edward Elgar, pp. 11–36.

Bernick, M. and Cervero, R. (1997). *Transit villages for the 21st century*. New York: McGraw-Hill.

Bertolini, L. and Spit, T. (1998). *Cities on rails: the redevelopment of railway station areas*. London: E & FN Spon.

Campbell, J. L. (2004). *Institutional change and globalization*. Princeton: Princeton University Press.

Cervero, R. (1998). *The transit metropolis: a global inquiry*. Washington, DC: Island Press.

Cervero, R. and Murakami, J. (2008). *Rail + property development: a model of sustainable transit finance and urbanism*. Berkeley: UC Berkeley Center for Future Urban Transport.

Chow, C. K. (2008). Sustainable and integrated development of railway in Hong Kong. Presentation at the MTR seminar on sustainable and integrated planning for railway, Hong Kong, November 2008.

Graham, S. and Marvin, S. (2001). *Splintering urbanism: networked infrastructure, technological mobilities, and the urban condition*. London: Routledge.

Harvey, D. (1973/1988). *Social justice and the city*. Oxford: Blackwell.

Hirshleifer, J. (2001). *The dark side of the force: economic foundations of conflict theory*. Cambridge: Cambridge University Press.

Ling, K. K. (2014). Innovating the DNA for sustainable development in Hong Kong. Presentation at the Planning and Development Division Annual Conference 2014 of the Hong Kong Institute of Surveyors: Review of Strategies for Planning of Development Space and Renewal, Hong Kong, 18 October 2014.

MTRC (2006). *MTRC 2006 annual report*. Hong Kong: MTRC Limited.

MTRC (2012). *Moving experience: the MTR's first 36 years*. Hong Kong: MTRC Limited.

MTRC (2014). *MTRC 2014 annual report*. Hong Kong: MTRC Limited.

North, D. C. (1981). *Structure and change in economic history*. New York: W.W. Norton & Company.

North, D. C. (1990). *Institutions, Institutional Change and Economic Performance*. Cambridge: Cambridge University Press.

North, D. C. (1997). Prologue, in: John N. Drobak and John V. C. Nye. (eds). *The frontiers of the new institutional economics*. San Diego, US: Academic Press, pp. 3–12.

Tang, B. and Wong, S. (2008). A longitudinal study of open space zoning and development in Hong Kong. *Landscape and Urban Planning*, 87, 258–268.

Tang, B., Chiang, Y. H., Baldwin, A. N., and Yeung, C. W. (2004). *Study of the integrated rail-property development model in Hong Kong*. Hong Kong: The Hong Kong Polytechnic University.

UITP (2003). *The Metro: an opportunity for sustainable development in large cities*. Brussels: International Association of Public Transport.

UITP (2009). *Integrating public transport and urban planning: a virtuous circle*. Brussels: International Association of Public Transport.

World Bank (2013). *Planning, connecting, and financing cities now: priorities for city leaders*. Washington DC: World Bank.

Yeung, R. (2008). *Moving millions: the commercial success and political controversies of Hong Kong's railway*. Hong Kong: Hong Kong University Press.

Zhang, M. (2007). Chinese edition of transit-oriented development. *Transportation Research Record*, 2038, 120–127.

3

Teardowns and reinvestment in the inner-ring suburbs of Chicago

Suzanne Lanyi Charles

Abstract

Teardowns – the redevelopment process in which an older single-family house is demolished and replaced with a larger single-family house – are a conspicuous form of residential real estate development with the potential to transform the physical, economic, and social character of American suburbs. Teardowns are a widespread and highly visible form of residential real estate development in suburban neighborhoods, but they are not ubiquitous. This chapter examines where and why teardown redevelopment occurs using data on teardown redevelopment in the 128 inner-ring suburbs of Chicago, located in Cook County, Illinois, between 2000 and 2014. The chapter begins with an overview of the nature and magnitude of teardowns across inner-ring suburban Chicago, Illinois, followed by an exploration of the variety of inner-ring suburbs in which redevelopment takes place. Then, the chapter reveals the property and neighborhood characteristics associated with teardown redevelopment, examines the spatial clustering of teardowns, and describes how the physical form of teardowns varies across different types of suburban neighborhoods. The chapter concludes with a discussion of the potential for suburban gentrification brought on by teardown redevelopment.

Introduction

Teardowns – the incremental redevelopment of single-family housing in postwar suburban neighborhoods – are a conspicuous form of residential real estate development, with the potential to transform the physical, economic, and social character of American suburbs. Although infill development is a common occurrence in urban areas, it has only recently begun to occur in postwar suburban neighborhoods. Unlike previous waves of suburban housing construction, suburban infill redevelopment takes place incrementally on individual single-family lots, upon which houses were originally built during the post-World War II period. Many of these same postwar suburbs are now undergoing a second housing construction boom. As the original residents of postwar suburbs age, their suburban single-family homes, the ownership of which was once seen as the epitome of the American Dream, are being demolished, and new, larger homes are being built in their place.

In the teardown redevelopment process an older single-family house is demolished and replaced with a larger single-family house (Charles, 2013; Charles, 2014; Pinnegar et al., 2010; Wiesel et al., 2013; Randolph and Freestone, 2012). Figure 3.1 presents an example of this type of redevelopment. Teardowns may be undertaken by property owners who demolish and rebuild a home for their own use, as exemplified in the "knock-down-rebuild" (KDR) process in Sydney, Australia (Pinnegar et al., 2010). Alternatively, a developer may purchase a house with the intent to demolish and rebuild it and then sell it either speculatively or for a previously identified client. Teardowns are widespread in many metropolitan regions and have garnered substantial media attention (Nagourney, 2015; Gopal and Perlberg, 2015; Van Voorhis, 2012). Although media accounts and reports by groups such as the National Trust for Historic Preservation focus on the demolition of historically significant houses in affluent areas (Fine and Lindberg, 2002; Podmolik, 2008; Hirshey, 2008; Spula, 2015), this chapter demonstrates that teardowns take varied forms and occur in a geographically and socio-economically diverse range of suburban neighborhoods. It is important to examine the spatial pattern of teardown redevelopment, as the newly built houses are significantly larger than their predecessors, sell for significantly higher prices, and attract higher-income households, contributing to physical, economic, and social change in suburban neighborhoods.

Figure 3.1　A redeveloped single-family house adjacent to an original house in suburban Chicago, Illinois

Source: Photograph by Katherine Lanyi

This chapter examines suburban teardown redevelopment, beginning with an overview of the nature and magnitude of teardowns across inner-ring suburban Chicago, Illinois, followed by an exploration of the variety of inner-ring suburbs in which redevelopment takes place. Then, the chapter reveals the property and neighborhood characteristics associated with teardown redevelopment, examines the spatial clustering of teardowns, and describes how the physical form of teardowns varies across different types of suburban neighborhoods. The chapter concludes with a discussion of the potential for suburban gentrification brought on by teardown redevelopment.

Background

The development of land outside central cities began well before the twentieth century; however, it accelerated significantly after World War II (Warner, 1978; Jackson, 1985; Hayden, 2003; Teaford, 2008; Fishman, 1987; Gleeson, 2006). In the United States, federal policies such as the construction of interstate highways, loan programs for returning veterans, and the redlining of urban neighborhoods based on race, as well as new construction technologies and a period of postwar prosperity, led to an unprecedented period of housing construction (Jackson, 1985). In the first 20 years following the end of World War II, over 26 million single-family homes were constructed (Nicolaides and Wiese, 2006). In many places, this postwar suburban housing is now becoming functionally obsolete. As Pinnegar et al. (2010) observe, the expendability of the existing houses, which are reaching the end of their expected lives, results in their redevelopment.

The mechanics of the teardown redevelopment process in suburban neighborhoods are straightforward. Initially, the land was divided into parcels and developed to its most profitable use as single-family housing. The original houses reflected the household preferences of the target homebuyer market at the time. Over time, as the original house ages and depreciates and household preferences change, it may no longer represent the most profitable, "highest-and-best" use for that particular piece of land. The disparity between the economic return of a property's original development and that of its most profitable potential use is the basis of the "rent gap" (Smith, 1979). Smith (1979) contends that when the rent gap grows large enough for a developer to purchase the property, pay to redevelop it, bear carrying costs, and then sell it for a profit, redevelopment of the land will occur. The demolition of the original house and the rebuilding of a larger house that is more in keeping with the popular tastes of a new group of homebuyers may return the land to its most profitable use once again.

Often controversial, teardowns have positive and negative implications for municipalities (Langdon, 1991). Teardowns replace older housing with homes that are more in keeping with current household preferences in house size, features, and styles; they also attract new higher-income households, raise property values, and create additional municipal revenue through increased property tax assessments – all desirable outcomes for suburban municipalities. However, existing residents may complain that teardowns create inconveniences during construction, result in a loss of neighborhood character, and do not aesthetically fit within the existing neighborhood. From a regional perspective, teardowns may reduce the regional stock of smaller, affordable (or mid-priced) housing and result in shifts in the social and economic composition of neighborhoods.

Suburban teardowns are frequently reported as occurring in historically wealthy neighborhoods with older pre-war housing (Fine and Lindberg, 2002). Teardowns are indeed prevalent in neighborhoods with high property values (Pinnegar et al., 2010; Charles, 2014), and it is in these affluent places where teardowns are most likely to attract public attention.

However, as this chapter demonstrates, teardowns also occur in more modest postwar suburban neighborhoods. Although the mechanics of the process are somewhat universal, the resultant forms are quite varied in these very different types of suburban neighborhoods. While teardowns in many older affluent neighborhoods are heavily controlled by local zoning codes, which attempt to regulate the resultant physical form to ensure that it is in keeping with the overall neighborhood, teardowns in many other neighborhoods stand in stark contrast to the existing physical context. Figure 3.1 provides an example of a teardown in a moderate-income suburb in which a new, 3,300-square-foot house replaced a modest, 864-square-foot postwar house. Since a new and significantly larger house commands a higher sales price than the original house, the residents of the redeveloped house are likely to be wealthier than their predecessors.

Teardowns in the inner-ring suburbs of Chicago

This study examines teardown redevelopment in the 128 inner-ring suburbs of Chicago, located in Cook County, Illinois, the county that immediately surrounds and includes the city of Chicago. Between 2000 and 2014, 6,983 parcels – slightly more than 1.25 percent of the single-family detached housing stock – were redeveloped. Figure 3.2 presents a graph of the number of teardowns per year during this period. During the first two years of the decade, fewer than 400 properties per year were redeveloped. The relative magnitude of teardowns roughly parallels home prices, increasing steadily during the first half of the decade and reaching its highest point in 2005 (Standard & Poors, 2015). At the peak, over 1,200 properties were redeveloped annually. The numbers then declined rapidly; in 2010 only 116 teardowns took place. Teardowns began to make a modest resurgence in some areas in 2013, albeit to an

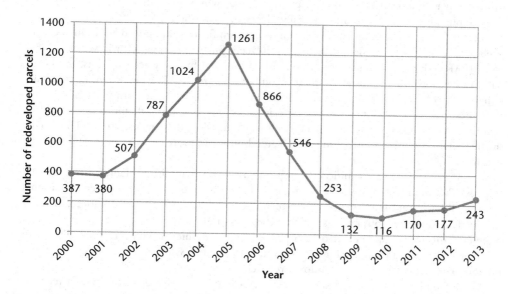

Figure 3.2   Graph of redeveloped single-family residential parcels, 2000–2014
Source: Author

annual total that is less than 20 percent of the high reached in 2005 and by 2016 not yet returned to pre-housing crisis levels.

Teardowns are not confined to the most affluent suburbs. The municipalities with the 20 highest rates of teardown redevelopment are socio-economically diverse. Table 3.1 presents the municipalities with the 20 highest rates of redevelopment. Within the 20 municipalities, the median house value falls within a wide range from US$137,600 to US$976,400, according to the 2000 US Census. Similarly, the median household income ranges from US$43,288 to over US$200,000, the highest census category. While four of the suburbs with the highest rates of teardown redevelopment had median household incomes in excess of US$100,000, five had household incomes less than US$50,000. Moreover, differences in the overall median property values do not have a large effect on the likelihood of a house's redevelopment (Charles, 2013).

The diversity in the median household income in neighborhoods with high rates of teardown redevelopment is reflected in the diversity of residents' occupations. In some suburbs, a large majority of residents are employed in white-collar occupations; in others, over half of the residents are employed in blue-collar occupations. Thus, teardown redevelopment is not restricted only to affluent neighborhoods comprised of white-collar professionals. It occurs with similar frequency in neighborhoods populated by residents working in service, construction, production, and transportation occupations.

Race and ethnicity have a significant effect on the likelihood of teardown activity in a neighborhood; houses located in neighborhoods with greater proportions of black and Hispanic residents are less likely to be redeveloped (Charles, 2013). Not surprisingly, the suburbs with the 20 highest rates of redevelopment were predominantly white, with all of them having 74 to 98 percent white residents. However, several municipalities are somewhat ethnically diverse. Of the suburbs with the 20 highest rates of teardowns, three have Hispanic populations of over 6 percent (with one having a Hispanic population of over 27 percent), and two have Asian populations of over 20 percent. Thus, although redevelopment is more likely to occur in neighborhoods with high populations of white residents, instances of teardown redevelopment do occur in more ethnically diverse neighborhoods.

Several physical characteristics of a house are significantly associated with an increased likelihood that it will be demolished and replaced. Smaller houses with lower floor-area-ratios (FARs) that are closer to commuter rail stations and highway access points, and houses with values lower than those of their neighbors are more likely to become teardowns (Charles, 2013). Characteristics such as masonry construction and finished basements – attributes that make a house more desirable as a residence or more difficult to demolish – are significantly less likely to lead to a house being redeveloped (Charles, 2013). Thus, older, smaller, relatively lower valued and well-located houses are more likely to be demolished and rebuilt.

Prewar housing is not significantly more likely to be demolished than postwar housing (Charles, 2013). In fact, postwar housing – built between 1945 and 1970 – experienced much of the redevelopment that took place between 2000 and 2014. One postwar suburb, in which the median year the housing was built is 1967 was the location of 649 teardowns, which represents the redevelopment of over 6 percent of the municipal housing stock. Teardowns occur at high rates in prewar suburbs; however, older suburbs with vintage architecture do not predominate as the locations of teardown redevelopment.

While the discussion above demonstrates that teardown redevelopment occurs in relatively diverse types of places, neighborhoods with high rates of teardowns are similar in that they are located in very highly regarded school districts (Charles, 2013). The quality of the public school district within which a property is located is a significant determinant of teardown activity;

Table 3.1 Cook County suburbs with the highest rates of teardown redevelopment, 2000–2014

	Municipality	Total number of redeveloped single-family detached houses, 2000–2014	Total number of single-family detached houses, 2000	Percentage of redeveloped single-family detached houses, 2000–2014	Median house age, 2000	Median household income, 1999	Median single-family detached house value, 2000	Percentage white residents, 2000	Blue-collar occupation (service, construction, production, transportation), 2000
1	Winnetka	430	3,709	11.6%	before 1940	$167,458	$736,800	96.3%	6.7%
2	Western Springs	399	4,180	9.5%	1954	$98,876	$322,000	98.3%	10.0%
3	Glenview	649	10,626	6.1%	1967	$80,730	$302,200	85.6%	18.2%
4	Wilmette	489	8,017	6.1%	1952	$106,773	$424,800	89.7%	8.4%
5	Glencoe	154	2,831	5.4%	1951	$164,432	$642,400	95.1%	7.9%
6	Kenilworth	42	779	5.4%	before 1940	$200,001	$976,400	97.3%	5.5%
7	La Grange	216	4,078	5.3%	1950	$80,342	$255,500	91.0%	16.3%
8	Northfield	80	1,603	5.0%	1959	$91,313	$380,200	92.5%	17.0%
9	Countryside	61	1,250	4.9%	1971	$45,469	$172,200	92.3%	34.5%
10	Harwood Heights	87	1,804	4.8%	1963	$43,288	$189,800	92.1%	41.5%
11	Northbrook	451	9,578	4.7%	1969	$95,665	$353,100	89.2%	10.6%
12	Norridge	197	4,381	4.5%	1959	$47,787	$198,000	94.8%	42.1%
13	Lincolnwood	158	3,517	4.5%	1957	$71,234	$288,200	74.5%	18.5%
14	Park Ridge	434	11,077	3.9%	1956	$73,154	$286,000	95.4%	20.2%
15	Palos Park	48	1,385	3.5%	1968	$78,450	$274,100	97.2%	22.4%
16	Morton Grove	222	6,962	3.2%	1958	$63,511	$208,200	74.0%	25.9%
17	Niles	203	6,567	3.1%	1963	$48,627	$198,000	83.2%	36.4%
18	Burbank	213	7,669	2.8%	1962	$49,388	$137,600	90.7%	52.7%
19	Prospect Heights	62	2502	2.5%	1973	$55,641	$196,800	77.4%	40.1%
20	Indian Head Park	11	499	2.2%	1978	$63,250	$188,500	95.9%	17.4%

houses located in an elementary school district with standardized test scores above 90 out of 100 (87th percentile) are 2.5 times more likely to be redeveloped than those that are located in lesser-quality districts (Charles, 2013). Thus, school quality is an important determinant of whether teardown redevelopment occurs, suggesting that the market for redeveloped suburban houses may include households with school-age children.

The spatio-temporal pattern of teardown redevelopment

Although teardowns occur in a wide variety of suburbs in terms of their socio-economic characteristics, the spatial distribution of teardowns in suburban Cook County is highly uneven. The percentage of single-family detached houses redeveloped per census place is presented in Figure 3.3. Twenty suburbs – 16 percent of suburban municipalities – account for over 75 percent of all teardown activity. Teardowns are having a substantial effect on the built environment in some places. In one municipality, over 11 percent of the single-family housing stock was redeveloped. Ten suburbs experienced the demolition and rebuilding of over 5 percent of the total single-family detached housing stock. The mapped data at the census place (municipal) level indicate that redevelopment is primarily confined to suburbs located north, north-west, and south-west of the city of Chicago. Variation within municipalities begins to emerge when teardown activity is mapped at a finer-grain spatial resolution. Figure 3.4 presents the percentage of single-family residential parcels redeveloped per census block group. The percentage of redeveloped parcels ranges from zero to over 21 percent, suggesting that in smaller neighborhoods within municipalities, teardowns are more concentrated and consequently have a greater impact on the physical form of the built environment.

Teardowns appear to act like a contagion; in neighborhoods where one teardown occurs, others are likely to follow (McMillen, 2009). To test whether teardowns are spatially clustered within the inner-ring suburbs of Chicago, there was a calculation of global and local measures of spatial autocorrelation, the Moran's I and local indicator of spatial association (LISA) statistic, respectively. The Moran's I statistic measures whether teardowns are generally located in close proximity to each other (Anselin, 1988; Getis, 2010). The LISA statistic provides a measure of a geographic area's tendency to have a rate of teardown redevelopment that is correlated with that of nearby areas (Anselin, 1995). LISA statistics are used to identify the spatial location of statistically significant spatial clusters of teardown activity. Using the rates of teardown activity per census block group, I first computed the global Moran's I statistic to test for global spatial autocorrelation; then I created LISA significance maps of cumulative redevelopment for each year during the study period to examine changes in the locations of statistically significant clusters of teardown activity. The LISA maps were overlaid upon maps illustrating socio-economic variables obtained from the 2000 US Census. Figure 3.5 presents a map of the statistically significant spatial clusters of teardowns from 2000 to 2014.

Findings reveal that teardown activity is significantly spatially clustered. Spatial clusters of teardowns encompass relatively diverse neighborhoods in terms of house prices, household incomes, resident occupations, and school district quality. Although the spatial clusters expanded over the course of the decade, the overall clustering of redevelopment increased each year – the Moran's I statistic increased from 2000 to 2014 and was statistically significant at the 1 percent level each year. Thus, as redevelopment became more geographically diverse, it was also becoming more clustered. As many more teardowns started to occur, they began to appear in neighborhoods in which they had not previously been seen. However, teardowns are not ubiquitous in the inner-ring suburbs of Chicago. In neighborhoods where teardowns occurred, they occurred with great frequency.

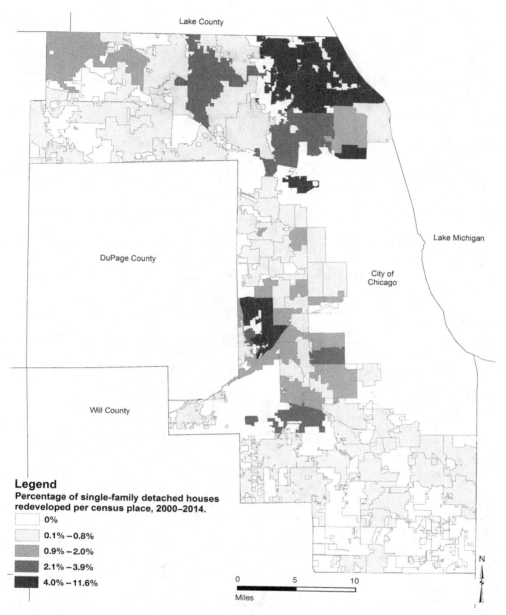

Figure 3.3 Map of the percentage of single-family detached houses redeveloped per census place

Source: Author

Over the course of the decade, clusters of teardowns generally first appeared in places with the highest incomes and house values. With one exception, these places all have high median household incomes and are located in highly ranked school districts. The exception to this pattern is the redevelopment cluster in two blue-collar, middle-income suburbs, which was statistically significant from the beginning of the decade. Clusters began in very or moderately

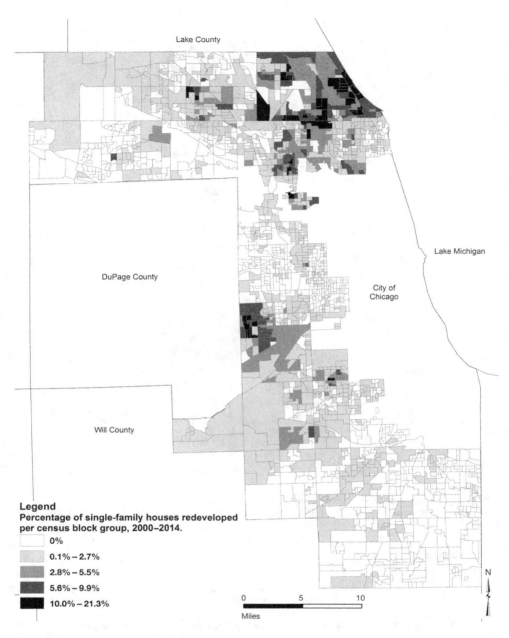

Figure 3.4 Map of the percentage of single-family houses redeveloped per census block group

Source: Author

affluent neighborhoods and in highly ranked school districts. Teardown activity then progressively extended into adjacent, less affluent neighborhoods. Teardowns began to appear in neighborhoods with median household incomes similar to the average for the whole of

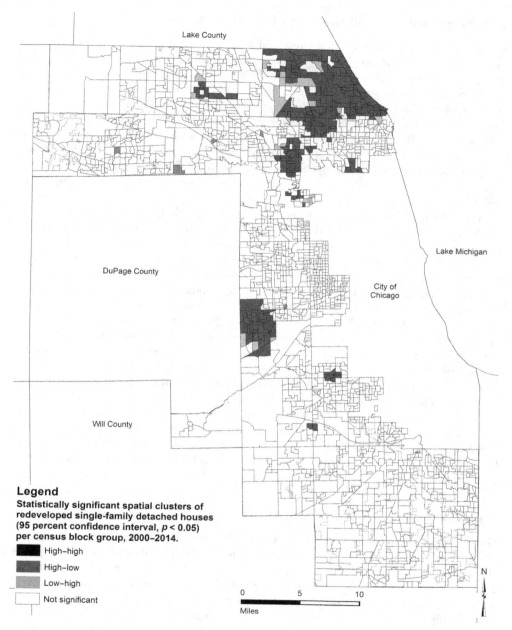

Figure 3.5   Map of statistically significant spatial clusters of redeveloped single-family houses
(95 percent confidence level, p < 0.05) per census block group, 2000–2014

Source: Author

suburban Cook County, albeit within highly ranked school districts. Teardown activity
retracted during the housing crisis and then restarted toward the end of the study period in the
places with the highest incomes and home values.

Differing physical manifestations of teardowns

Teardowns have had a large physical effect – in terms of the increase in house size – on the built environment. The floor area of the average demolished house is 1,500 ft^2, while the average size of the new house is greater than 3,700 ft^2. However, the physical change due to teardowns varies across inner-ring suburbs. In some areas the new house stands out among a neighborhood of smaller houses. In others the redeveloped house is more similar in size and bulk to its neighbors. Figure 3.6 presents the average ratio of the new to the original house floor area per census tract, separated into percentiles. Parcels with the lowest ratios (the 20th percentile ranges from a ratio of 1.1 to 2.1) are primarily located in older prewar neighborhoods where municipal officials are explicitly concerned with conserving their older housing stock and maintaining the character of residential streets. In some of these cases, large and expensive housing is being demolished and replaced with larger and more expensive housing. The relatively small ratio of new to original house floor area is due in part to the relatively large size of the houses that were demolished, but the floor area, volume, and lot placement of the new houses are also constrained by local zoning regulations and original parcel configurations. And even when a house is replaced with one twice its floor area, local regulations may limit the difference in appearance as viewed from the street.

Areas with high ratios of post- to pre-redevelopment house floor area occur throughout the remaining part of the county. However, clusters of areas with ratios in the 80th percentile – areas where the redeveloped house is over 3.4 times larger than the original house – are located in many places with moderate property values and household incomes. This indicates that redevelopment has a larger and more conspicuous physical impact in these middle-income neighborhoods. For example, in neighborhoods with ratios of the new house to the old house above the 80th percentile, the average demolished house floor area is approximately 1,030 ft^2. In these areas houses have been replaced by houses with an average floor area of approximately 3,700 ft^2. The bulk and size of the new houses stand in stark contrast to those of neighboring properties. In these neighborhoods, relatively small and moderately priced housing is being permanently removed from the housing stock and replaced with housing that is significantly larger and more expensive. And high rates of redevelopment of this magnitude may be reflected in a substantial shift in the overall physical, social, and economic character of neighborhoods.

Teardown redevelopment and gentrification

Gentrification is usually considered an urban phenomenon. Consequently, gentrification research has focused on the particular motivations for redevelopment in urban neighborhoods (Lees et al., 2008; Helms, 2003; Rosenthal and Helsley, 1994). Factors such as the proximity of cultural institutions and entertainment, racially and ethnically diverse neighborhoods, historic architecture, and the dynamism of city living are cited as explanations for urban residential redevelopment (Bridge, 2001; Hamnett, 1991; Ley, 1996). In contrast, many of the suburban neighborhoods where clusters of teardown redevelopment occur do not offer quaint streetscapes of distinctive architecture, nor are they particularly racially and ethnically diverse; however, suburban teardowns do offer new, larger housing in neighborhoods with convenient access to public transportation and highways and high-quality public schools. Urban redevelopment and gentrification are often attributed to the preferences of childless couples and empty nesters (Kern, 1984; Schill and Nathan, 1983), but suburban teardown redevelopment responds to a very different set of household preferences, including those of larger households with school-age children.

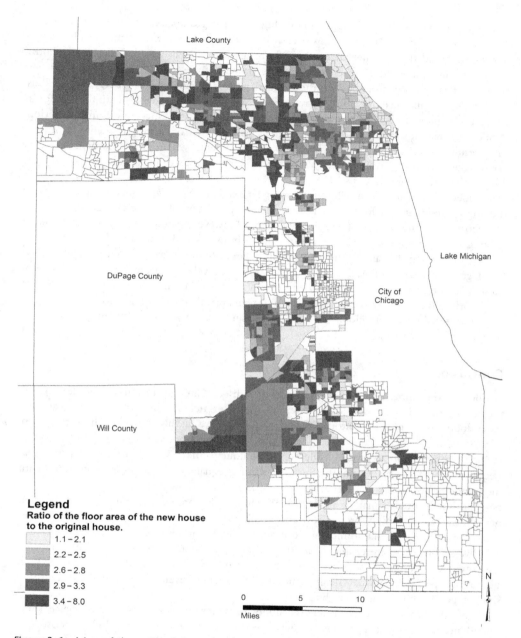

Figure 3.6   Map of the ratio of the new to original house floor area per census block group
Source: Author

Given the findings of the inner-ring suburban Chicago case study, it is important to consider the possibility that teardown redevelopment may result in suburban gentrification. Lees et al. (2008) argue that any definition of gentrification must include: (1) the reinvestment of capital,

(2) landscape change, (3) social upgrading by high-income groups, and (4) the displacement of low-income groups. Suburban teardowns clearly meet the first two criteria: they involve capital reinvestment and the commensurate physical change in the landscape of inner-ring suburbs. They may also fulfill the third criterion: when properties are redeveloped, the sale price is typically at least three times that of the original property (Fine and Lindberg, 2002). Thus, teardowns result in social upgrading by replacing lower-income households with those that can afford the higher price of the new house. Finally, teardowns may meet the fourth criterion. While they may not cause the eviction or direct displacement of existing residents, teardowns may result in market-driven, exclusionary displacement, when redevelopment prevents households with incomes similar to that of the residents of the original house from buying or renting the new house (Marcuse, 1985).

Residents of teardown redevelopments must earn significantly higher incomes than those of the original residents to afford to live there, which may result in substantial changes in the social and economic composition of the neighborhood. However, teardowns may also respond to existing residents who wish to upgrade in place – households that acquire the wealth or the access to financing that allows them to rebuild a larger house in their current neighborhood (Pinnegar et al., 2010). Thus, the process of teardown redevelopment presents the possibility of substantial neighborhood change in terms of the social and economic character of neighborhoods in addition to the obvious physical changes that occur. The association between suburban teardown redevelopment and suburban gentrification is an important topic warranting further study.

Conclusion

Teardowns are numerous in the inner-ring suburbs of Chicago, incrementally altering the physical, economic, and social character of suburban neighborhoods; however, teardown redevelopment is not ubiquitous. While teardowns are highly spatially clustered north, northwest, and west of the city of Chicago, the geographic boundaries of teardown clusters include within them a diverse collection of municipalities. Teardowns occur with great frequency in the wealthiest suburbs, but they occur at equally high rates in municipalities with more modest household incomes and house values. And teardowns are not strictly relegated to neighborhoods predominantly made up of white-collar professionals; teardowns occur in neighborhoods with high rates of residents employed in blue-collar jobs as well as in white-collar jobs. The racial and ethnic composition of neighborhoods in which teardowns occur is less diverse, but there are examples of ethnically diverse suburbs having equally high rates of teardown redevelopment activity. Moreover, the study finds that teardowns occur in both prewar and postwar suburban neighborhoods. Teardown redevelopment is not confined to a single type of suburban municipality.

While the types of suburban neighborhoods in which teardowns occur are diverse, some underlying patterns to the teardown redevelopment process emerge. Several characteristics of the property and the neighborhood are significantly associated with redevelopment. Properties with smaller houses, lower floor-area to lot-size ratios, and lower ratios of their value to that of the neighborhood's average home value are more likely to be redeveloped. Moreover, properties located near commuter rail stations and highway access points are also more likely to be redeveloped. The median property value of a neighborhood does not have a large effect on whether a property is developed. However, neighborhoods with high proportions of black and Hispanic residents were significantly less likely to be redeveloped. Most significantly, properties in the highest-quality school districts are more likely to be redeveloped.

An examination of the spatio-temporal pattern of teardown clusters reveals that in general, clusters first appeared in places with the highest incomes and house values and the most highly ranked school districts. As house prices rose rapidly during the first six years of the decade, teardowns expanded into neighborhoods that had not previously seen redevelopment. Teardowns began to spread from wealthy neighborhoods into adjacent, less affluent neighborhoods. As the housing boom cooled, teardowns retracted from these middle-income neighborhoods. And as the housing market began to recover, so too did teardown activity. It started again in the most affluent inner-ring suburban neighborhoods.

Teardown redevelopment is a straightforward process, but its physical impacts vary across the inner-ring suburban landscape. Teardowns with the lowest ratio of the new house floor area to that of the original house are located primarily in very affluent suburbs. Teardowns with the highest ratios – those that result in the greatest physical change to the properties – occur in many places with modest property values and household incomes. In middle-income neighborhoods the substantial physical change in the housing stock may be accompanied by commensurate economic and social changes in the neighborhoods. These changes may be a harbinger of a new phenomenon – suburban gentrification. As moderately priced housing in well-located, desirable inner-ring suburban neighborhoods is demolished and replaced with significantly more expensive housing, the neighborhoods may no longer be affordable to the middle-income households that once lived there. Understanding the teardown redevelopment process is important in that teardowns have the potential to transform household location patterns and the spatial structure of American suburbs.

Acknowledgments

This chapter draws from Charles (2013; 2014) but uses a significantly updated and expanded version of the dataset used in the two papers.

References

Anselin, L. 1988. *Spatial Econometrics: Methods and Models.* Boston, MA: Kluwer Academic Publishers.

Anselin, L. 1995. Local indicators of spatial association – LISA. *Geographical Analysis*, 27, 93–115.

Bridge, G. 2001. Bourdieu, rational action and the time–space strategy of gentrification. *Transactions of the Institute of British Geographers*, 26, 205–216.

Charles, S. L. 2013. Understanding the determinants of single-family residential redevelopment in the inner-ring suburbs of Chicago. *Urban Studies*, 50, 1505–1522.

Charles, S. L. 2014. The spatio-temporal pattern of housing redevelopment in suburban Chicago, 2000–2010. *Urban Studies*, 51, 2646–2664.

Fine, A. S. and Lindberg, J. 2002. *Protecting America's Historic Neighborhoods: Taming the Teardown Trend.* Washington, DC: National Trust for Historic Preservation.

Fishman, R. 1987. *Bourgeois Utopias: The Rise and Fall of Suburbia.* New York: Basic Books.

Getis, A. 2010. Spatial autocorrelation. In: Fischer, M. M. and Getis, A. (eds) *Handbook of Applied Spatial Analysis: Software Tools, Methods and Applications* (1st edn). New York: Springer, pp. 255–278.

Gleeson, B. 2006. *Australian Heartlands: Making Space for Hope in the Suburbs.* Sydney: Allen & Unwin.

Gopal, P. and Perlberg, H. 2015. Mansions replacing old homes in suburban makeovers across US. *Chicago Tribune*, May 30. Available at: www.chicagotribune.com/business/ct-home-teardowns-suburbs-20150530-story.html

Hamnett, C. 1991. The bind man and the elephant: the explanation of gentrification. *Transactions of the Institute of British Geographers*, 16, 173–189.

Hayden, D. 2003. *Building Suburbia: Green Fields and Urban Growth, 1820–2000.* New York: Pantheon Books.

Helms, A. 2003. Understanding gentrification: an empirical analysis of the determinants of urban housing renovation. *Journal of Urban Economics*, 54, 474–498.

Hirshey, G. 2008. Today a house, tomorrow a teardown. *The New York Times*, November 19. Available at: www.nytimes.com/2008/11/23/nyregion/connecticut/23Rhome.html

Jackson, K. T. 1985. *Crabgrass Frontier: The Suburbanization of the United States*. New York: Oxford University Press.

Kern, C. 1984. Upper income residential revival in the city: some lessons from the 1960s and 1970s for the 1980s. In: Ebel, R. and Henderson, J. (eds) *Research in Urban Economics*, Vol. 4. Greenwich, CT: JAI Press, pp. 79–96.

Langdon, P. 1991. In elite communities, a torrent of teardowns. *Planning*, 57, 25-27.

Lees, L., Slater, T., and Wyly, E. K. 2008. *Gentrification*. New York: Routledge.

Ley, D. 1996. *The New Middle Class and the Remaking of the Central City*. New York: Oxford University Press.

Marcuse, P. 1985. Gentrification, abandonment, and displacement: connections, causes, and policy responses in New York City. *Washington University Journal of Urban and Contemporary Law*, 28, 195–240.

McMillen, D. P. 2009. Teardown Clusters: GMM Spatial Probit with Sample Selection. Working paper presented at the Federal Reserve Bank of New York, June 1.

Nagourney, A. 2015. In Los Angeles, vintage houses are giving way to bulldozers. *The New York Times*, February 6. Available at: www.nytimes.com/2015/02/08/us/classic-or-ramshackle-old-homes-in-los-angeles-are-being-bulldozed-into-history.html

Nicolaides, B. M. and Wiese, A. 2006. *The Suburb Reader*. New York: Routledge.

Pinnegar, S., Freestone, R., and Randolph, B. 2010. Suburban reinvestment through "knockdown rebuild" in Sydney. *Suburbanization in Global Society*, 10, 205–229.

Podmolik, M. E. 2008. Taming of the monster home. *Chicago Tribune*, September 19. Available at: www.chicagotribune.com/chi-local-scene_chomes_0919sep19-story.html

Randolph, B. and Freestone, R. 2012. Housing differentiation and renewal in middle-ring suburbs: the experience of Sydney, Australia. *Urban Studies*, 49, 2557–2575.

Rosenthal, S. and Helsley, R. W. 1994. Redevelopment and the urban land gradient. *Journal of Urban Economics*, 35, 182–200.

Schill, M. and Nathan, R. 1983. *Revitalizing America's Cities: Neighborhood Reinvestment and Displacement*. Albany, NY: State University of New York Press.

Smith, N. 1979. Toward a theory of gentrification: a back to the city movement by capital, not people. *Journal of the American Planning Association*, 45, 538–548.

Spula, I. 2015. Teardowns are rising in Chicago. *Chicago Magazine*, February 11. Available at: www.chicagomag.com/Chicago-Magazine/February-2015/Teardowns-Are-Rising-in-Chicago/

Standard & Poors. 2015. Standard & Poors Case-Shiller Home Price Index, Chicago metropolitan area [Online]. Standard & Poors Financial Services. Available at: http://us.spindices.com/index-family/real-estate/sp-corelogic-case-shiller

Teaford, J. C. 2008. *The American Suburb: The Basics*. New York: Routledge.

Van Voorhis, S. 2012. Teardowns on the rise in some communities inside Route 128. *The Boston Globe*, May 6. Available at: www.bostonglobe.com/metro/regionals/west/2012/05/05/teardowns-rise-some-communities-inside-route-teardowns-rise/cs09s6pf2iw73ZHsBpduLJ/story.html

Warner, S. B. 1978. *Streetcar Suburbs: The Process of Growth in Boston, 1870–1900*. Cambridge, MA: Harvard University Press.

Wiesel, I., Pinnegar, S., and Freestone, R. 2013. Supersized Australian dream: investment, lifestyle and neighbourhood perceptions among "knockdown-rebuild" owners in Sydney. *Housing Theory & Society*, 30, 312–329.

4

Public real estate development projects and urban transformation

The case of flagship projects

*Nadia Alaily-Mattar, Johannes Dreher, Fabian Wenner,
and Alain Thierstein*

Abstract

Analysis of real estate development should not limit its considerations to the direct and indirect effects associated with physical or functional spatial development but rather encompass also catalytic effects in addition to the spatial distribution of benefits and costs. The development of public cultural facilities is increasingly legitimized as primarily flagship architectural projects, which are capable of drawing a large number of visitors not just to the facility in question but also to the city at large. Beyond city beautification and urban renewal, such public investments in the development of physical space aim at re-positioning cities on regional and international economic circuits. This chapter argues that a multi-scalar approach is needed to understand the range of impacts associated with such projects. In the context of increasing competition between cities and the struggle of particularly small- to medium-sized cities to remain on the radar of competitive places, cities will increasingly be pressured to resort to spectacular measures to attract and retain attention to their localities. This will include architectural measures and dramatic changes to the physical spatial environment. Huge financial investments are associated with such risky endeavors, whose impacts are speculative. If we become more knowledgeable about how flagship architectural projects "work" and what drives their impacts, we can use increasingly scarce public financial resources more effectively. Adopting a multi-scalar analytic approach assists in identifying a wider net of potential winners and losers in the game of urban transformation.

Introduction

This chapter discusses the role that public real estate development projects play in steering urban transformation. Urban transformation is not just change, but rather fundamental change; it is about spatial relationships and how changes in these relationships transform spatial characteristics and the plethora of specificities – social, economic, functional, and so on – associated with them. We specifically focus in this chapter on flagship architectural projects that have been developed by public authorities. By flagship architectural projects, we are referring to cultural facilities that are conceived as urban generators and for the design of which star architects have been commissioned. Beyond city beautification and urban renewal, such public investments in the development of physical space aim at re-positioning cities on regional and international economic circuits. We argue that while such initiatives are frequently criticized for aiming at clandestine short-term real estate development gains – resulting in gentrification and so on – we have to understand such initiatives beyond local urban regeneration to include notions of urban competitiveness as well. We argue that a multi-scalar approach is needed to understand the range of impacts associated with such projects. In the context of increasing competition between cities and the struggle of particularly small- to medium-sized cities to remain on the radar as competitive places, increasingly cities will be pressured to resort to spectacular measures to attract and retain attention to their localities. This will include architectural measures and dramatic changes to the physical spatial environment. Huge financial investments are associated with such risky endeavors, whose impacts are speculative. If we become more knowledgeable about how flagship architectural projects "work" and what drives their impacts, we can use increasingly scarce public financial resources more effectively. This is not a plea for or against the development of cultural facilities as flagship architectural projects, but rather an effort to highlight the limitations and effectiveness of such particular real estate development projects in steering urban transformation.

In the following, we provide first a short review of some of the prevailing academic perspectives that investigate urban transformation through flagship projects. Second, we present a framework with which to analyze and assess the role of flagship architectural projects in urban transformation. Third, we discuss the externality effects associated with real estate development in general on a neighborhood and city scale and the real estate market. In the fourth section, we elaborate on spatial incidence analysis as a method that captures the spatial distribution of benefits and costs. The chapter concludes that analysis of real estate development should not limit its considerations to the direct and indirect effects associated with physical or functional spatial development but rather encompass also catalytic effects. Adopting such an approach assists in identifying a wider net of potential winners and losers in the game of urban transformation.

Urban transformation through flagship projects

Hiring the best contemporary architects for the design of public cultural facilities is not a new phenomenon. Although the majority of architects who have designed buildings, including public buildings, throughout history are anonymous, the history of the architectural profession is ample, with references to individual architects whose names rose to fame because of exceptional public buildings they designed. In more recent history, public recognition of architecture and the built environment as identity-generating elements that help transmit cultural identities from one generation to another, legitimizes the importance of public investment in contemporary architecture (Heynen, 1999). What is new though is the legitimization of the conception of public cultural facilities first, as primarily flagship projects and second, the association of "flagshipness" with the fame of "star" architects. Flagshipness refers to the ability to draw a large number of visitors not only to the facility but also to the city at large. "Star" architects are recognized as leaders in their field, not necessarily for their professional merit, but for their "mass appeal" (Adler, 1985: 208) and celebrity status; they are public personas associated with artistic genius, aesthetic innovations, and public quirks (McNeill, 2009). This undoubtedly relates to "media's newfound consciousness of architecture"

(Klingmann, 2007: 246) and architecture's endorsement of an "aesthetic of expendability" (Banham, 1981), in a "world in which the difference between mass culture and high culture, good and bad taste, popular culture and avant-garde has almost become irrelevant" (Klingmann, 2007: 246). Architects' increasingly "flirt with fame" (Bayley, 2005: ix) and the development of the image of the architect (Saint, 1983) in a "winner-take-it-all society" (Frank and Cook 1995) is well documented (Davies and Schmiedenknecht, 2005).

Patterson (2012) argues about an affinity between public institutions and star architects by drawing attention to the fact that the top employers of winners of the prestigious Pritzker Architecture Prize and RIBA awards are indeed public authorities. This points to a new relationship between the state and architecture. On the one hand, architecture is abandoning its social engagement and becoming fully instrumentalized in the pursuit of capitalist urbanization. On the other hand, the state is transforming its conception of public cultural buildings from purely public goods – to be primarily collectively consumed – into positional or status public goods – to primarily boost competitiveness. The transformation of public cultural buildings into status goods is linked to the shift of governance of cities from managerialism to entrepreneurialism and the emergence of the "entrepreneurial city" (Hall and Hubbard, 1998; Harvey, 1989). It begs us to interrogate the logics that drive public sector engagement with star architecture by investigating the roles that these projects play in urban transformation.

Within this context, the Guggenheim Museum in Bilbao designed by architect Frank Gehry has been a well-discussed example. Building on its supposed success, the promise of urban transformation through star architecture's capacity to boost tourist numbers and local optimism has been coined as the "Bilbao or Guggenheim effect" (Plaza et al., 2009). Plaza et al. note that there are clearly positive effects triggered by the Guggenheim museum in Bilbao, namely, 12 years after its inauguration, boosting the number of visitors and tourists, contributions to the development of landmark and economic restructuring of the region. However, Plaza et al. (2009) warn that effects of flagship architectural projects cannot be guaranteed. Klingmann states that "strategic and deliberate use of architecture as a catalyst to set off economic and social transformations … cannot be attained by a mindless repetition of a formula (that only worked once, anyway) … it would be a great mistake to think that the Guggenheim effect is a generic recipe that can be duplicated" (2007: 250). This risk is accentuated if we take into consideration that effects are related to international attention, which as Thierstein cautions, is a highly volatile good whose use is short-lived (Thierstein, 2013). Hence, the "Bilbao effect" is a controversial concept. Indeed, some argue that such interventions have hardly created any jobs for those who have been hit hard by de-industrialization. In a twist of words this has been termed the "Bilbao defect" (Girgert, 2011).

Critical literature questions whether the intended impacts of flagship architectural projects are ever declared or publicized. For example, Patterson (2012) states that organizations in charge of public cultural institutions need to attract outside support and donations necessary to support themselves. This dependency motivates these organizations to hire star architects to capitalize on "perceived notions of success and acceptability among clients of architecture and their external stakeholders" (Patterson, 2012: 3301). Such motivations to establish public legitimacy will hardly be made explicit or publicized. They need to be uncovered by qualitative research involving interviews, discourse analysis, and so on.

In the context of urban transformation, we focus on declared and intended impacts to which public officials can be held accountable. Hence, the more pressing question would be whether the declared intended impacts are really delivered. In other words, if the commission of a star architect for the design of a public cultural facility is intended to result in long-term fundamental changes of spatial relationships and the social, economic, and functional specificities associated with them, are these changes really delivered? Do these projects really become urban generators contributing to urban transformation? Beyond pitting the declared intended impacts against some socio-economic indicators of change proving thus the degree of (in)-effectiveness of these projects, the full spectrum of effective impacts is rarely presented. Indeed, investigations of effective impacts remain in the domain of government-funded reports that rarely make it outside the bureaucratic drawers. Evans (2005) argues that lack of access to "hard evidence" and decision-making processes limit scientific analysis. Even when this evidence is made available, frequently, "the underlying assumptions, subtleties, and potential error sources associated with economic-impact studies" (Crompton, 2006: 79) pose another layer of haze for the comprehensibility of these impacts. Therefore, in order to understand urban transformation through star architecture it is important to uncover the (lack of) consistencies between intended and effective impacts of flagship architectural projects. To do so, it is imperative, first, to open the bureaucratic drawers to get "hard facts" and second, to scrutinize these facts beyond their "political payoffs" (Crompton, 2006: 79).

To assist us in identifying what we should be looking for it is important to structure our understanding of what these impacts might be and how they are connected to each other and to flagship architectural projects in a causal relationship. The following section presents an impact model that organizes our understanding of how flagship projects "work."

A conceptual framework of the impacts of flagship projects

A defining characteristic of flagship projects designed by star architects is their particular capacity to attract media attention and to become identity (re-)generating agents that disrupt or reorient the evolution of the inward and/or outward image of their respective cities. Public authorities' declared rationale for the commission of star architects to design public buildings is based on the hypothesis that due to these particular capacities, buildings designed by star architects have significant long-term effects on the economic and social performance of their cities. We propose an impact model that describes the hypotheses that underlie the interconnected workings of complex effects associated with flagship architectural projects. These effects include urban economic impacts and social effects. These effects are linked to the accentuation of media exposure and the drama associated with a changed urban morphology.

We use an impact model in order to identify the most important inter-relationships between driving or influencing factors (Figure 4.1). The impact model starts from a premise that the development of flagship projects is related to a city's underperformance and the intentions of a city's decision-makers to address this underperformance. Underperformance is meant as a situation in which a city is experiencing either levels of decline or unsatisfactory levels of growth. Levels of decline can be related, for example, to loss of an employment base due to urban restructuring and deindustrialization. On the other hand, unsatisfactory levels of growth relate to the idea that a city is believed to have the capacity to perform better. Pressure groups' awareness of this state of underperformance triggers initiatives for change. Champions of these projects are the spearheads of pressure groups to whom they are accountable. In democratic political contexts, these pressures could emanate from the party's members, city managerial apparatus, or an electoral base. Hence, to understand the underlying intentions of the champions of these projects requires going beyond the declared intentions, uncovering the origin of the pressures to (pro-)act, and identifying to whom decision-makers are accountable.

At the onset of these projects, and as a response to pressures, city officials take a decision to allocate public resources for the development of a flagship architectural project. The declared

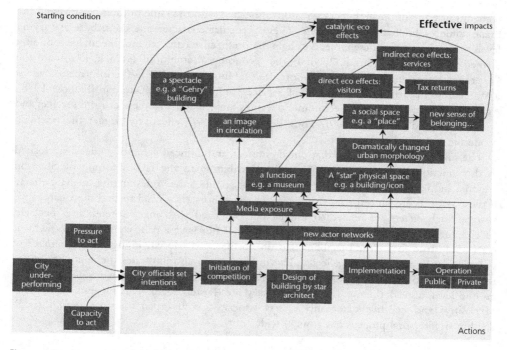

Figure 4.1 Impact model of how flagship architectural projects "work"

rationale is that such public investment is needed because cities are in competition for residents, businesses, and tourists. These have pre-mediated tastes, representational needs (Ronneberger et al., 1999) and "thematic predispositions" (Sternberg, 1997: 957). Sklair (2006) links today's unprecedented accentuated architectural iconicity to the search of uniqueness or difference by the agents and institutions of an emergent transnational capitalist class.

The process of developing these projects usually takes many years; it is accompanied by media exposure from its onset. Unlike most buildings that are simply ignored, these buildings "arouse public attention causing disputes and controversies" (Yaneva, 2009: 8), an attention that accentuates along this process. In addition, the "social life" (Arjun Appadurai, quoted in Yaneva, 2009) of these buildings yields dynamic identities, which shape and are shaped by dynamic actor networks (Yaneva, 2012); they are "not a static object but a moving project" (Latour and Yaneva, 2008: 80) in a state of continuous flow. This means that even after being implemented and appropriated, the dynamic nature of these projects demands their continuous management. For these projects to sustain effective impacts, they must be conceived as continuously ongoing projects and not as finished products.

The decision of public authorities to allocate public funds for these projects is weighed against other alternative investments. Hence, uncovering the particular capacities of these projects related to their particular characteristics is important for understanding why these projects are chosen and how they are instrumentalized to achieve a purpose. The impact model identifies these particular characteristics of star architectural projects. They include, first, the iconicity of their architectural form, second, their spectacularity associated with the avant-gardism of their architectural style, their costs and surrounding controversies, and third, the recognition value of the architect's signature associated with the fame of their persona and the brand value of his/her other artefacts. These particular characteristics contribute at varying levels to the capacity of these projects to help their respective cities tap into the image economy, termed the "iconomy" (Smith, 2008: 4), the "economy of attention" (Franck, 1998) and the "experience economy" (Pine and Gilmore, 1999). Human attention in the information age is a scarce resource that the providers of services or products compete for (Franck, 1998); experiences have become new economic offerings for which demand exists as for services and goods (Pine and Gilmore, 1999). Thus, these projects will effectively trigger the intended economic and social impacts.

The impact model illustrates that direct, indirect, and induced economic effects are related to the function of these projects. Towards contributing to the re-positioning of cities on regional and international economic circuits, what counts is the wide-ranging catalytic effects. If these projects are indeed conceived as urban generators, they need to become catalysts for change. Such effects are related to soft factors associated with image, brand, and spectacle. Indeed, for such wide-ranging effects, function does not have a primary importance. This is particularly true for cultural facilities that do not draw an overwhelmingly large number of visitors, such as museums or galleries. The impact model also illustrates the centrality of the role allocated to media exposure in driving impacts. Indeed, these projects cannot be read at ground level, instead, the media is their primary site (Foster, 2008). This suggests that the net of positive and negative externality effects associated with urban transformation through flagship architectural projects has a much wider cast.

Externality effects of real estate development through flagship architectural projects on a neighborhood and city scale

In economics, "externality" or "external effect" describes a situation in which social costs differ from private costs, i.e. economic actors do not fully take into account the consequences of their actions for third parties (cf. Laffont, 2008). Such externalities can be positive or negative, depending on whether the third parties derive advantages or disadvantages from the actions.

Urban development is rife with such externality situations. The classic example is a newly constructed polluting enterprise that does not compensate nearby residential landowners for noise, smell, or toxic emissions. Were the enterprise to include financial compensation for the landowners in its budget calculations to "internalize" the effect, its production might become unviable – in this case, the combined private and public costs are higher than the overall benefits. Since the establishment of market mechanisms for the resolution of spatial externality situations (such as "cap-and-trade") suffers from high transaction costs, municipalities often resort to land use planning as an instrument to mitigate unwanted negative or encourage positive externalities (Klosterman, 1985: 7–8).

In the case of merit goods, externalities are also closely linked to another market deficiency. The characteristics of public goods, non-rivalry in consumption, non-excludability of users, and in turn non-chargeability for its use, lead to an under-provision by private agents in market equilibrium. Hence, public authorities use this justification to tax market agents and provide public goods themselves, if they deem the overall social merit greater than the cost (Musgrave, 1987). However, the provision of public goods and services has side effects, and is often localized; hence, advantages and disadvantages always have a spatial component.

Publicly funded flagship architectural projects are a good example for the mechanisms outlined above. The direct benefits of these public buildings' use – such as educational or recreational – only reflect a small share of the overall benefits, and the costs incurred by the operator do not account for external costs. Typically, municipal administrations deliberately try to place these flagship projects in economically deprived neighborhoods to encourage local "revitalization" besides city- or region-wide effects. Locally, visitors and tourists attracted by the project mean increased pedestrian frequency and spending in the surrounding streets, or the main street between the flagship building and other existing pedestrian destinations, such as a traditional core city, a transport hub, etc. This creates a positive local externality for shopkeepers and landowners, but not necessarily for the current residents (e.g. higher rents, dangers of gentrification, and so on). On the other hand, more people mean more noise and pollution, leading to disadvantages for residents and owners at the main access roads. On a city scale, added consumption of the visitors, e.g. as hotel and restaurant guests, creates a benefit for the local economy, resulting in employment and income effects. Flagship architectural projects are also aimed at disrupting established narratives about the city or region in which they are located and creating positive image effects. These yield effects that are wide ranging.

The market value of land typically reflects the discounted future income stream generated by the best theoretically possible use. A real estate project with positive external effects will thus lead to a rise in land values of the surrounding plots, since their use prospects for retail or residential purposes are improved and the demand is increased, while negative external effects will lead to a decline. Owners of land that happens to be in the vicinity of a project generating added social value enjoy windfall profits without investments. Hence, landowners are one of the driving forces behind "growth machine" coalitions identified in studies of local power distributions (Molotch, 1976: 311). Despite recent attempts at value capturing taxation or levies, externalities created by public goods are rarely fully internalized.

While political solutions to internalize spatial externalities on the land market still leave room for improvement, they are actively exploited by a growing strand of research in urban economics and planning. Using a hedonic pricing model – under the premise that at any point in time a spatial equilibrium of land values exists, i.e. that welfare is equalized across space (Glaeser and Gottlieb, 2009) – it becomes possible to decompose the land values according to the locational characteristics and quantify an individual externality. This method of compensating differentials has long been used in cost–benefit analyses for infrastructure projects. It is now also applied to identify the value that the "iconicity" of a building creates for its surroundings by comparing their land values before and after the intervention with a similar area ("control group") with equal starting characteristics (e.g. Ahlfeldt and Holman, 2015; Ahlfeldt and Kavetsos, 2014; Ahlfeldt and Maennig, 2009; Ahlfeldt and Maennig, 2010; Ahlfeldt and Mastro, 2012; Franco and Macdonald, 2015). These studies find a significant premium in the willingness to pay that landowners enjoy for land close to iconic buildings, such as sports stadia or conservation areas. For the difference-in-differences method, it is crucial to define a space where the changes in land values are expected to occur prior to the assessment, as well as a carefully chosen control space. We have seen, however, that different externalities occur on various scales, and they often occur spatially close to positive externalities. A simple treatment and control group approach might be too crude.

Capturing changes in land value, however, falls short in providing evidence about how such projects can indeed become urban generators. Land value might go up near the project and down in another area of the city. In other words, there might be a redistribution but not a net change in the overall performance of a city. Hence, it is important to investigate external effects that are not reflected in the immediate land market – this relates to social and ecological externalities, such as the added utility of the image effect. This utility is associated with the iconicity and narratives that accompany "signature buildings" and "iconic landmarks." The building's iconography and the sign/brand value of its architect both contribute to the circulation of its images and narratives. The brand value of the architect relates to his/her established professional standing and reputation and its contribution to the hype and marketability of the building and by extension the setting in which the building is erected. These brands are detached from their use value and linked to image, symbolism, and experience produced by the media (Klingmann, 2007). Hence, there is a "marketing weight" (Fuerst et al., 2011: 167) associated with these projects inducing economic effects that are not captured immediately in the land market. Overall utility and overall land values are therefore not the same. Most importantly, some external effects occur immediately, while others only appear in a long-term view. Thus in order to increase transparency and ultimately equalize misallocations, it is necessary to utilize tools that take into account the fine-grained nature and differences in scale of externalities, differences between payment flows and utility, as well as the change of effects over time.

Spatial incidence analysis

As described in the previous section flagship architectural projects can cause various positive and negative impacts. The impacts can be diverse and are not always quantifiable or measurable in a monetary sense. Therefore, it is necessary to consider which effects – that contribute to the urban regeneration of a city – are actually caused through these public real estate development projects and how these can be measured or assessed. It is also important to clarify where these effects occur in space in order to measure their contribution to urban regeneration, that is, do the impacts occur where the regeneration is intended. Investigating the spatial distribution of effects is important because public real estate development projects are public goods. An important characteristic of public infrastructures in a federalist state is a distribution of infrastructure investments between nation state, the county, and the local community. Thus, for the investigation of the spatial distribution of these effects it is important to answer the question relating to which institutional level profits through these projects and who pays for them. This question is politically charged, especially against the background that flagship projects do not guarantee benefits or use; they also create enormous costs and are associated with critical public discourse. As public real estate development projects with public functions and utilizing public funds or public subsidies, the question of use and costs, beneficiaries and benefactors is relevant for the public.

The confusion as to the spatial distribution of the economic impacts produced by flagship architectural projects fuels much of the controversy over effects or defects associated with them. For example, Girgert (2011) argues that they might create jobs but not necessarily for the local population. Flagship architecture projects also cause positive local externalities and benefits, while the costs are not fully paid locally, but include partial state compensation. Hence, understanding the spatial distribution of effects, the spillovers in terms of costs and benefits, is important. A simple cost–benefit analysis does not capture the spatial distribution of these benefits and disadvantages. We argue that spatial incidence analysis can be an appropriate method to investigate a wider range of effects related to flagship architectural projects on various spatial scales, which considers more than just the direct neighborhood or the district scale, and at the same time takes account of the time component. The spatial incidence analysis also contributes to the question of effectiveness of public real estate development projects as urban regenerators as mentioned earlier, by comparing the intended and the effective impacts.

Spatial incidence analysis is a method first used in the finance sector to analyze the impacts of public infrastructure (Musgrave et al., 1977; Musgrave et al., 1993) and was adopted for the analysis of spatial impacts of regional infrastructure projects and their operation (Frey and

Häusel, 1982). This method has been applied in regional economic studies (Thierstein and Wilhelm, 2000). Incidence analysis is a systematic analysis of benefits and disadvantages for different sections of the population, caused by fiscal regulations, a project or an action (Frey and Häusel, 1982; Frey, 1990; Musgrave et al., 1993; Zimmermann and Henke, 2001). Spatial incidence analysis is based on a multiplier model and the Keynesian multiplier theory expanded by the basic idea of the regional economic export-base theory. The fundamental expectation is that a region receives payment flows from outside due to the realization of a publicly funded project, that is an infrastructure, event, or public facility. Caused by the implementation of a publicly funded project, the demand for local goods and services increases and entails an increase in production and rising employment. These are the primary effects of such projects. If the project is infrastructure, then the project itself can contribute to a rise in production to the region. For example, in the case of traffic infrastructure, the implementation of the project results in enhanced accessibility, which in turn can generate a rise in production in the region. The above-mentioned primary effects cause further effects, termed secondary effects. These secondary effects are usually captured by a specific regional multiplier, which is based on input–output calculations and the analysis of economic interdependencies. On the one hand, primary effects raise the demand of enterprises for intermediate inputs; on the other hand, additional incomes are created and spent, they raise the demand for goods and services again. This process, caused by additional income in the region, is continuous, but the demand diminishes with every repetition, because of drains on purchasing power in other regions, imports, taxes, or money that is not used for commerce but for saving, for example. In the end, these effects can by far exceed the primary effects, depending on the regional multiplier (Giuliani and Berger, 2009; Kronthaler and Franz, 2003; Zimmermann, 1981).

Hence, spatial incidence analysis is similar to a cost–benefit analysis but with an important distinction in that it regards the spatial distribution and redistributions (also described as "external effects" or "spillovers") of different impacts, which are not only monetary impacts of public action. It systematically investigates spillover effects in the form of benefits – that is, use – and disadvantages – that is, costs (Fischer and Wilhelm, 2001; Frey and Häusel, 1982; Scherer et al., 2002; Scherer et al., 2012; Zimmermann and Henke, 2001). Distributional effects always occur if the spatial distribution of beneficiaries does not match the spatial distribution of the benefactors (Scherer et al., 2002; Scherer et al., 2012). These spillovers can be cost spillovers or use spillovers. In the case of cost spillovers, regions bear costs but do not claim benefits. In the case of use spillover, a region claims use without bearing costs (Frey and Häusel, 1982; Frey, 1990).

Besides the regional analysis of payment flows, spatial incidence analysis investigates the spatial distribution of goods and services as well as the utility of these public investments. Through mapping these impacts the balance sheet is drawn, highlighting which institutional level profits from these three categories of effects and which institutional level pays more than it receives (Frey and Häusel, 1982; Frey, 1984; Giuliani and Berger, 2009; Scherer et al., 2002; Scherer et al., 2012; Thierstein and Wilhelm, 2000; Zimmermann, 1981).

The strength of the spatial incidence analysis is based on the registration and spatial mapping of payment flows. The analysis of the effects on the use level – for example image effects caused through a spectacularly designed museum by a star architect – is much more difficult because these effects are mainly intangible, meaning not measurable. They have to be described and valued qualitatively and the underlying expectation, hypothesis, and consideration have to be revealed (Frey and Brugger, 1984; Thierstein and Wilhelm, 2000). These intangible effects have to be investigated with methods for the special case and questions (e.g. personal interviews, network analysis, and media analysis).

A comprehensive spatial incidence analysis includes all institutions, persons, and actions that are linked inseparably with the public investment project; hence, at the very beginning of the research it is vital to define what should be included. Furthermore, the temporal and the regional frame of investigations have to be determined (Frey and Häusel, 1982; Frey and Brugger, 1984; Thierstein and Wilhelm, 2000). Typically these impacts are examined at different spatial scales ranging from, first, the local municipality of a project – that in a federalist, decentralized nation state takes the largest chunk of related infrastructure investments – to second, the wider defined region – the agglomeration, the county – third, the space of the nation state, and fourth, the rest of the world (Thierstein and Wilhelm, 2000).

An analysis undertaken by the University of St. Gallen in 2012 (Scherer et al., 2012) to study the regional economic impact of the Culture and Congress Centre in Lucerne (KKL) utilized a spatial incidence analysis in an exemplary way to demonstrate the initiated payment flows caused through a private–public flagship project. Commissioned by the Kultur- and Kongresszentrum Luzern Management AG, the study concluded that a purchasing power inflow of CHF50–52 million in the region is created by the KKL and the events held there. It is assumed that this purchasing power effects increase by externalities (such as intermediate consumption, increased tourist numbers, increased income, and in turn, increased consumption) by another CHF20–24 million, so that the whole region accrues CHF72–74 million in total (Scherer 2012: 25-26). Thus, the measurable economic effects go far beyond the direct environment of the project (Scherer 2012). Additional to that, the KKL generated a media value calculated as a derivative of the accounts in the media of CHF6.6 million in 2011.

Conclusion

Pointing to a gap between public rhetoric and effective impacts, Kunzmann declares that "Each story of regeneration begins with poetry and ends with real estate" (Kunzmann in Evans, 2005). In another take at the danger of mistaking the activity of building physical structures with achieving effective impacts, Glaeser cautions that

> for centuries, leaders have used new buildings to present an image of urban success … General Grigory Potemkin created a prosperous-looking fake village to impress Empress Catherine the Great. Today urban leaders love to pose at the opening of big buildings that seem to prove that their municipality has either arrived or come back … The tendency to think that a city can build itself out of decline is an example of the edifice error, the tendency to think that abundant new building leads to urban success … Building is the *result*, not the cause, of success.
>
> *(Glaeser, 2011: 61–62, emphasis added)*

This calls for an urgency to evaluate the increased involvement of the public sector in the development of cultural facilities as flagship architectural projects. It will be a waste of much needed public resources to continue unabatedly what Rocco (2014) calls "museum madness" and keep the rush to develop even more cultural facilities worldwide. In order to identify the successes or failures of such projects we need to compare their intended and effective impacts. To do so, we need to dig deeper into the internal value chain of ambitions, interests, and objections. We have to open the archives of these public facilities and conduct interviews with key stakeholders who were involved with these projects at their very inception. We also need to cast the net much wider than the direct spatial vicinity of these projects. If indeed these projects were intended to contribute to urban competitiveness and urban regeneration, we need to identify the short-term and long-term winners and losers beyond the direct and indirect impacts associated with these projects.

Acknowledgments

This work was supported by the German Research Foundation (Deutsche Forschungsgemeinschaft, DFG) under grant number TH 1334/11-1, entitled, "Star architecture and its role in re-positioning small and medium sized cities."

References

Adler, M. 1985. Stardom and talent. *The American Economic Review*, 75, 208–212.

Ahlfeldt, G. M. and Holman, N. 2015. *Distinctively Different: A New Approach to Valuing Architectural Amenities.* SERC Discussion Papers. London: Spatial Economics Research Centre (SERC).

Ahlfeldt, G. M. and Kavetsos, G. 2014. Form or function? The effect of new sports stadia on property prices in London. *Journal of the Royal Statistical Society: Series A (Statistics in Society)*, 177, 169–190.

Ahlfeldt, G. M. and Maennig, W. 2009. Arenas, arena architecture and the impact on location desirability: the case of "Olympic Arenas" in Prenzlauer Berg. *Urban Studies*, 46, 1343–1362.

Ahlfeldt, G. M. and Maennig, W. 2010. Impact of sports arenas on land values: evidence from Berlin. *Annals of Regional Science*, 44, 205–227.

Ahlfeldt, G. M. and Mastro, A. 2012. Valuing iconic design: Frank Lloyd Wright architecture in Oak Park, Illinois. *Housing Studies*, 27, 1079–1099.

Banham, R. 1981. A throw-away aesthetic. In: Sparke, P. (ed.) *Design by Choice*. London: Rizzoli, pp. 90–93.

Bayley, S. 2005. Foreword. In: Davies, P. and Schmiedeknecht, T. (eds) *An Architect's Guide to Fame*. Oxford: Elsevier, pp. ix–xiv.

Crompton, J. L. 2006. Economic impact studies: instruments for political shenanigans? *Journal of Travel Research*, 45, 67–82.

Davies, P. and Schmiedeknecht, T. (eds) 2005. *An Architect's Guide to Fame*. Oxford: Elsevier.

Evans, G. 2005. Measure for measure: evaluating the evidence of culture's contribution to regeneration. *Urban Studies*, 42, 959–983.

Fischer, G. and Wilhelm, B. (eds) 2001. *Die Universität St. Gallen als Wirtschafts- und Standortfaktor. Ergebnisse einer regionalen Inzidenzanalyse*. Bern: Haupt.

Foster, H. 2008. Image building. In: Vidler, A. (ed.) *Architecture Between Spectacle and Use*. Williamstown, Massachusetts: Sterling and Francine Clark Art Institute, pp. 164–179.

Franck, G. 1998. *Ökonomie der Aufmerksamkeit: ein Entwurf*. Munich: Hanser.

Franco, S. F. and Macdonald, J. 2015. The Effects of Historic Amenities on Residential Property Values: Evidence from Lisbon. FEUNL Working Paper Series No. 603. Available: http://dx.doi.org/10.2139/ssrn.2776207

Frank, R. H. and Cook, P. J. 1995. *The Winner-Take-All Society: How More and More Americans Compete for Ever Fewer and Bigger Prizes, Encouraging Economic Waste, Income Inequality, and an Impoverished Cultural Life*. New York: The Free Press.

Frey, R. L. 1984. Die Inzidenzanalyse: Ansatz und Probleme der Erfassung von Spillovers. Infrastruktur, Spillovers und Regionalpolitik. In Frey, R. L. and Brugger, E. A. (eds) *Infrastruktur, Spillovers und Regionalpolitik. Methode und praktische Anwendung der Inzidenzanalyse in der Schweiz*. Diessenhofen: Rüegger, pp. 37–55.

Frey, R. L. 1990. *Städtewachstum Städtewandel. Eine ökonomische Analyse der schweizerischen Agglomerationen*. Munich: Helbing and Lichtenhahn.

Frey, R. L. and Brugger, E. A. (eds) 1984. *Infrastruktur, Spillovers und Regionalpolitik. Methode und praktische Anwendung der Inzidenzanalyse in der Schweiz*. Diessenhofen: Rüegger.

Frey, R. L. & Häusel, U. 1982. *Infrastruktur, Spillovers und regionale Disparitäten Zwischenbericht: Fragestellung und methodische Grundlagen*. Bern: Institut für Sozialwissenschaften Universität Basel.

Fuerst, F., McAllister, P., and Murray, C. B. 2011. Designer buildings: estimating the economic value of "signature" architecture. *Environment and Planning*, 43, 166–184.

Girgert, W. 2011. Der Bilbao-Defekt. german-architects.com [Online]. Available: www.german-architects.com/pages/23_11_bilbao [Accessed 08.06.2011].

Giuliani, G. and Berger, S. 2009. Leitfaden für die regionalwirtschaftliche Beurteilung von Entwicklungsstrategien und -projekten. Theoretischer Grundlagenbericht. Zürich.

Glaeser, E. 2011. *Triumph of the City: How our Greatest Invention Makes Us Richer, Smarter, Greener, Healthier, and Happier.* New York: Penguin.

Glaeser, E. L. and Gottlieb, J. D. 2009. The wealth of cities: agglomeration economies and spatial equilibrium in the United States. *Journal of Economic Literature, American Economic Association*, 47, 983–1028.

Hall, T. and Hubbard, P. 1998. *The Entrepreneurial City: Geographies of Politics, Regime, and Representation.* New York: Wiley.

Harvey, D. 1989. *The Condition of Postmodernity.* Oxford: Blackwell.

Heynen, H. 1999. Petrifying memories: architecture and the construction of identity. *The Journal of Architecture*, 4, 369–390.

Klingmann, A. 2007. Beyond Bilbao. In: Klingmann, A. (ed.) *Brandscapes, Architecture in the Experience Economy.* Cambridge, Massachusetts: The MIT Press, pp. 237–255.

Klosterman, R. E. 1985. Arguments for and against planning. *The Town Planning Review*, 56, 5–20.

Kronthaler, F. and Franz, P. 2003. Methoden und Probleme der Abschätzung regionalökonomischer Effekte großer Sportveranstaltungen. *Tourismus Journal (Lucius & Lucius)*, 7, 439–455.

Laffont, J. J. 2008. Externalities. In: Durlauf, S. N. and Blume, L. E. (eds) *The New Palgrave Dictionary of Economics* [Online]. Available at: www.dictionaryofeconomics.com/extract?id=pde2008_E000200

Latour, B. and Yaneva, A. 2008. Give me a gun and I will make all buildings move: An ANT's view of architecture. In: Geiser, R. (ed.) *Explorations in Architecture: Teaching, Design, Research.* Basel: Birkhäuser, pp. 80–89.

McNeill, D. 2009. *The Global Architect: Firms, Fame and Urban Form.* New York: Routledge.

Molotch, H. 1976. The city as a growth machine. *American Journal of Sociology*, 82, 309–330.

Musgrave, R. A. (ed.) 1987. *Merit Goods.* London: Macmillan.

Musgrave, R. A., Musgrave, P. B. and Kullmer, L. 1977. *Die öffentlichen Finanzen in Theorie und Praxis / 3.* Tübingen: J.C.B. Mohr

Musgrave, R. A., Musgrave, P. B. and Kullmer, L. 1993. *Die öffentlichen Finanzen in Theorie und Praxis / 2.* Tübingen: J.C.B. Mohr

Patterson, M. 2012. The role of the public institution in iconic architectural development. *Urban Studies*, 49, 3289–3305.

Pine, B. J. and Gilmore, J. H. 1999. *The Experience Economy: Work is Theatre & Every Business is a Stage.* Boston, Massachusetts: Harvard Business School Press.

Plaza, B., Tironi, M., and Haarich, S. N. 2009. Bilbao's art scene and the "Guggenheim effect" revisited. *European Planning Studies*, 17, 1711–1729.

Rocco, F. 2014. Museum Madness. LSE Arts and Breese Little public lecture. London, November 24.

Ronneberger, K., Lanz, S. and Jahn, W. 1999. *Die Stadt als Beute.* Bonn: Dietz.

Saint, A. 1983. *The Architect as Hero and Genius. The Image of the Architect.* New Haven, London: Yale University Press.

Scherer, R., Strauf, S., and Bieger, T. 2002. *Die wirtschaftlichen Effekte des Kultur- und Kongresszentrums Luzern (KKL).* Schlussbericht. St. Gallen: Universität St. Gallen.

Scherer, R., Strauf, S., Riser, A., and Gutjahr, M. 2012. *Regionalwirtschaftliche Bedeutung des Kultur- und Kongresszentrums Luzern.* St. Gallen: Universität St. Gallen.

Sklair, L. 2006. Iconic architecture and capitalist globalization. *City*, 10, 21–47.

Smith, T. 2008. Spectacle architecture before and after the aftermath: situating the Sydney experience. In: Vidler, A. (ed.) *Architecture between Spectacle and Use.* Williamstown: Sterling and Francine Clark Art Institute, pp. 3–24

Sternberg, E. 1997. The iconography of the tourism experience. *Annals of Tourism Research*, 24, 951–969.

Thierstein, A. 2013. The creative economy and the knowledge economy. In: Wang, W. (ed.) *Culture: City.* Zürich: Lars Müller Publishers, pp. 43–46.

Thierstein, A. and Wilhelm, B. 2000. Hochschulen als Impulsgeber für die regionale Entwicklung. In: Thierstein, A., Schedler, K., and Bieger, T. (eds) *Die lernende Region. Regionale Entwicklung durch Bildung.* Zürich: Rüegger, pp. 9–35.

Yaneva, A. 2009. *The Making of a Building. A Pragmatist Approach to Architecture.* Oxford: Peter Lang.

Yaneva, A. 2012. *Mapping Controversies in Architecture.* Farnham: Ashgate Publishing.

Zimmermann, H. 1981. *Regionale Inzidenz öffentlicher Finanzströme. Methodische Probleme einer zusammenfassenden Analyse für einzelne Regionen*. Baden-Baden: Nomos.

Zimmermann, H. and Henke, K. D. 2001. *Finanzwissenschaft*. Munich: Franz Vahlen.

Part II

Markets and economics of real estate development

5

Investment and development behavior following the Great Recession

Raymond G. Torto

Abstract

Commercial real estate is capital intensive, and managers operate their businesses to serve the expectations of capital: to invest in and develop commercial real estate. This chapter investigates the operations of the global commercial real estate (hereafter, CRE) investment management business within the context of the changed environment of investing and developing into CRE since the global financial crisis and the subsequent Great Recession.

The chapter documents the changes in the market along with changes in the operations of CRE investment management. These changes have been observed by the author and have been corroborated with interviews with senior professionals in the CRE investment management business.

The chapter draws out the implications for market performance from the changes in the investment management business operations. While it is clear to this writer that the industry has moved from a deal-making culture to a business operational culture, aligning its interests with its institutional clients, targeting to minimize risk rather than reach for performance, it is ironic that the pricing in the market is pushing these same managers to take on more risk today.

The interesting and unanswered question, from the point of view of the market for commercial real estate, is what will be the amount of supply coming to market due to this change in investment strategy. Will it follow the traditional path of a building/construction boom which was the case when development was financed by lender capital? Or will equity capital, provided through the vehicle of CRE investment management firms, be more tempered in financing new development? Of course, time will tell.

Introduction

In an industry that is capital intensive, such as commercial real estate, the quantity, expectations, and behavior of the capital sources are paramount. "Capital sources" in this chapter are institutional capital, which is usually sourced from pension funds, sovereign wealth funds, insurance companies, etc. The dimensions of this behavior are usually captured in metrics such as transaction volume, asset pricing, investment structures, return expectations, holding period, investment strategy, etc. Generally lost in these details is how investment managers operate their businesses, as opposed to their choice of investments, to serve the expectations of capital: to invest in and develop commercial real estate. Importantly, investing in and developing commercial real estate have different return and risk profiles, and therefore are not one and the same goal.

This chapter investigates the operations of the global commercial real estate (hereafter, CRE) investment management business within the context of the changed environment of investing and developing into CRE following the global financial crisis and the subsequent Great Recession. The chapter is divided into several parts. The first section compares the commercial real estate market as of the end of 2015 to the market pre-2007. This section is followed by several sections that provide a comparison of the management of assets or investment management of CRE and how investment management has changed since the Great Recession. The final section pulls the lessons together to analyze the prospects of an emerging "bubble" and a subsequent "bust" in CRE. As we entered 2016, the issue on most minds in the industry was whether CRE was priced too high, and therefore, the market on the edge of a bursting price bubble, or as many phrase the question "will we do it again?" In other words, will the CRE industry bid prices too high leading to a price bubble that will subsequently burst!

The commercial real estate market

By the beginning of 2016, the CRE market had recovered significantly from the Great Recession. As shown in Figure 5.1, global transaction volumes, as measured in dollars, had almost recovered to the level last seen in 2007, with some shift in underlying volume by region (Real Capital Analytics 2015a). Volume was very strong in North America (Canada, United States, and Mexico) while the dollars invested in Asia were down significantly, reflecting the fact that the number of development sites being sold or bought in Asia (particularly China) was down significantly. European investment volume was up 23% in 2015 over 2014, with more caution reflected in the market as the year ended. The fourth quarter volume was about 6% lower than in the fourth quarter of 2014. However, London, particularly Central London, which is the top market in volume and pricing in Europe had an increase of 30% in 2015 (Real Capital Analytics, 2015b, p. 9).

This decline or shift as to where transactions were being globally undertaken mirrored the changes seen since 2007 in the economic prospects of the regions/countries and also the degree to which investors decided to make defensive decisions: judgements to invest in existing properties versus develop new properties.

In addition to volume having rebounded since 2007, prices in the CRE sector were generally above the peak levels of the earlier period. In addition, prices were growing faster than rents (Figure 5.2), and did so with pronounced strength in 2010 and 2011 when investors eschewed development in favor of buying existing assets. The faster growth in prices, in contrast with market rents, reflects the volume of capital in the market for existing properties. In the United States several indices reached new heights. The Moody's/RCA Commercial Property Index increased 14.7% year on year (YOY) in 2015 and extended its streak of double-digit annualized gains to 34 months. The CoStar value weighted index increased 12.2% through November 2015 and by the beginning of 2016 was 18.4% above its 2007 prior peak. The equal weighted index, which reflects pricing for smaller properties, gained 11.7%, but remained 4% below the 2007 peak.

This phenomenon reflects the investment priority of capital in the early part of this recovery. Capital was invested in "safe" or core assets, defined as assets with attributes such as credit

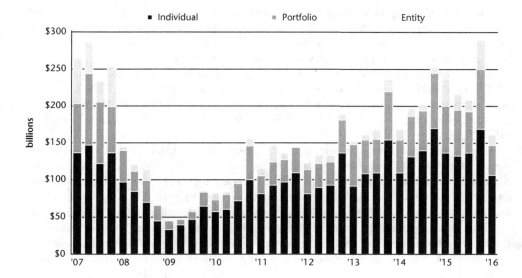

Figure 5.1   Global transaction volume

Source: Real Capital Analytics, 2015a

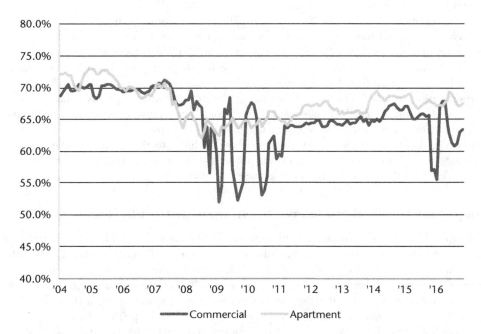

Figure 5.2   Global capital and rent indices

Source: Real Capital Analytics, 2015a

tenants and good locations. With many investors having the same priorities, prices of these core assets rose quicker than assets in general as many investors were motivated by safe assets, i.e. looking for downside protection. This was reflected in comparative yields or capitalization rates for core vs non-core assets, where rates for core assets were lower.

With volume and prices above 2007 peak levels, some observers have raised concerns that CRE is in a "bubble" again, i.e. that prices are unsustainable, and that this bubble is about to burst and prices will decline precipitously. These observers maintain that the CRE industry has not learned the lessons of the Great Recession and, hence, will repeat the mistakes of the past (Allen 2015).

There is no doubt that the amount of capital looking to invest in CRE has ballooned over the last several years. There were 492 funds in the market as of the end of 2015, and in 2016 Andrew Moylan, head of real assets products for Preqin, suggested "investor demand for real estate remains considerable, and the strong fundraising seen in 2015 is expected to continue well into 2016," although he predicted there would be some "softening" in 2016 (Preqin 2016).

This international money has multiple motivations: higher returns in CRE, diversification for the portfolio, money into another currency or country, among others. Some in the industry are concerned that this money/capital will bid up prices, in order to gain/acquire a long position in the CRE market, and at prices that will be unsustainable. Prices will become frothier and debt on these assets will not be supported by the operating income of the assets, nor by the ultimate values after the expected decline.

Prices have been bid higher. But if it is frothy, it is happening due to an abundance of equity capital, not debt capital. For instance, a closer look at the debt side of the market does not show a high level of loan to value ratio. For example, the data for the USA – the largest CRE market in the world – shows that loan to value ratios are in the 66% range (Real Capital Analytics 2015a) versus 70% in the 2006/7 time period (Figure 5.3).

Also supportive of pricing in 2016 were the fundamentals of the market – the relationship between the demand and supply of space. Because of the Great Recession, all supply channels for new property were halted for several years following 2007/8. It took time for the demand side of the market to recover and even more time for supply to respond as excess capacity needed to be reduced first. At the beginning of 2016, the supply/demand relationship in the leasing market was fairly balanced with vacancy levels stabilizing and rents rising in many markets. The fundamentals implied steady or rising net operating incomes (NOI) and suggested support for the high prices in the market. Of course, this statement is not true universally. As one example, the Houston market was hit with the oil supply glut, and the reduction in that business's demand for space, just as several new office buildings were entering the market. In 2016 Houston vacancy rates were 16.5% up from 12% in 2015.

The commercial real estate investment management business

The overall investment management business has shifted over the years as documented by Matthew H. Lynch, CEO of UBS at the Advanced Management Program in Real Estate at the Harvard Graduate School of Design, July 24 2015. Overall, portfolios have grown but there is even faster growth in alternative asset classes, of which CRE is one alternative. The reason for the shift to alternative assets is that the returns were better in alternative assets: in 2014 alternative assets were 30% of industry revenue and 12% of assets (Lynch 2015, p.8) (see Figure 5.4).

According to the Tower Watson Global Alternative Survey 2015 (Tower Watson, 2015), direct real estate funds accounted for about 33% of alternative assets under management and were projected to grow quite handsomely through 2020. With strong direct CRE investments, the

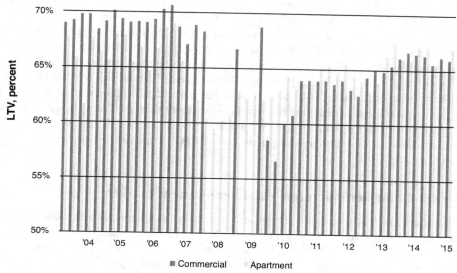

Figure 5.3 US CRE loan underwriting

Source: Real Capital Analytics 2015b

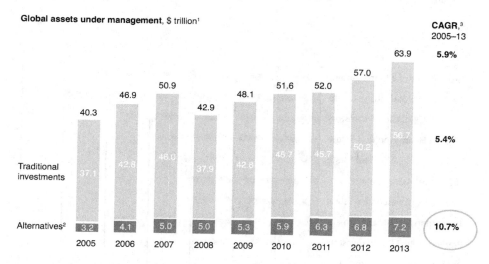

Figure 5.4 Alternative investments have grown twice as fast as traditional investments since 2005

Source: ANREV/INREV/NCREIF 2015

large fund managers turn out to be larger, capturing more investors and funds. The top ten managers saw growth of 12.9% YOY for 2014 and they control US$790 billion or 36.5% of assets under management (AUM). The top 10 CRE managers as of 2014 are listed in Figure 5.5.

Not only are CRE managers growing in size, but also their client base is changing. Based on a 2014 survey the majority of the client base was institutional and pension fund orientated, as shown in Figure 5.6, and this trend continued in 2016. A 2016 survey by BlackRock, who polled over 170 of its largest institutional clients, representing US$6.6 trillion in assets under management, found that 47% of their investors were increasing their allocation to real estate in 2016 while only 9% were reducing allocations (Duffell 2016). The authors of the survey noted that investors were shifting into "illiquid assets" as they reduced their allocations to equities and were looking for "alpha generating opportunities that match their liabilities." It is of note that this shift into liabilities is long term. As we will argue in the last section, this shift in focus is having an effect on the management philosophy of CRE firms, and by implication on the market for CRE.

The post-Great Recession real estate investment market

Following the Great Recession, the investment market in CRE saw early risk takers return to the market in earnest and volume, beginning in 2011. The focus at that time was on buying existing assets, not new development. The early focus on existing assets meant that real estate development across all property types and all regions was limited, especially in North America and Europe. There was little excess supply of existing space but nevertheless there was limited equity or lending capital for development. Some limited development was undertaken in parts of Asia, where some emerging economies were recovering fastest from the Great Recession and prime quality assets were in short supply. In such markets, only through development could

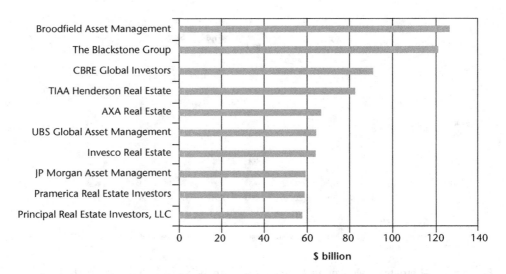

Figure 5.5 Top 10 fund managers worldwide by AUM as of December 31, 2014

Source: ANREV/INREV/NCREIF 2015

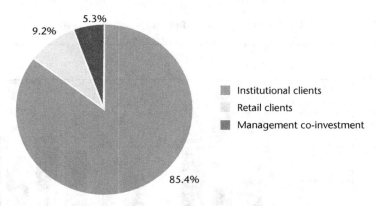

Weighted % breakdown by client base as of 31 December 2014

- Institutional clients
- Retail clients
- Management co-investment

Figure 5.6 Client base of CRE managers

Source: Real Capital Analytics 2015b

tenant demand be met, but even here, much foreign equity capital was repatriated for several years following the Great Recession.

They invested mainly into existing core assets, which are thought to have a combination of good location and credit tenants and are therefore safe assets, investment structures, and locations. As a result, asset values rose in urban areas, but not in the outer locations as equity capital avoided the less popular and less populated cities. This pattern was particularly evident in the USA and the UK as cities such as New York, San Francisco, and London had a plethora of buyers that did not venture to the outer regions of these metros. As prices of core assets reached record levels, and yields were low, equity capital began investing in 2015 in areas outside the larger metros and in assets with more risk than found in core assets.

One data point to support the above conclusion is from London and the rest of the UK (Real Capital Analytics 2015b) where the data on development site sales in the UK is shown in Figure 5.7. The data shows that as we moved into 2015, more development sites were transacted outside London than in earlier post-recession years. Additionally, development site sales in the UK, both in London and outside London, were equal to or above pre-2008 levels.

As the focus on buying existing assets was driving prices higher, equity capital and investment managers recognized that the price of existing core assets was higher than estimated development costs – i.e. replacement costs – and these managers started to ask: "why buy it when we can build it cheaper?" Slowly, investment managers began shifting their strategy from core existing assets to adding value, or put differently, from buying existing assets to developing new assets.

The development incentive today is not only driven by the price of buying existing assets versus building assets but also by tenants shifting their preferences for space amenities and locations. This shift makes some existing assets and existing locations obsolete (Peiser and Torto 2015). Most particularly, tenant preferences are changing with regard to location and design. Users want fundamentally different environs and appear less price sensitive. In the former category is the desire of many housing tenants to find areas of density and proximity to work and play, and in the latter category are the changes dictated by how work is defined. As one

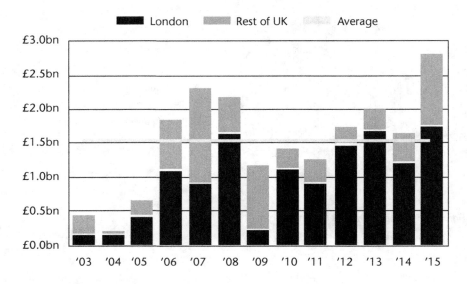

Figure 5.7 Development site sales in the UK

Source: Real Capital Analytics 2015b

observer has noted: "work is not where you go, but what you do." Both of these factors are changing what is a viable investment for the long term and are changing land use policies and obsolescence rates for commercial real estate assets.

As we have shown above, following the Great Recession, the CRE investment management industry grew, with growth coming from a client base of new institutional investors, whose investment horizon is long term, not short term.

Investment management

There have been changes in the market along with changes in the operations of the CRE investment management business since the Great Recession of 2008/9. The above sections note some of the changes in the CRE market, while this section will identify the changes within the business operations of the CRE investment management firms. These changes have been observed by the author and have been corroborated with interviews with senior professionals in the CRE investment management business. There are implications for market performance from the changes in the investment management business operations, which will be discussed in the next section.

We would categorize the changes in the operations of the investment management business as follows:

Globalization – prior to 2008, probably 10% of the clients of large investment managers were global capital. Now, over 50% are global. This requires investment managers to go beyond their country borders as global capital sees fewer borders and boundaries for investing their funds.

Additionally, we are seeing global capital "clubbing-up" with other large scale investors or investment management firms to buy scale and diversity. This means they are considering bigger investment deals as a means to deploy large capital blocks with like-minded partners. Investment management firms need to keep pace and this is reflected in further growth of the largest firms.

Due diligence – there is more requirement for a thorough due diligence as regulations require that new clients as well as the seller or buyer of properties go through a due diligence background check. For example, in the USA, the Office of Foreign Assets Control requires a representation and warranty from the investment management firm that the client is not on the terrorist list. This requirement has certainly added to costs through overheads and time and is another factor leading to larger sized investment managers.

Fees – the fees that investment management firms can charge – assets under management (AUM) and performance-based fees – continue to drop. Fifteen years ago, asset management fees were around 200 basis points (bps). Now they are closer to 50bps and for large investors fees are in the 35bps range. Investment manager's payments put more weight on the "back-end" performance instead of the annual AUM fees.

Returns – there has been a significant change in client return expectations. Internal-rate-of return-driven capital has seen expectations forced steadily downward since 2009. And with yields compressed over the last few years due to central bank policies, there seems to be acceptance of "low yields for longer." This has changed investment strategies, which we will discuss in the next section.

Technology – there continues to be dramatic technological change in the investment management operations making the business more efficient. Reports and offering memoranda are prepared electronically for clients and accessible on the internet, and it is no exaggeration to say that "everything is in the cloud now." Fast networks and wireless connections, video conferencing and large screen videos allow group reviews of deals and more interactions. As one interviewee noted: "Not sure we are better investors or sellers, but definitely faster and can use less people to do more."

Governance – CRE firms now have a Head of Compliance, who has to review many aspects of the business operations, including all external presentations. There also are strict rules and regulations on tracking internal information. For example, investor reports must itemize affiliate fees and there is an abundance of footnotes! There are also independent members on the Investment Committee whose vote on investments is theoretically not biased or based on fees earned. All together these changes have added significant staff to compliance and legal (not to review deals but to make sure things are done correctly and engage outside counsel properly).

Conversations with investment management professionals indicate that the above items are unblemished changes in the operations of the investment management business. However, a number of observers noted that the institutional changes, the search for good governance, among others, were in the air prior to the Great Recession and just continued and intensified subsequent to it. We do not think it matters when this shift in the industry started, but rather, what impact has or will this shift have on the current and future CRE market?

Evidently, investors are looking for good government, transparency, and long-term investments. This is different from earlier periods when investment managers were measured only on performance. Today's capital sources expect several professional, distinguishing attributes in their investment managers. For instance, they expect their managers to have policies for succession planning, policies for maintaining depth on their bench, and compensation procedures that align interests and motivate staff. Along these lines, the Global Real

Estate Standards Board was brought up by interviewees, and it was noted that Environment, Social and Government (ESG) policies and data supporting such are requested in all client Requests for Proposals. It is "good governance and performance today, that is important, not just performance" noted one interviewee. Another noted that the strategy of his firm, which has been very successful, is threefold: "to de-risk, de-risk, de-risk" implying that less risk is paramount to higher returns.

Conclusion

The above begs the question as to why the industry has changed. Some argue that it is transitioning from "deals" to "structure" because of the nature of the clients and their demand for an "institutionalization" of the industry. But others, while acknowledging that institutionalization has played a role, emphasize that "regulation" has also been an important factor. As one observer put it: "Partly the institutionalization of the business, but to a large extent the vast array of new government/regulatory mandates. The new rules have definitely had an impact."

While it is clear to this writer that the industry has moved from a deal-making culture to a business operational culture, aligning its interests with its institutional clients, targeting more to minimize risk than reach for performance, it is ironic that the pricing in the market is pushing these same managers to take on more risk today. Initially, these managers served their clients by focusing on existing core strategies, looking for "safe/core" investment assets. But since this is a crowded strategy, prices of these assets have moved above replacement costs, and investment managers realize that they can achieve better performance by developing property, rather than buying assets. One example of this shift is the actions of the California Public Employees Retirement System (CalPERS) in the spring of 2016 to formalize a "build to core" strategy, which would allow "for the creation of new core assets in markets that offer limited opportunities to acquire existing assets" (Peterson 2016). But, of course, development has more risks – construction risk, leasing risk – associated with it than does buying an existing asset.

The interesting and unanswered question, from the point of view of the market for commercial real estate, is what will be the amount of supply coming to market due to this change in investment strategy. Will it follow the traditional path of a building/construction boom, which was the case when development was financed by lender capital? But in today's market the lending capital is sparse, and the lending regulators are raising capital requirements and are watching CRE markets very closely, remembering well the past boom/bust cycles in commercial real estate. Will equity capital, provided through the vehicle of CRE investment management firms, be more tempered in financing new development?

Of course, time will tell, but the view in the CRE industry is that CRE has a lot more equity backing it than during the last cycle where high leverage was the norm. Additionally, loan underwriting today is not nearly as aggressive as was the case during the last cycle. Regulatory pressures are increasing, capital requirements for banks are making it harder to lend in the secondary market – all of these factors are making lending a bit harder and more expensive as well. And together these factors should keep development in check as well. Furthermore, the CRE market is a lot more transparent than it was in earlier cycles. There are more independent research firms following pricing and fundamentals, and more media following CRE and asking tough questions. If the market shows weakness it will be fully reported and I would expect this will feedback quickly to the investment management boards and clients. In this environment, firms will not be chasing deals as much as managing risk, which is foremost on the minds of the long-term investor clients, and therefore on the minds of those managing CRE investment firms.

Acknowledgments

I am indebted to a number of CRE professionals for their time, viewpoint and the sharing of information with me. These include: Peter Donovan, Senior Managing Director, Capital Markets, Multifamily CBRE; Blake Hutchinson, CEO and President, Oxford; Christopher Ludeman, President, Capital Markets, CBRE; Brian Stoffers, Global President, Debt and Structured Finance, CBRE; Jeffrey Torto, Managing Director and Portfolio Manager, CBREGI; Gary Whitelaw, CEO and President, Bentall Kennedy. Omissions and errors of fact and interpretation are my responsibility.

References

Allen, Kate (2015). "Aggressive Pricing Raises Property Bubble Fears." *The Financial Times*. October 3, 2015.

ANREV/INREV/NCREIF (2015). Fund Manager Report, 2015.

Duffell, Thomas (2016). "Investors Fleeing Fixed Income for Real Estate." PERE January 28. Available at: www.perenews.com/news/global/2016-01-26/blackrock_-investors-fleeing-fixed-income-for re/ ?utm_source=Sailthru&utm_medium=email&utm_campaign=PERE%20DD%20012616&utm_ term=PERE%20Daily%20Smart%20List#sthash.VWskbQMX.dpuf

Lynch, Matthew (2015). Advanced Management Program in Real Estate at the Harvard Graduate School of Design, Presentation, July 24, 2015.

Peiser, Richard and Torto, Raymond (2015). "The Changing Demand for Office Space: A Boston Tale." unpublished manuscript.

Peterson, Jon (2016). "CaLPERS to Ramp Up Real Estate Development." *IPE Real Estate*, April 14. Available at: http://realestate.ipe.com/news/investors/calpers-to-ramp-up-real-estate-development/ 1001

Preqin (2016). *Preqin Spotlight*, 10(4).

Real Capital Analytics 2015a. *The US Capital Trends. The Big Picture. 2015 Year in Review*. New York: Real Capital Analytics.

Real Capital Analytics 2015b. *Europe Capital Trends. 2015 Review. January, 2016*. New York: Real Capital Analytics.

Tower Watson (2015). Global Alternatives Survey 2015. Tower Watson. Available at: www.towerswatson.com/en-GB/Insights/IC-Types/Survey-Research-Results/2015/07/Global-Alternatives-Survey-2015

6

Land use policies and markets in some selected African countries

Moses M. Kusiluka, Sophia M. Kongela, and Karl-Werner Schulte

Abstract

Land use policies and markets in Africa are significantly different from those operating in many other continents. Many countries in Africa are characterized by reforming property rights institutions and burgeoning property markets. This chapter gives an overview of land use policies and markets in Africa as a whole before focusing on Tanzania and Kenya for a detailed examination of the evolution of land use policies and markets and other developments in the property sector. It will be shown that many African countries share a common land policy and property market evolution path that involves four periods, namely pre-colonial, colonial, post-independence, and post-Cold War. During the first half of the twentieth century, colonial governments introduced formal systems of land administration although they focused much on allocating productive land to settlers and plantation agriculture. Land and property markets started developing during the post-independence period and gained much ground after the end of the Cold War when many countries embarked on institutional reforms. The reforms have significantly enhanced property rights and promoted land and property markets. Market transparency is improving in many countries, thereby attracting international and regional investors. With more reforms of land policies and property markets, Africa will continue attracting more investors from around the world.

Introduction

Land use policies and markets in Africa are significantly different from those operating in many other markets. Many African countries are characterized by reforming property rights institutions and burgeoning property markets (Kusiluka, 2012). Changes in demographic and economic trends in many African countries have increased pressure on the need to reform and formalize many institutions. African cities are urbanizing at an average rate of 1.1 percent, young consumers are increasingly dominating economic life, and a middle-class income group is emerging (von Gaertringen, 2014; AfDB et al., 2016). Apart from giving an overview of the continent as a whole, this chapter uses Tanzania and Kenya as case examples to examine land use policies and markets, and other developments in the property sector.

An overview of land use policies and markets in Africa

Demographic and economic facts

Africa, consisting of 54 countries and measuring about 30,221 million square kilometers, is the third largest continent. Africa's population stands at about 1.2 billion, with an annual growth rate of 2.6 percent, which is the highest growth rate amongst the continents (AfDB et al., 2016). With about 40 percent of the population living in urban areas, Africa is the least urbanized but one of the fastest urbanizing continents. Africa's average GDP growth since 2001 has stood at over 5 percent (AfDB et al., 2016). The economy is largely commodity driven with agriculture, mining, and oil and gas sectors being the major drivers. Tourism is another fast-growing sector.

Land use rights and the evolution of land markets

Owing to the large number and diversity of countries, the continent has a considerable diversity in land ownership and administration institutions. Land use rights differ from one country to another. In many countries, rights to land are constitutional matters. With the relatively low level of urbanization, many people still directly depend on land for their livelihoods. Rapid population growth and urbanization further increase pressure on land resources, which underscores the need to put in place efficient land administration systems. In the absence of good policies and administration systems, land has been a cause of many conflicts for many countries. It is also noteworthy that most of the land parcels in many countries are not formally demarcated and registered. This limits their transferability and acceptability as collateral for loans (Cotula et al., 2006; Deininger, 2003).

Another notable aspect about land tenure in Africa is the dominance of customary rights. A large portion, especially in the rural areas, is owned according to traditional arrangements. In most cases land is owned along patriarchal norms, which sometimes deny women equal access to land (Kusiluka et al., 2011). However, these discriminatory traditions are increasingly being abolished as many countries enact new land policies and laws. The evolution path of African land and property markets is summarized in Figure 6.1.

Land policies and markets for many African countries have gone through three distinct periods; namely colonial, post-colonial, and post-Cold War. Before the colonial period most of the countries were characterized by very low levels of urbanization as most of the people lived in small villages or family groupings. There are no detailed records on individual countries' land markets during the pre-colonial period. Between the late 1800s and the end of the first half of the twentieth century, many African countries were under European colonial regimes. It is during this period that many of the existing cities started taking the shape of modern urban centers.

Urbanization increased significantly when countries achieved their independence as many people started moving into urban centers in search of jobs and education. Many urban centers started being overwhelmed with rapid urbanization. Different countries adopted different political systems, ranging from socialism to dictatorship. Military coups characterized many governments, undermining democratic institutions. The end of the Cold War and economic hardships forced many countries to embark on reforming their economies. This entailed creating a conducive environment for private sector growth and secure property rights. It was during this period that many countries recorded noticeable progress in attracting private investment.

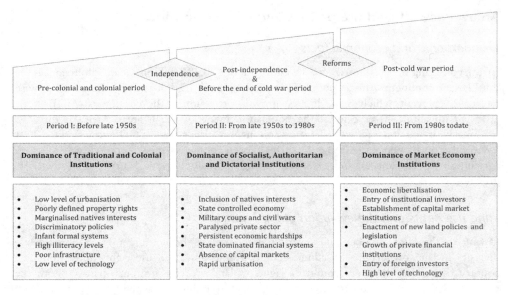

Figure 6.1 Evolution of African land policies and markets

Source: Adapted from Kusiluka, 2012

From Figure 6.1, it is clear that over the past century land ownership and utilization systems have evolved with time and with the influence of colonial institutions. Different interests in land were introduced during colonial times and modified during the post-colonial period. The most notable interests in land found in many African countries today include customary tenure, which is still dominant, various kinds of leaseholds, and few freeholds. Despite the evolution of land tenure arrangements, restrictions to land ownership by foreigners are consistently notable in many countries (Austin, 2010; Toulmin, 2009).

One of the main challenges facing Africa is land-grabbing, which is reported in many countries on the continent (Cotula et al., 2009). This is partly due to a lack of comprehensive land policies or the existence of weak institutional frameworks for some countries (EUC et al., 2010). Land-grabbing is both endogenous and exogenous in that both well-off individuals amongst Africans and large multinational companies tend to buy large tracts of productive land, leaving many indigenous people landless and destitute. Large-scale commercial agriculture, industrial uses, and land speculation are some of the main drivers of land-grabbing in Africa (GLP, 2010; Kusiluka et al., 2011).

Institutional reforms and property markets

Although institutional reforms are still ongoing, notable progress has been recorded, especially in developing market-oriented economic environments. Most of the recorded achievements can be largely attributed to the reforms introduced following the end of the Cold War. Despite the achievements, most of the property markets are still at their early stages of development. Apart from South Africa and a few other countries that are regarded as having stable markets, land and property markets in many countries are highly speculative, mainly due to lack of market transparency. Average land prices in prime locations range between US$1,500 and US$2,500 per square meter (Knight Frank, 2015). Rentals for prime properties differ significantly from one country to the other. For instance, prime office monthly rent in Angola could go as high as US$150 per square meter while in Nairobi prime office monthly rents are around US$20 per square meter (Knight Frank, 2015).

Land policies and markets in Tanzania and Kenya

Countries' overview

Tanzania and Kenya are both located in the East African region. The two countries together with Uganda are the founding members of the East African Community (EAC). Tanzania and Kenya have fundamental differences in terms of their land tenure systems and the historical institutions underpinning the evolution of the two land markets. There are many countries in Africa whose evolution path is similar to that of Kenya; there are also many countries with many similarities to Tanzania. The discussion of the two countries should therefore give an overall picture of land policies and markets found in many African countries.

The two countries have experienced rapid urbanization and have a young population, growing middle class, and a fast-growing private sector. These attributes are set against the backdrop of an environment that is defined by weak property rights institutions, highly imperfect property markets, and acute shortage of decent accommodation for various uses. Real estate business in these countries is predominantly defined by new development projects rather than trading of the existing stock of buildings. Besides, a large number of land parcels in these countries are unregistered, which inhibits land-based investment, productivity, and sometimes attracts land-related conflicts. Informal land and property markets accounting for more than 70 percent of the market are much more active, but remain opaque.

Agriculture remains the traditional sector of both economies. Some non-traditional sectors such as mining, tourism, and services are also growing fast. The two countries are among the African countries that have been attracting foreign investment. Each has relatively large populations and population growth. Favorable demographic details and trends for these countries coupled with their political stability increase their attractiveness to investors. Additionally, both countries have some of the famous African tourist destinations: the Serengeti, Ngorongoro Crater, Mt Kilimanjaro and Zanzibar in Tanzania; the Masai Mara, Tsavo, Mt Kenya and Mombasa in Kenya. Some of the key demographic and economic facts for the two countries are summarized in Table 6.1.

Table 6.1 Key facts for Tanzania and Kenya

Tanzania	Kenya
Country size: 946,265km²	Country size: 582,650km²
Population: 53.5 million	Population: 46.1 million
GDP growth rate: 7.0%	GDP growth rate: 5.5%
Inflation rate: 5.6%	Inflation rate: 6.0%
Urbanization level: 30.9%	Urbanization level: 26%

Source: Compiled from a range of sources: Ministry of Lands 2009; AfDB et al., 2016; NBS, 2016

As summarized in Table 6.1, Tanzania is larger, more populous and more urbanized than Kenya. Both countries have relatively low inflation rates compared to many African countries. Similarly, both countries have high GDP growth rates, with Tanzania being slightly ahead but Kenya having a higher GDP. In each of the two countries, some significant progress has been made following economic reforms that are ongoing. The reforms have resulted in new land policies being formulated or significantly reviewed to promote and protect property rights. During the reforms, a number of policies and pieces of legislation have been enacted. Reforms have promoted formal land markets, which are attracting a number of private investors, both local and foreign. Although informal land markets are still dominant, efforts are directed towards developing land markets as an instrument to enhance investment and productivity in land, including improving its collateral value for securing credit (EUC et al., 2010).

Land policies and markets in Tanzania

As stated in the National Land Policy of 1995, all land in Tanzania is public land and vested in the president as the trustee on behalf of all citizens. All occupiers of land must do so by way of having a right of occupancy. A right of occupancy may be granted or deemed. Whereas certificates of granted right of occupancy can only be issued on surveyed lands, deemed right of occupancy is mainly for unsurveyed lands. Foreigners in Tanzania can only own land through certificates of granted right of occupancy or rights derivative of granted right of occupancy. Derivative rights are offered to foreign investors by the Tanzania Investment Centre (TIC). Foreigners can also lease land from holders of certificates of right of occupancy (Kongela, 2013). The maximum term of a right of occupancy is 99 years. The law provides for the possibility of term renewal upon expiry. A deemed right of occupancy, on the other hand, has no term limit.

Formal land ownership was introduced during German rule, prior to which land was owned according to different institutional arrangements of different tribes (Kusiluka, 2012). By the Imperial Decree of 1895, the Germans declared all lands in Tanzania (then Tanganyika), whether occupied or not, to be Crown land. Private individuals started owning land formally by way of leaseholds or freeholds. These were mainly foreigners who were engaged in plantation agriculture. After the end of World War I, Tanganyika became a Trust Territory under British Administration. In 1923, the British came up with the Land Ordinance, which provided that all land was public land under the control of the governor, and that no title to the occupation and use of any such lands was valid without the consent of the governor (Shivji, 1998). The legislation further introduced a new system of land ownership known as right of occupancy. Titles and interests created prior to the enactment of the Ordinance were not affected by the provisions of the Ordinance (James, 1971).

When Tanzania attained its independence in 1961 there were four types of interests in land: freeholds, leaseholds, granted right of occupancy, and deemed right of occupancy. In 1962, Government Paper No. 2 was issued to abolish freeholds. In 1969, Government Leaseholds (Conversion to Rights of Occupancy) Act No. 44 was enacted and all leaseholds were converted to right of occupancy. Since then, land ownership in Tanzania has been in the form of granted and deemed right of occupancy. In 1967, through the Arusha Declaration, a major ideological shift towards socialist policies was announced. The Enactment of Acquisition of Buildings Act in 1971 led to the nationalization of private rental properties whose market value was above Shs. 100,000 (then equivalent to US$14,000) or whose rental value was above Shs. 833.3 per month (Meredith, 2006).

Institutional reforms in the land sector in Tanzania have significantly improved tenure security and property rights, which is essential in attracting private investment. Tanzania has a number of land use related policies and pieces of legislation dealing with different aspects of land management and administration. The newly enacted and revised land related policies and legislation have enhanced access to land thereby promoting land based investment. Some of the principal land use and markets related policies and pieces of legislation in Tanzania enacted or amended in the process of reforms include the National Land Policy of 1995, Human Settlements Policy of 2000, Land Act of 1999, the Village Land Act of 1999, and the Urban Planning Act of 2007.

As a result of the reforms, land and property markets have been fast evolving. The private investment environment has improved significantly, thereby attracting a considerable number of private investors. Besides, the burgeoning real estate markets reflect economic growth, rapid population growth, and the growing demand that is created by the expanding middle class. Both public and private investors have been increasing their investments in the sector. In Tanzania, pension funds and the National Housing Corporation (NHC) dominate the market. Real estate accounts for about 21 percent of pension funds' portfolio allocations in Tanzania. High demand for office space and retail space have attracted some commercial banks to provide partial short-term funding for some commercial property projects. Large-scale retail projects are also emerging in all large cities. However, in the absence of long-term financing, private developers still have difficulties raising funds.

Housing is one of the sectors that is also growing very fast. Pension funds and the NHC have attempted to undertake affordable housing projects but not on a scale that is capable of making a significant difference. Lack of housing finance has forced many people to build their homes on an incremental basis using family savings. Building a house may take more than ten years, and in many cases families move into semi-finished houses and continue with the construction while occupying the houses (Alananga et al., 2015; Kongela, 2013). Private developers have mainly focused on upmarket residential units fetching prices of above US$150,000 or renting at above US$500 per month.

Land policies and markets in Kenya

In 1895 Kenya became a protectorate under the British East African Protectorate with the powers to enact policies and laws under the Crown. It is from this time that the indigenous community land ownership and administration system started changing in favor of colonial master systems. The effects of this have been far-reaching and are still observable today. In Kenya this is the period during which settlers amassed large tracts of land. Unlike Tanzania, after attaining its independence in 1963 Kenya kept some of the land policies and laws that were enacted during the colonial period (McAuslan, 2013). By so doing, many native and poor Kenyans were denied access to prime or productive land. For many years before and after independence, demand for the restitution of land in Kenya has been unwavering. Even after independence, no adequate efforts were made to correct the earlier mistakes. A few individuals – settlers and influential citizens – continued to own large tracts of productive land.

After many years of debate, in 2009 Kenya enacted a comprehensive national land policy (Harbeson, 2012). The policy, among other things, sought to redress both historical and contemporary land related injustices and problems. The policy provided for the enactment of several statutes and the establishment of land administration institutions. Some of the laws that were enacted in the course of the implementation of the National Land Policy and the new Constitution of Kenya that was promulgated in 2010 include the Land Act of 2012, the Land Registration Act of 2012, and the National Land Commission Act of 2012.

All land in Kenya is vested in the people of Kenya collectively as a nation, as communities and as individuals (Manji, 2014). Three main categories of land exist in Kenya, namely public land, community land and private land. The National Land Policy of 2009 clearly distinguishes the three categories of land. Public land refers to all land that is not private land or community land and any other land declared to be public land by an Act of the Parliament. Community land refers to land lawfully held, managed, and used by a given community as defined by the Land Act. Private land refers to land lawfully held, managed, and used by an individual or other entity under statutory tenure. The National Land Policy further provides that ownership of land in Kenya will, apart from customary land tenure, be through freeholds and leaseholds. However, freeholds are increasingly being replaced by leaseholds as foreigners are no longer allowed to have freeholds. Foreigners can own land through leaseholds of a term of up to 99 years.

Kenya is one of the few countries in sub-Saharan Africa that have attracted many private investors. It is estimated that more than 100 multinational companies have their offices in Nairobi (JLL, 2016). The country has attracted international equity investors and private real estate funds, which have mainly focused on acquisition of prime properties. Some of the real estate funds investing in Kenya include Actis, Sanlam, Stanlib, and Fusion Capital. Nairobi remains the most advanced real estate market in the Eastern African region and Kenya is the only country that has attained a semi-transparent status (JLL, 2016). Most of the large international real estate consulting firms such as Jones Lang LaSalle, CBRE, Cushman and Wakefield, and Knight Frank have their regional offices in Nairobi.

Other market developments in Tanzania and Kenya

Speculative land prices and urban sprawl

Land speculation is rife in Kenya and Tanzania. In many urban centers land prices are very high, which denies many people access to such lands. For instance, bare land price in prime areas in Dar es Salaam ranges between US$1,000 and US$1,500 per square meter. For secondary residential locations the price per square meter ranges between US$15 and US$30. This is one of the reasons for sprawl in most urban centers in the two countries. Urban sprawl presents a problem in providing basic services such as clean water, sanitation, electricity, paved roads, and security to scattered neighborhoods (Kiunsi, 2013).

Despite having legal provisions discouraging land speculation, the problem is still pervasive. For instance, in Tanzania a holder of an urban land title is required to complete development of the land within 36 months from the date of the commencement of the lease term. The main limitation of this legal requirement are the difficulties that the majority of the people have in raising funds for developing their land. Besides, this condition applies only to land parcels that have titles or letters of offer while in practice most of the land parcels are held without titles.

Development of capital market-based property investments

Broadly, institutional reforms in African countries have seen the development of capital markets institutions. Many countries have stock markets, although most of the stock markets are still at the early stages of development. Nairobi Stock Exchange (NSE) is the largest stock market in East Africa. Until 2015, the real estate sector had not taken advantage of the local capital markets. Both Tanzania and Kenya introduced REITs in 2015. The first REIT in Kenya, namely Stanlib Fahari I-REIT, launched its IPO in November 2015. It was followed by Fusion Capital in June 2016. Stanlib Fahari I-REIT is listed at NSE, which enhances its governance and accessibility to a wider range of investors. Watumishi Housing Company is a private REIT and the only one in Tanzania; the future plan is to go public and get listed at Dar es Salaam Stock Exchange (DSE).

Development of mortgage markets

Property market sector growth is supported by the emergence of primary mortgage market institutions. Kenya's mortgage market is the third most developed in sub-Saharan Africa after South Africa and Namibia (IBRD, 2011). While mortgage lending in Kenya accounts for about 2.5 percent of the GDP, mortgage lending in Tanzania accounts for only 0.46 percent (IBRD, 2011; BoT and TMRC, 2016). Mortgage lending rates in the two countries have varied from time to time but they have mostly reflected treasury bills and bonds rates. The lending rates in Tanzania have ranged from 14 percent to 19 percent reflecting treasury bills rates, whose yields reached their highest level, 17.8 percent, in 2015 (BoT and TMRC, 2016). Low levels of financial literacy for the majority of the people is another reason for low mortgage take up, especially in Tanzania. To address some of the problems, some steps have been taken including establishing a mortgage liquidity facility for primary lenders and increasing public education and awareness on mortgage literacy.

Secondary property markets

Secondary formal property markets in both countries are still developing, but Kenya has a more vibrant market when compared to Tanzania. Both countries have land registries, however they only cover transactions involving surveyed and registered lands. It is noteworthy though that informal property markets are much more active than formal markets. It is thus important to consider them because they play a much more important role as market movers. Tracking of transactions in the informal market is more complicated. However, both countries have a well-established real estate valuation practice. Ardhi University in Tanzania and the University of Nairobi in Kenya have been offering undergraduate programs in real estate valuation since the 1970s. Each country has a specific law catering for valuation practice. Recently, the Royal Institution of Chartered Surveyors (RICS) has opened its regional office in Nairobi. Moreover, Kenya has a law catering for estate agency business; Tanzania is in the process of enacting a similar law. Tanzania is also in the process of setting up an institutional framework that will oversee the entire real estate sector including collecting, processing and publicizing key sector data.

Emergence of satellite towns

The tendency is emerging for some large developers to embark on satellite city development. The satellite town concept seems to be sensible considering the exorbitant land prices within the cities. Besides, many of the city centers have obsolete infrastructure and are characterized by traffic jams. Satellite cities seem to give a solution to some of these problems, as they are located at the outskirts and have adequate infrastructure provisions. Tatu city and Konza city, both in Kenya, are the largest satellite city projects in East Africa. Kawe city is one of the satellite cities being developed in Dar es Salaam. Two satellite cities are planned to be developed on the outskirts of Arusha.

However, satellite towns in these countries are still a new phenomenon. There are no success stories yet from which one could derive optimism. The main criticism for the satellite city concept is that their designs are copycats reflecting Western lifestyles and many of them focus on expensive units, making them exclusive to the rich. Even the pre-sales are apparently based on speculative prices and projections. Some of the satellite city projects are likely to end up as non-performing projects due to their disregard for consumer taste.

China's influence in Tanzania and Kenya

As Western governments and institutional investors from those countries are hesitant to engage in large-scale investment projects in Africa, the government of China and some Chinese companies are heavily engaged in different sectors, including construction and real estate (Rothenberger, 2010). Pre-dating the recent flow of Chinese funds to Africa, East Africa, and especially Tanzania, has for many years enjoyed cooperation with China. This dates back to the 1960s when Tanzania decided to follow socialist policies. The first large project undertaken by China in Tanzania was the construction of a railway connecting Dar es Salaam in Tanzania and Kapri Mposhi in Zambia (TAZARA). The railway, which is 1,860km long, was completed in 1975 (Monson, 2009). Over recent years, China has been involved in many of the large infrastructure projects, including the construction of a natural gas pipeline, upgrading railways to standard gauge, and construction of roads. Some of the future projects in which China is heavily involved include the Mombasa to Nairobi railway upgrading in Kenya and the planned construction of a more than US$10 billion port in Bagamoyo, Tanzania (Knight Frank, 2015).

Recently, Exim Bank of China has agreed to provide a concession loan to the government of Tanzania to finance the construction of a railway that will connect Dar es Salaam and the hinterlands, including Rwanda, Burundi, and the Democratic Republic of Congo. Exim Bank also financed the construction of a 532km gas pipeline from Mtwara to Dar es Salaam, which was completed in 2015; the project is estimated to have cost US$1.2 billion. China also financed and undertook the construction of Tanzania's new football stadium and an international conference center, both located in Dar es Salaam. AVIC International and CJRE Estate are among the Chinese real estate developers in Tanzania.

Summary

Many African countries share a common land policy and property market evolution path. Prior to the colonial period, most of the countries did not have formal property markets and land administration institutions. Besides, very little is documented about land issues during the period. The colonial governments introduced formal systems of land administration, but in most cases customary systems were retained, especially for rural lands. As such, many African countries still maintain a dual land ownership system, consisting of statutory and customary land tenure systems.

Land policy and legal reforms that have been taking place in many countries around the continent in the process of institutional reforms have, among other things, significantly enhanced property rights and thus promoted land and property investment markets. Market transparency is improving, which has attracted international and regional investors and real estate consulting firms. Chinese firms are investing heavily in many countries in Africa. With more reforms of land policies and property markets, African countries are likely to attract many more investors from around the world.

References

AfDB, OECD, UNDP (2016). *African Economic Outlook: Sustainable Cities and Structural Transformation*. Paris: OECD Publishing.

Alananga, S., Lucian, C., and Kusiluka, M. (2015). Significant Cost-Push Factors in Owner-Built Incremental Housing Construction in Tanzania. *Construction Management and Economics*, 33(8), 671–688.

Austin, G. (2010). African Economic Development and Colonial Legacies. *International Development Policy*, Series 1, 11-32.

BoT and TMRC (2016). Tanzania Mortgage Market Update, 31 March 2016. Dar es Salaam: The Bank of Tanzania.

Cotula, L. Toulmin, C., and Quan, J. (2006). *Better Land Access for the Rural Poor. Lessons from Experience and Challenges Ahead*. Rome: IIED/FAO.

Cotula, L., Vermeulen, S., Leonard, R., and Keeley, J. (2009). *Land Grab or Development Opportunity? Agricultural Investment and International Land Deals in Africa*. London/Rome: IIED/FAO/IFAD.

Deininger, K. (2003). Land Policies for Growth and Poverty Reduction. A World Bank Policy Research Report No. 26384. Washington, DC: World Bank.

EUC, AfDB, and ECA (2010). *Land Policy in Africa: Southern African Region Assessment*. Addis Ababa: EUC Publications.

Gaertringen, C. H. von (2014). *Africa is the New Asia: A Continent in the Upswing* [in German]. Hamburg: HoCa-Sachbuch.

GLP (2010). *Land Grab in Africa: Emerging Land System Drivers in a Tele-Connected World*. Copenhagen: GLP International – University of Copenhagen GLP.

Harbeson, J. W. (2012). Land and the Quest for a Democratic State in Kenya: Bringing Citizens Back. *African Studies Review*, 55(1), 15–30.

IBRD (2011). *Developing Kenya's Mortgage Market*. Washington: The World Bank.

James, R. W. (1971). *Land Tenure and Policy in Tanzania*. Dar es Salaam: East Africa Literature Bureau.

JLL (2016). *Real Estate Transparency Improves in Sub-Saharan Africa*. London: Jones Lang LaSalle Inc.

Kiunsi, R. (2013). A Review of Traffic Congestion in Dar es Salaam City from the Physical Planning Perspective. *Journal of Sustainable Development*, 6(2), 94–103.

Knight Frank (2015). *Africa Report 2015–2016*. London: Knight Frank.

Kongela, S. (2013). *Framework and Value Drivers for Real Estate Development in Sub-Saharan Africa: Assessment of the Tanzanian Real Estate Sector in the Context of the Competitiveness Model*. Cologne: Immobilien Manager Verlag.

Kusiluka, M. M. (2012). *Agency Conflicts in Real Estate Investment in Sub-Saharan Africa: Exploration of Selected Investors in Tanzania and the Effectiveness of Institutional Remedies*. Cologne: Immobillien Manager Verlag.

Kusiluka, M. M., Kongela, S., Kusiluka, M. A., Karimuribo, E. D., and Kusiluka, L. J. M. (2011). The Negative Impact of Land Acquisition on the Indigenous Communities' Livelihood and Environment in Tanzania. *Habitat International*, 32(1), 66–73.

Manji, A. (2014). The Politics of Land Reform in Kenya 2012. *African Studies Review*, 57, 115–130.

McAuslan, P. (2013). *Land Law Reform in Eastern Africa: Traditional or Transformative? A Critical Review of 50 Years of Land Law Reform in East Africa, 1961–2011*. New York: Routledge.

Meredith, M. (2006). *The State of Africa: A History of Fifty Years of Independence*. New York: Free Press.

Ministry of Lands (2009). *Sessional Paper No.3 of 2009 on National Land Policy*. Nairobi: Government Printer.

Monson, J. (2009). *Africa's Freedom Railway: How a Chinese Development Project Changed Lives and Livelihoods in Tanzania*. Indianapolis: Indiana University Press.

NBS (2016). *Tanzania in Figures*. Dar es Salaam: National Bureau of Statistics.

Rothenberger, S. (2010). *Foreign Real Estate Investment in Sub-Saharan Africa: A Behavioural Approach in Countervailing the Investment Stigma by Image Analysis and Exploration of the Tanzanian Real Estate Market*. Cologne: Immobilien Manager Verlag.

Shivji, I. G. (1998). *Not yet Democracy: Reforming Land Tenure in Tanzania*. Dar es Salaam: Hakiardhi.

Toulmin, C. (2009). Securing Land and Property Rights in Sub-Saharan Africa: The Role of Local Institutions. *Land Use Policy*, 26(1), 10–19.

7

Residential prices, housing development, and construction costs

Paloma Taltavull de La Paz and Raul Pérez

Abstract

This chapter covers the relationships between residential market and housing development focusing on the role that construction costs play as an incentive for housing prices. Attention is paid to the existing links between construction costs and price changes. A specific database is built from real development projects seeking to analyse several cost groups and selling prices. Two non-linear models (average cost model and new supply model) are defined to explore the cost–price relationship using panel regression techniques. The first one identifies components that have negative effects on average costs, highlighting scale economies in those inputs. The second one estimates individual impacts on selling prices of nine cost groups measured in euros and pooled from real projects' accounting data in a new supply model framework. Results support the existence of scale economies in the cost of pavements and labor, the strong effects of foundation costs on prices, and the existence of sunk costs, that is parts of construction costs which are not reflected in house prices (such as carpentry or painting). It also identifies direct and indirect effects of costs on prices, as in the case of green areas, suggesting that extra-construction costs incurred by creating housing amenities increase housing prices through affecting demand tastes and the housing quality perception.

Introduction

House prices are the key determinant of market equilibrium. What distinguishes the roles of prices in the housing market from those played in other markets is the bundle of influences on house prices that make those prices fluctuate far from the expected path sometimes. The literature has broadly focused on identifying housing price fundamentals and the sources of their dynamics, as a way to explain how housing prices actually evolve. However, the role of construction costs in housing price formation has received less attention from the literature.

It is widely agreed that housing prices depend upon two main demand-driven groups of factors: socioeconomic fundamentals, and financial innovations (Hwang and Quigley, 2006; Himmelberg et al., 2005; Haffner and de Vries, 2009; Mikhed and Zemcik, 2009). The bulk of literature (see Smith et al., 1988 as a good summary; and Di Pasquale and Wheaton, 1992 for an explanation of the model) clearly explains that the increases (decreases) in fundamental variable (demography, income or purchasing power) movements press house prices to grow (or diminish) in the market. Financial flows and capital market conditions facilitate housing investment or increase the household purchase capacity as well (see seminal papers such as those by Meltzer 1974, or Kearl 1979). A summary can also be found in Smith et al. (1988), and for more recent works see Miles and Pillonca (2008), especially as regards homeowner markets directly affecting prices.

In accordance with the economic theory of market principles, changes in prices coming from changes in demand sources constitute the determining factor for supply responses, insofar as the total quantity supplied usually rises when prices exceed the average unit cost, through the establishment of a new equilibrium. It is assumed that the housing market follows the law of supply that is valid for any other good, but, unlike other markets, housing products remain in the market for a long time (Glaeser and Gyourko, 2001), which means that the housing market supply combines both existing and new units to satisfy all the housing needs. The present chapter delves into the area of new unit constructions as the response to market signals on the part of the developers' sector. New units (or housing investment) have been examined by the literature from two perspectives associated with the reaction to market incentives, both of them directly linked to building decisions: as a response to market price changes; and as a reflection of direct investment plans from capital flows into the housing sector.

Rising prices consequently lead developers and landlords to react, either offering an increasingly high number of existing units or developing new ones. In the latter case, it is also agreed that the development process may be affected by several constraints, two of which are clearly identified by the literature: (a) the shortage of land supply; and (b) the delays resulting from the administrative permissions system. The former could appear for multiple reasons, such as lack of land, oligopoly situations in which the land market is controlled by few landlords, or restrictions concerning land regulation, all of which results in higher-than-normal land prices. Land is a key input for development. The lack of available land or the absence of land market flexibility reduces the possibility for developers to build, since these problems lengthen the building period and increase the degree of uncertainty with regard to expectations of builders' profits.

On some occasions, excessively high housing prices (or land prices) discourage house construction because the price level does not match the existing purchase capacity amongst the potential household demand, thus making houses unaffordable. When this happens, the housing demand diminishes, reducing the pressure on prices, and then on new construction, which in turn shows the limits of growth and contributes to the end of the boom-cycle phase (Meen, 2000). The effect of house prices on house development additionally differs depending on time (based on the strength of demand sources) and space (based on local markets) factors. The new development reaction to price changes varies across local markets due to idiosyncratic local features (Mayer and Somerville, 2000b; Malpezzi and Vandell, 2002) where aspects such as land market, administrative regulations, or developer sector structure play a relevant role in the definition of housing construction levels.

From the developers' perspective, the final new unit price has a lot to do with construction costs. This mainly comes as a result of uncertainty in relation to future selling prices when a development is starting, and the need exists to set a selling price based on the (estimated) total construction costs. In fact, what developers take into account is the traditional economic "maximum benefits" rule of thumb, according to which they should start developing new houses when it becomes clear to them that housing prices exceed their average total cost per unit. High risk exists in such situations because prices may change with the housing cycle during that period.

The present chapter focuses on the price formation mechanism in new units. It is structured in four sections. The second explains market equilibrium and summarizes the literature devoted to explain the house price dynamics from several economic perspectives, after which the following section focuses on the role that construction costs play in housing prices and development, providing evidence of the effects that construction costs have on housing prices. The chapter finishes with a summary of the most important conclusions drawn from our research work.

Housing market equilibrium: a view from the new housing supply perspective

The key issue in this work is the extent to which house price influences the housing supply mechanism by inducing new construction. Regarding the supply as a whole (new and existing units), the total available units in the market are considered fixed in the short run (Arnott, 1987) as a result of the special characteristics of the construction process, because increasing new building takes time and requires administrative procedures that delay the development process (Topel and Rosen, 1988; Quigley, 1997; Arnott, 1987). Since the stock cannot increase instantaneously (all of a sudden), any change in the demand component leads to an immediate change in housing rents. Figure 7.1 illustrates this mechanism, which consists in determining the rent in the property market where demand changes affecting rents can be seen by the shift on the demand curve due to changes in the economic conditions that affect demand sources. Rents go from r_1 to r_2 in response to a positive shift in housing demand.

New rent levels induce new construction through the impact on house market prices. The so-called cap-rate relates rent prices to property prices following a "predictive" financial rule – the capitalization rate (net rent/price ratio, r = rent/price; price = rent/r) – i.e. being considered as an indicator of the yield that makes the investor hold the real estate asset.

When rent rises, housing price should also increase to maintain the cap-rate constant. By way of example, when r_1 rise to r_2 (in Figure 7.1), the cap-rate ($Cr_0 = r_1/P_1$) changes ($Cr_1 = r_2/P_1$)

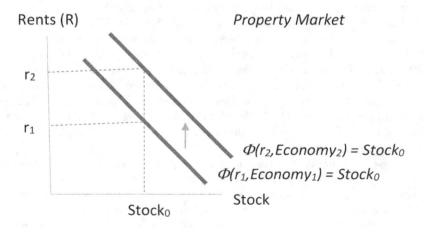

Figure 7.1 Rent determination in property markets

Source: Adapted from Di Pasquale and Wheaton, 1992: 188

making investors react and affecting house prices in the same direction as rent movements bringing the cap-rate to the equilibrium ($Cr_0 = r_2/P_2$). The cap-rate cannot be observed but is influenced by four variables: the economy's long-term interest rate; the expected growth in rents/comparable returns in capital markets; the risks associated with the rental income stream; and the tax treatment (Di Pasquale and Wheaton, 1992).

The new price level affects new development through the real estate market defined in Figure 7.2, where the impulse on starts can be seen as a result of the price incentive or market signal (prices rising from P_1 to P_2 in the curve P). The development response will depend on the new supply elasticity (e), that is, the development industry's sensitivity to price changes in each particular housing market, represented by the new supply curve slope. Such sensitivity reflects the existence of any limitation stemming from inputs such as land scarcity, the permits system, or market power, amongst others. As new construction increases the stock, the total supply rises to meet a larger demand. This process, which extends over several periods to reach the equilibrium, constitutes one of the origins of housing cycles.

The supply curve may change (Figure 7.2, function P★) with the increase of any cost component – such as interest rate, land price, construction materials, or wages – which in turn makes the cost of new units rise and leads to a reduced development level at the current prices. In other words, there will be less construction at a price level P_2. The new cost combination has made the new construction costs rise thus establishing a new equilibrium between costs and prices that inevitably reduces the incentives to build (due to problems such as lower profitability or financial losses).

The key mechanism from the developers' perspective can be found in Figure 7.2, which shows how construction reacts to property price movements that could change due to, amongst other reasons, demand source variation, changes in the financial/asset market, or other investor-related variables (including risk perception), all of which will affect the capitalization rate. Any shock increasing property prices will encourage new supply towards a larger level of

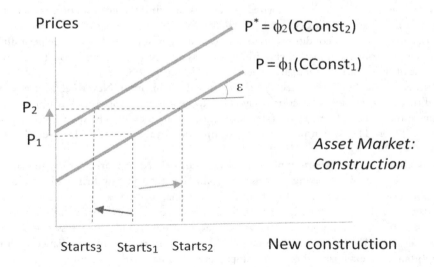

Figure 7.2   Asset market, construction

Source: Adapted from Di Pasquale and Wheaton, 1992: 188

construction (the reverse is also true). In this case, larger new supply flows enter the real market raising the available supply and reducing rents, matching the new house price level (see Di Pasquale and Weaton, 1992 and 1996 for greater detail). Furthermore, when any shock affects construction materials or land prices (either scarcity or prices), the new supply curve moves to fit its cost–price equilibrium. An increase in costs is seen as a negative effect on new supply that causes the curve to shift to the left and reduce the construction level (for a deeper explanation see Rosenthal, 1999; Di Pasquale, 1999; or Mayo and Sheppard, 2001).

The role of construction costs in development

Following the Di Pasquale and Wheaton model (1992), new housing supply reaction firstly depends on prices and construction costs, the response scope of these two factors being closely linked to a number of other institutional and market variables. Several studies have found that construction costs (costs of material and labor), the cost of land and land availability (Goodman, 2005; Malpezzi and Vandel, 2002), along with the cost of finance, are the main determinants of house building within a market-oriented equilibrium framework (Blackley, 1999; Somerville, 1999a; Di Pasquale, 1999; Mayer and Somerville, 2000a; Somerville, 1999b). Taltavull de La Paz and Gabrielli (2015) checked that construction materials constitute a key house-building determinant in Italy, but are insignificant in Spain's new housing supply. Taltavull de La Paz (2014) also found empirical evidence that some markets do not react strongly to changes in prices to explain new development.

Furthermore, the construction literature generally argues that developers determine the selling price for new units basing their calculations on total construction costs despite the effect of other visible exogenous variables. For instance, Akinci and Fischer (1998: 70–72) define two groups of variables that affect final construction costs: the early technical estimate of project costs (structures, foundations, enclosures, façades, as well as others related to climate location and the project as such); and other costs (related to the general economy and risks of a different kind). Akintoye (2000: 77) uses a factor analysis to identify seven areas affecting final costs, while Warsame (2006: 10–16) summarizes the factors determining construction costs as four: project factors (size, complexity, quality); client/developer factors (size of parts, type of contract); competitiveness conditions (market power and development level in the period); and finally, macroeconomic and market conditions (such as interest rates and inflation, to quote just two). Most of those variables actually depend on non-construction-related facts, which points towards the role of the market mechanism in price determination. Nevertheless, it is widely believed that most development costs really come to be known before the development starts (Warsame estimated that 95 percent of development costs are identified before the construction phase starts (2006: 11)), which permits the determination of the unit price for every house to be sold.

The micro-economic theory explains how costs are related to market prices through the widely accepted decision-making process for production theory: Any firm will establish the production level when its marginal cost equals its marginal income as a general rule, or when the market price (P) equals (or exceeds) the total average costs. That is, prices are compared to the cost by unit ($P \geq CTAv$) or to the variable average cost (minimizing losses: $P \geq CVAv$, Krugman et al., 2013: 208) in order to take the decision to produce. In accordance with this, the importance of exploring the relationships between final market (selling) price and cost directly has to do with what the economic theory supports. A need exists to reconcile the approaches to explain how costs affect prices, identifying how prices reflect costs in house development with the micro-economic principles.

Construction costs and new house prices: empirical evidence

This section explores the relationships between construction costs and prices hypothesizing whether or not construction costs determine final house prices. A real-cost-based model is presented to that end using a database specially collected for this purpose. It comes from the accounting of 16 house development projects from which the main construction cost components have been extracted and analyzed.

The data were collected directly from accountancy information at the companies' headquarters. It comes from 16 projects built during the 1998–2012 period by three different construction companies operating in Alicante. The documentation provided included both cost-related and technical information, which permits us to describe these 16 projects from an architectural and economic perspective. The database was built analyzing the whole project, including construction techniques and materials used, thus allowing us to account for total constructed size, soil type and foundation features, façade types, or other technical characteristics. The latter include the execution costs list by type of material controlled by the foremen (building engineer responsible for that process under Spanish regulation) as well as the transaction price (from the notarial document also included in the package).

Some initial data descriptions suggest consistency in the analyzed projects. There are 16 apartment projects, three large ones (with 92 to 116 apartment-units built), seven medium-sized ones (between 14 to 21), and six small ones (between one house and six units). All of them show different amenities but the cost by unit (of one apartment) is consistent across them (Figure 7.3) showing a sign of scale economies as the project becomes larger.

Selling prices also give consistent average values according to the type of house built, ranging from € 200,000 in isolated or very low-density housing, to around € 100,000 in the rest of the cases (Figure 7.4, panel a). When applying the theoretical principle of decision, the

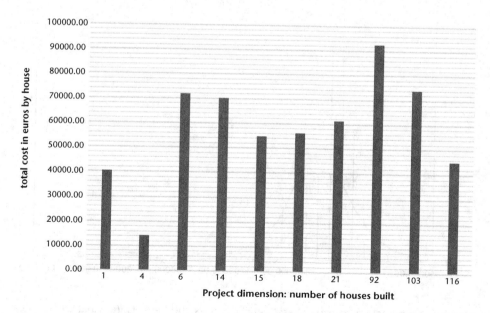

Figure 7.3   Cost per house in euros

Source: Author

a) Final selling prices

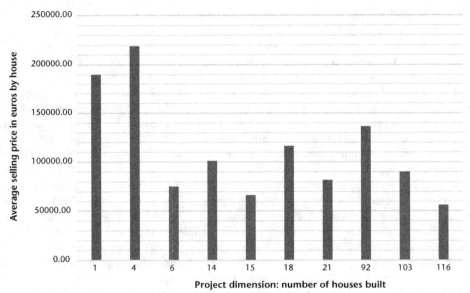

b) Price–average costs difference

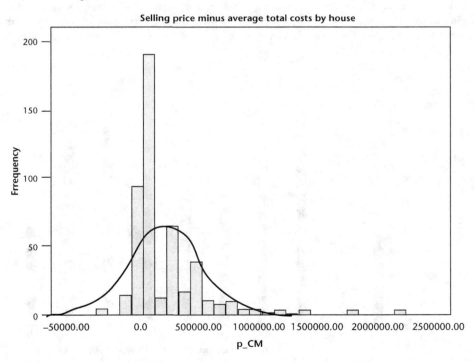

Figure 7.4 Construction costs database: detail of prices and benefits. (a) Final selling price, (b) price–average costs difference

Source: Author

differences between price and average costs (not including land costs) give normal benefits
($P \geq AvCt$; $P - AvCt \geq 0$), the normally distributed benefits being consistent with the general
theory.

Two relationships were tested in order to explore the cost component linkages: first, the
sensitivity of average costs to each different cost group; and second, the price sensitivity to cost
groups through a new housing supply model framework.

The specific groups of construction costs were pooled into nine sets of those sharing
common characteristics (from C1 to C9) and are those used in the empirical exercise described
below.

Average cost sensitivity to construction-cost groups

A log–log cost model relates the average cost per unit to the groups of costs explained above,
finding consistent evidence of the role played by each set of inputs used in project construction.
The model estimated can be seen in Equation 7.1.

$$AvC = \alpha + \Gamma \star C + \Omega \star K + \varepsilon \qquad (7.1)$$

Where AvC is the average cost per house/apartment, C is a matrix of costs including the nine
sets of aggregated costs, K a matrix of four control variables which captures basic project
features affecting cost (Akinci and Fischer, 1998: 70–72), such as size, existence of green
common areas, type of basement built, and number of rooms. Γ and Ω are matrices of the
parameters to be estimated and ε is the error term. Since costs could present non-linear
relationships, the role played by each set of costs affecting the average total cost per house is
estimated here by means of a log–log model specification, as shown in Equation 7.2.

$$lAvC = \alpha + \sum_{i=1}^{9} \beta_i \, lc_i + \sum_{j=1}^{4} \gamma_j \, lk_j + \mu \qquad (7.2)$$

Where lAvC is the log of AvC, lc_i is the log of cost 'i', lk_j is the log of the 'j' control variable
and α, β and γ are the parameters to be estimated, with μ as the error term in the equation.

In this model, the non-linear relationships between costs approach their cross-contribution
to average cost changes through the estimated parameters that may be interpreted as elasticities
showing the responses (in %) of average costs to a change in 1% of every cost set. Table 7.1
provides a two-panel result: panel one is for a pure cost-model (showing the estimated betas in
Equation 7.2); and panel two contains results of the cost model controlled by project-specific
features, mainly defined as the project size affecting the total cost in some accounting items
(showing both beta and gamma parameters also represented in Equation 7.2).

Results in both models consistently show how changes in cost groups are transmitted to
average prices. For instance, groups 1, 2, 3, 5, and 7 cost increases have the effect of reducing
average costs, showing a kind of scale economies in those inputs. Most of those groups refer to
common production processes. For instance, a 1% increase in the cost of installations results in
a 0.57% decrease in the average cost per house. Furthermore, cost-set numbers 4, 6, 8 (mainly
manufactured goods to be installed, labor, and painting), and 9 (other expenses, project fees, and
taxes) positively affect the total average cost showing the extent to which it increases.
Regarding full model results, an increase in taxes in group 9 elastically affects the average cost
in such a way that a 1% increase in this item makes the average income rise by 1.23%. The
remaining positive parameters are inelastic – therefore an increase of labor costs by 1% results
in an 0.1% increase of the average cost per house.

Table 7.1 Average cost and construction cost blocks: Sensitivity to changes in project characteristics

Dependent variable: log of average costs	β	Standard error	β	Standard error
	Panel 1		Panel 2	
Constant	5.838***	0.296	4.673***	0.258
LogC1_Earth movements and foundations	−0.283***	0.041	−0.295***	0.066
LogC2_Installations	−1.486***	0.138	−0.571***	0.076
LogC3_Pavements and tiling	−0.694***	0.068	−0.089***	0.033
LogC4_Sanitation, kitchen furniture and lift	1.987***	0.103	0.565***	0.104
LogC5_Construction materials	−0.429***	0.062	−0.254***	0.064
LogC6_Labor	1.177***	0.094	0.102**	0.052
LogC7_Carpentry (wood and others)	−1.609***	0.067	−0.229***	0.043
LogC8_Painting and false ceilings	0.842***	0.113	0.082*	0.049
LogC9_Other expenses, tax, project fees	0.913***	0.131	1.236***	0.117
			γ	
Existence of green common areas			0.452***	0.074
With basement wall and reticular wrought			−0.090**	0.036
Project size (m²*1000)			−0.051***	0.000
Number of pieces by project			0.000	0.000
R^2	0.828		0.972	
Adj R^2	0.825		0.971	
Equation standard error	0.129		0.052	
F	247.84***		1231.336***	
DW	0.421		1.558	

Notes: *** p-value < 0.01, ** p-value < 0.05, *p-value < 0.1

Source: Author

Results in Table 7.1 also indicate the effect of having common green areas, which increases the average cost per house by 0.45%. As for negative parameters, they quantify the marginal reduction of unitary cost as the project grows.

New housing supply model

This section estimates the new housing supply model as presented in Figure 7.2, using information from the construction cost database. The housing prices registered in the database are transaction based and collected from the notarial document of transmission included in the accounting documents corresponding to each project. Seeking to control for heterogeneous houses, as the database is supported on micro-data observations, price is quality controlled by means of a hedonic specification in the supply equation. The new supply model definition reads as follows:

$$H_i = \alpha + \Gamma \star C + \beta Ph_i + \Phi \star X_i + \nu \tag{7.3}$$

Where H is the number of start units (new supply), C is a matrix of the nine sets of construction costs that have labor costs as a separate variable, Ph is the selling price of each house and X is a matrix with a set of {x} housing characteristics for control purposes in the model. Fixed effects by each of the three cities are considered including dummy variables. Cost components will be disaggregated into the nine cost sets used in the cost model in order to capture the separate contribution on prices of each group (following Somerville, 1999b; and Coulson, 1999), while Cost_09 includes the financial costs paid in the project. Hence our decision to exclude interest rates from the model (as Cost_09 already includes those costs).

Along the lines of hedonic model methodology traditions, the specification is non-linear with a semi-log functional form in which the continuous variables are expressed by means of natural logarithms while categorical variables appear in levels. The general expression of Equation 7.3 can be found in Equation 7.4. a, G, b,F are the parameters to be estimated and υ is an error term.

$$lH_i = \alpha + \sum_{i=1}^{8} \beta_i lC_i + \beta_9 lC_{lab} + \beta_{10} lPh_i + \sum_{j=1}^{8} \rho_j lx_j + \upsilon_i \tag{7.4}$$

Rearranging the terms:

$$lPh_i = \frac{lH_i - [\alpha + \sum_{i=1}^{8} \beta_i lC_i + \beta_9 lC_{lab} + \sum_{j=1}^{8} \rho_j lx_j]}{\beta_{10}} + \upsilon_i \tag{7.5}$$

$$lPh_i = \vartheta + \sum_{i=1}^{8} \gamma_i lC_i + \gamma_9 lC_{lab} + \psi_i lH_i + \sum_{j=1}^{8} \varphi_j lx_j + \varepsilon_i \tag{7.6}$$

Where $\gamma_i = b_i/b_{10}$, $\psi_1 = 1/b_{10}$, $\varphi_j = r_j/b_{10}$; expressed differently, the parameters are inverse elasticities amongst characteristics, costs and prices. Since one of the variables in 'x' is house size in m², Equation 7.6 could become Equation 7.7:

$$lPh_i - lm_i^2 = lPh/m_i^2 = \vartheta + \sum_{i=1}^{8} \gamma_i lC_i + \gamma_9 lC_{lab} + \psi_i lH_i + \sum_{j=1}^{8} \varphi_j lx_j + \varepsilon_i \tag{7.7}$$

Equation 7.7 is estimated using panel-OLS in three steps. First, the hedonic house price model per m² is calculated on the basis of housing attributes (x). Second, residuals are calculated from the former to represent quality-controlled prices (cqP), and to serve as a dependent variable for the housing new supply model, defined in this step as an inverse new supply function cqP = f (cost, new supply), i.e. approaching the cost role in the unobserved price components. And third, the full model as Equation 7.7 is estimated.

It is assumed here that parameter j for log m² is close to the unity. Any value $\neq 1$ refers to non-linearity between housing price and size and is partially captured by the price per m².

Table 7.2 provides the results of all three models in panels. Panel 1 shows the parameters and test results in the hedonic model. The houses observed in this sample are main residence apartments where aspects such as height, having three or more bedrooms, having more than four external rooms, being closer to the city center, proximity to Elche, having common green areas, being a non-public house, forming part of lower density buildings (constructions), with two or more bathrooms or toilets and being new, are highly valued. The model explains 95% of price-per-m² variability according to such attributes. Panel 2 explains the reduced inverse supply equation based on residuals from the hedonic model, that is, quality-controlled prices. Only a small number of cost-set components are statistically significant to explain the unobserved component of price per m² in this model. Some components can affect prices negatively and positively. Those with negative parameters are cost of foundations (C1) and labor costs (C6) (at 5%) suggesting scale economies with size; construction materials (C5 at 5%) and other expenses (C9) show positive parameters, which suggests that any increase in these

Table 7.2 Semilog Inverse New Supply Model

Dependent variable	Log of real house price/m²		Unstandardized residual		Log of real house price/m²	
	β	St. error	β	St. error	β	St. error
	Panel 1: Hedonic regression		Panel 2: Hedonic residual		Panel 3: Full new supply model	
Constant	4.835***	0.192	−0.46***	0.17	−0.70	0.77
D_ South-East orientation (best)	0.019	0.011			0.01	0.01
D_ 1 bedrooms (house)	−0.075	0.044			0.03	0.04
D_ 2 bedrooms (house)	0.034	0.028			0.06**	0.02
D_ 3 bedrooms (house)	0.040***	0.015			0.07***	0.01
D_ 2 external rooms	−0.252***	0.041			−0.16***	0.04
D_ 3 external rooms	−0.049***	0.013			−0.06***	0.01
D_ 4 external rooms	0.093***	0.011			0.09***	0.01
Log Distance to Town Hall (m)	−0.159***	0.014			−0.38***	0.03
Torrellano (Excluding Elche)	0.141**	0.057			−0.42***	0.14
Alicante (Excluding Elche)	−0.318***	0.039			−1.50***	0.16
D_ With urbanization common green area	0.309***	0.040			1.54***	0.22
D_ Free house/(public house)	0.159***	0.024			0.01	0.04
D_ 4 Floors (including basement)	−0.428***	0.045			−0.46***	0.09
D_ 5 Floors (including basement)	−0.379***	0.035			−0.44***	0.05
D_ 6 Floors (including basement)	−0.311***	0.040			−0.17***	0.06
D_ 7 Floors (including basement)	−0.046**	0.021			−0.18***	0.04
D_ 1 bathrooms in the house	−0.095***	0.030			−0.13***	0.03
D_ > 2 bathrooms in the house	0.002	0.020			0.00	0.02
Log Years since built	−0.748***	0.058			0.41**	0.18
LogC1_Earth Movements and Foundations			−0.05**	0.02	0.36***	0.06
LogC2_Installations			−0.11	0.08	0.47**	0.21
LogC3_Pavements and tiling			−0.06	0.04	−0.51***	0.09
LogC4_Sanitation, kitchen furniture and lift			−0.02	0.06	Exclud	
LogC5_Construction materials			0.07**	0.03	0.81***	0.10
LogC6_Labour			−0.11**	0.05	−0.43***	0.06
LogC7_Carpentry (wood and others)			0.05	0.05	Exclud	
LogC8_Painting and false ceilings			0.08	0.06	Exclud	
LogC9_Other expenses, tax, project fees			0.19***	0.07	−0.19	0.23
Log New houses			−0.05**	0.02	−0.45***	0.05
R²	0.954		0.06		0.97	
Adj R²	0.952		0.04		0.97	
Equation standard error	0.066		0.06		0.05	
F	494.53***		2.97***		526.10***	
DW	1.450		1.54		1.79	

Notes: D_ = Dummy variable
       *** p-value < 0.01, ** p-value < 0.05, *p-value < 0.1

Source: Elaborated by the authors

cost items contributes to raise differences in prices. Panel 3 presents the full model. Results match with the previous two in signs and some additional parameters become significant this time. For instance, the most significant differences are the elastic responses of existing green areas in the house (parameter shift from 0.309 to 1.54 in the presence of costs) and the elasticity of Alicante location, which also becomes highly elastic (from -0.318 to -1.50), suggesting that price/m² dramatically diminishes for houses located within the Alicante city limits, and when costs are taken into account.

Some of the cost-sets become statistically significant in the full model. For instance, foundation and installation costs (C1 and C2) increase their influence on final price per m² in an inelastic way (a 1% increase in foundation costs leads to an increase by 0.36% in the final house price per m²). Construction materials also show a similar effect (an 1% increase results in a 0.81% growth of prices/m²). Two construction cost components strongly support the existence of scale economies as the project becomes larger, namely: costs of pavements and labor (the higher the labor costs increase, is associated to the lower the house price per m² increase), consistently with other model results in the literature. Finally, four cost components are excluded from the model or regarded as non-significant to explain changes in house prices.

The results above suggest that the influence of construction costs on final prices not only directly but also indirectly affects housing prices. For instance, it can be said that the better the amenities, the higher the price, because the demand values amenities not because of construction costs but through the price increase due to the availability of amenities as such.

The aforementioned results additionally suggest that not all construction cost components affect house prices. Some costs are clearly transmitted to an increase in house prices, such as foundation costs, installations or construction materials; instead, others transmit their influence through final quality, attributes or amenities in the house, like green area; in other words, indirectly. This is almost the same as accepting that part of the construction costs are sunk ones (they remain unseen) which can hardly be perceived by the demand, and therefore do not affect the final price at all.

In summary, some cost components do directly affect house prices, whereas others do not; and some show the existence of scale economies according to size. Those findings suggest the need for careful costs analyses in separate groups when a developer evaluates the effects that incurring construction costs has on housing prices. After all, the relationships between costs and final housing prices are far from clear and direct.

Conclusion

The literature has broadly focused on identifying housing price fundamentals. Consensus exists on the fact that housing prices depend on two main demand-driven groups of factors: socio-economic fundamentals, and financial innovations. Changes in prices arise as the main determinant(s) of supply responses, insofar as the total quantity supplied usually rises when prices go up. The literature justifies that development reacts heterogeneously to price changes across regions, showing different responses reflecting regional idiosyncratic and behavioral characteristics. This chapter explains the economic rationality as well as the market mechanisms behind developers' decisions to start housing projects, and it equally provides empirical evidence about the relationship between various groups of construction costs and housing prices.

The existing literature supports the idea of prices and construction costs (drawing a distinction between those associated or related to materials, labor, and finance) directly affecting the decisions to build. This chapter explores price sensitivity to different construction costs using a new database especially built for this purpose. The database extracts the actual costs, pooling them in consistent production phases included in the whole development as cost components in a new-supply model. Results demonstrate that building processes show the existence of sunk costs, that is, parts of construction costs that are not reflected in house prices (e.g. carpentry or painting). Other costs have a (direct) influence on prices, in such a way that prices rise when those costs increase (installation costs, for instance). Finally, a number of relevant cost-accounting items show clear scale economies as project size grows (pavement and labor costs, amongst others). By way of conclusion, it can be stated that construction costs affect housing prices in a heterogeneous way, which suggests that further research needs to be undertaken so that the precise price effect of every group of costs can be estimated.

References

Akinci, B. and Fischer, M. (1998). Factors affecting contractors' risk of cost overburden. *Journal of Management in Engineering, 14*(1), 67–76.

Akintoye, A. (2000). Analysis of factors influencing project cost estimating practice. *Construction Management and Economics, 18*(1), 77–89.

Arnott, R. (1987). Economic theory and housing. *Handbook of Regional and Urban Economics, 2*, 959–988.

Bramley, G. (2003). Planning regulation and housing supply in a market system. In Tony O'Sullivan and Kenneth Gibb (eds), *Housing Economics and Public Policy*, RICS Foundation, Oxford: Blackwell Publishing, pp. 193–217.

Coulson, N. E. (1999). Housing inventory and completion. *The Journal of Real Estate Finance and Economics, 18*(1), 89–105.

Di Pasquale, D. (1999). Why don't we know more about housing supply? *The Journal of Real Estate Finance and Economics, 18*(1), 9–23.

Di Pasquale, D. and Wheaton, W. C. (1992). The markets for real estate assets and space: A conceptual framework. *Real Estate Economics, 20*(2), 181–198.

Di Pasquale, D., and Wheaton, W. C. (1996). *Urban Economics and Real Estate Markets*. Englewood Cliffs, NJ: Prentice Hall.

Glaeser, E. L. and Gyourko, J. (2001), Urban decline and durable housing, NBER working paper no. 8598. Available at www.nber.org/papers/w8598.pdf (accessed 01/10/2008).

Goodman, A. C. (2005). The other side of eight mile: Suburban population and housing supply. *Real Estate Economics, 33*(3), 539.

Haffner, M. and de Vries, P. (2009) Dutch house price fundamentals, Australian Tax Research Foundation Housing and Taxation Symposium, Melbourne, February 11. Available at http://repository.tudelf.nl/view/ir/uuid:cc125196-993-4d26-a5bb-ee74b171b5b4/ (accessed 02/01/2016).

Himmelberg, C., Mayer, C., and Sinai, T. (2005), Assessing high house prices: Bubbles, fundamentals, and misperceptions, Federal Reserve Bank of New York Staff Reports, no. 218, September.

Hwang, M. and Quigley, J. M. (2006), Economic fundamentals in local housing markets: Evidence from US metropolitan regions. *Journal of Regional Science, 46*(3), 425–453.

Kearl, J. R. (1979). Inflation, mortgage, and housing. *The Journal of Political Economy, 87*(5), 1115–1138.

Krugman, P., Wells, R., and Graddy, K. (2013). *Fundamentos de Economia*. Barcelona: Reverté.

Malpezzi, S. and Vandell, K. (2002). Does the low-income housing tax credit increase the supply of housing? *Journal of Housing Economics, 11*(4), 360–380.

Mayer, C. J. and Somerville, C. T. (2000a). Residential construction: Using the urban growth model to estimate housing supply. *Journal of Urban Economics, 48*(1), 85–109.

Mayer, C. J. and Somerville, C. T. (2000b). Land use regulation and new construction. *Regional Science and Urban Economics, 30*(6), 639–662.

Mayo, S., and Sheppard, S. (2001). Housing supply and the effects of stochastic development control. *Journal of Housing Economics, 10*(2), 109–128.

Meen, G. (2000). Housing cycles and efficiency. *Scottish Journal of Political Economy, 47*(2), 114–140.

Meltzer, A. H. (1974). Credit availability and economic decisions: Some evidence from the mortgage and housing markets. *The Journal of Finance, 29*(3), 763–777.

Mikhed, V. and Zemcík, P (2009). Do house prices reflect fundamentals? Aggregate and panel data evidence. *Journal of Housing Economics, 18*(2), 140–149.

Miles, D. and Pillonca, V. (2008). Financial innovation and European housing and mortgage markets. *Oxford Review of Economic Policy, 24*(1), 145–175.

Quigley, J. (1997). *The Economics of Housing.* Cheltenham: Edward Elgar.

Rosenthal, S. S. (1999). Residential buildings and the cost of construction: New evidence on the efficiency of the housing market. *Review of Economics and Statistics, 81*(2), 288–302.

Smith, L. B., Rosen, K. T., and Fallis, G. (1988). Recent developments in economic models of housing markets. *Journal of Economic Literature, 26*(1), 29–64.

Somerville, C. T. (1999a). Residential construction costs and the supply of new housing: Endogeneity and bias in construction cost indexes. *The Journal of Real Estate Finance and Economics, 18*(1), 43–62.

Somerville, C. T. (1999b). The industrial organization of housing supply: Market activity, land supply and the size of homebuilder firms. *Real Estate Economics, 27*(4), 669–694.

Taltavull de La Paz, P. (2014). New housing supply and price reactions: Evidence from Spanish markets. *Journal of European Real Estate Research, 7*(1), 4–28.

Taltavull de La Paz, P. and Gabrielli, L. (2015). Housing supply and price reactions: A comparison approach to Spanish and Italian markets. *Housing Studies, 30*(7), 1036–1063.

Topel, R. and Rosen, S. (1988). Housing investment in the United States. *The Journal of Political Economy, 96*(4), 718–740.

Warsame, A. (2006). Supplier structure and housing construction costs. Thesis. Division of Building and Real Estate Economics, Royal Institute of Technology, Stockholm. Available at www.kth.se/polopoly_fs/1.173389!/Menu/general/column-content/attachment/73_thesis.pdf (accessed on 04/02/2015).

8

Retail market analysis for development projects

Coping with new frontiers in retailing

Petros Sivitanides

Abstract

The successful development of a retail scheme in today's competitive and ever-changing retail marketplace presents a formidable challenge to real estate entrepreneurs. Market analysis is one of the most crucial stages in designing a retail development scheme that will successfully cope with the new challenges of today's retail marketplace. Within this context, this chapter discusses first the new frontiers in retailing as well as their implications in terms of planning new retail development projects and assessing their market potential. It then proceeds with the discussion of the four basic stages of a typical market analysis for a retail development project, while pointing out in each stage the relevance of the new trends in retailing. The first stage focuses on the different methodologies for defining the project's trade area, while the second stage focuses on the analysis of competing retail developments, which needs to take into account features that make them more attractive in the new retailing environment. In the third and fourth stages the focus shifts on consumer shopping patterns and the estimation of the project's sales potential and supportable square footage, with special emphasis on assessing as accurately as possible potential in-store sales as opposed to online sales.

Introduction

Over the last decade, the retailing landscape has been undergoing significant structural changes as a result of the rapidly growing use of the internet and mobile devices by consumers and businesses alike around the globe. These structural changes include the steadily and rapidly rising share of retail sales captured online; the emergence and implementation of multi-channel strategies that integrate e-commerce with physical store space; and the changing role of the physical store to better complement the online shopping experience (Piotrowitz and Cuthbertson, 2014; Platt Retail Institute, 2014). In addition to these developments, the consumer base has been undergoing significant structural changes in its composition and characteristics that have direct implications on shopping demands and preferences (Platt Retail Institute, 2014). These new frontiers in retailing have important implications in terms of optimizing the design and assessing the potential of a retail development project in today's environment. Thus, when carrying out market studies for retail development projects the implications of these trends need to be appropriately taken into account when quantifying demand for the specific project and assessing the competition.

Data from the USA confirms the rapid growth of online sales over the period 2000–2014. Online sales in the United States grew from about $5 billion in 2000 (representing 0.9 percent of total retail sales) to $299 billion in 2014 representing 6.4 percent of total retail sales. No arguments have been presented in the literature as to why this consistent and rapid growth of e-commerce's share should cease. On the contrary, the continued rise of this share seems unavoidable due to the rising percentage of the population that gains daily access to the internet worldwide; the vastly superior capabilities that the internet provides in terms of price comparison and product assortment; and the evolving e-commerce technologies that continue to upgrade and enrich the internet shopping experience.

As e-commerce continues to grow, many retailers are led in the adoption of multi-channel strategies integrating management of customers, pricing, and retail mix across online and offline sales channels (Verhoef, Kannan, and Inman, 2015). The need for such an integration is reflected in the rising popularity of click and collect. Click and collect allows consumers to take full advantage of the superior capabilities provided by the internet in terms of price and product search, and at the same time gain possession of the product quickly by picking it up from their closest store (Piotrowitz and Cuthbertson, 2014).

The seamless integration of online with in-store shopping is changing the role of the physical store, not only as the point at which consumers can have a three-dimensional interaction with the product, or pick it up after ordering it on the internet, but also as the point that integrates all channels of interface with the customer. According to Piotrowitz and Cuthbertson (2014) the new role of physical stores will also be characterized by increasing in-store technologies, such as interactive screens and augmented reality, as well as appropriate store layout design to organically incorporate them in the store functioning and customer shopping experience.

Another important dimension of the new frontiers in retailing is the changing structure of the consumer base. This new structure is characterized by a basic dichotomy in the consumer makeup, consisting of the aging baby-boomers with high buying power, and a new internet-addicted generation, born between the early 1980s and early 2000s and referred to as "Generation Y" or the "Millennials," with considerably smaller buying power (Platt Retail Institute, 2014). The other major characteristic of the changing consumer base is the increasing multiculturalism and diversity, manifested in an increasing number of diverse consumer groups seeking specialization, customization and niche interest products and services (Platt Retail Institute, 2014).

Implications of new frontiers for retail market analysis and shopping center design

The new frontiers in retailing have important implications in terms of planning new retail developments and assessing their viability. First of all, the increasing market share of internet retailing at the expense of bricks-and-mortar retailing, points to the risk of diminishing demand for store space in retail developments (Henderson Research, 2013). According to Platt Retail Institute (2014) there have been estimates that by 2025 40–70 percent of retail transactions might take place online. On the other hand, Damesick (2001) and Hortacsu and Syverson (2015) suggest that the rise of the internet shopping may not necessarily result in the demise of the physical store, but in a "bricks-and-clicks" hybrid retailing format. Along the same lines, AT

Kearney (2014) suggests that the two channels may end up complementing each other with e-commerce "specializing in product search and discovery" and physical stores in product testing, purchase, and return. Although there are varying opinions in terms of the size of the negative effect of the internet on demand for physical store space, this debate underscores the importance of a thorough market analysis for retail development projects, and the accurate assessment of the online and in-store shopping patterns of the consumer groups within the project's trade area.

The rise of internet retailing has important implications in terms of designing an appealing and attractive retail development concept and tenant mix. In particular, the fierce competition from the internet points to the need for transforming the shopping center from mostly a shopping destination to a destination of "meaningful lifestyle experiences – fun family destinations where all age groups can spend quality free time together" (Evans, 2014, p.37). Such transformation requires a considerable increase of the non-retail tenants and specifically food and beverages, leisure, entertainment, and services in order to provide a pleasant three-dimensional experience that cannot be replicated in the two-dimensional electronic space. According to Evans (2014), the optimal level of the leisure component is 25 percent of the tenant mix and he predicts that leading shopping centers will increase their leisure component to such levels in the short term.

Another implication of the recent trends in retailing in terms of tenant mix is the importance for new retail developments to include technology-savvy tenants that have embraced the new technologies and offer a seamless integration of physical and electronic space. The inclusion of such tenants will reduce the longer-term risk of the development, as they are more likely to withstand the challenges of the new environment and provide the shopping center owner with a more stable cash flow in the long term. Furthermore, the development will be more attractive as the consumers will be able to enjoy an integrated online and in-store shopping experience.

The new demographic frontiers also have important market analysis implications as they call for in-depth consumer research, so as to accurately identify the diverse consumer groups that compose the project's clientele and thoroughly map their shopping habits, both online and offline, as well as their leisure, entertainment, and other preferences. Such an analysis will provide a solid basis not only for designing an appealing tenant mix so as to best serve the project's clientele, and strengthen its "destination" character, but also for accurately assessing the supportable square footage for the different lines of trade that will be included in the project. Within this context, the analysis of consumer profiles and their shopping patterns in the development's trade area is of crucial importance in evaluating and refining the project concept so as to successfully cope with the new frontiers in retailing.

Market analysis for retail development projects

The market analysis process for evaluating a retail development concept focuses on the analysis of demand for the goods and services that are expected to be traded at the project and the analysis of the competition. The new trends in retailing discussed earlier are relevant in the analysis of both the demand and the supply side of the market. In order to properly analyze these two aspects for a particular development, the project's trade area needs to be defined first. Thus, the market analysis process includes four basic stages for estimating the project's supportable square footage in different product lines. The first stage includes the preliminary definition of the project's trade area and the second the refinement of its boundaries through the detailed analysis of the competition. The third and fourth stages include the analysis of the shopping patterns of the consumers residing within the trade area and the quantification of the project's potential sales and supportable square footage in the different product lines expected to be traded at the center. These four stages along with the relevant analytical implications of new frontiers in retailing, wherever applicable, are discussed below.

Stage 1: Preliminary definition of the boundaries of the development's trade area

A retail project's market or trade area is the area encompassing all locations from which the majority of its clientele is drawn. For this reason, it is important to define the geographical boundaries of the project's trade area before proceeding to the analysis of consumer demand and the competition. At this stage the new trends in retailing are not relevant, as a typical methodology for the preliminary definition of a project's trade area is based on reasonable maximum driving times that consumers are willing to drive depending on the typology of the retail cluster under consideration. In estimating driving times to the site under consideration the analyst needs to take into account road access, speeds, and the configuration of the area's transportation network, as well as public transportation routes. The driving time norms typically used in the industry for the different types of retail clusters are 15 minutes for a neighborhood center, 30 minutes for a community center, 45 minutes for a regional center, and 60 minutes for a super-regional center (Schmitz and Brett, 2001).

A more simplistic approach for defining the project's trade area is the concentric ring approach, which defines the boundaries of the development's primary trade area as a circle with a radius that varies depending on the type of the development considered. According to Schmitz and Brett (2001), the industry norm with respect to the length of this radius, measured in straight distances, is one and a half miles for a neighborhood shopping center, three to five miles for a community center and eight to 12 miles for a regional mall. The major pitfall of this approach though is that it ignores the uneven accessibility patterns as determined by the area's topographical characteristics and the existing transportation and public transit network.

A more complicated approach for a preliminary definition of the project's trade area is the "near twin" or "analog" approach, according to which the geographic capture of existing similar retail developments is used to guide the determination of the geographic capture of the planned development (Fanning, Grissom, and Pearson, 1994). "Near twin" developments are selected on the basis of their similarity to the planned development in terms of several factors, including development type and size, tenant mix, layout, and physical design, surrounding streets, location with respect to downtown and other competing retail clusters, access and distance to sources of demand, and volume of traffic on adjacent streets for stop-off business. Once the "near twin" development is identified, information is collected regarding consumer shopping patterns, such as frequency of visits by type of consumer and by location, in order to provide the basis for identifying the spatial extent of the project's area of influence and the delineation of the development's primary, secondary, and tertiary trade areas. This approach is expensive to implement though and may not be feasible or lead to inaccurate definition of the project's trade area and its potential clientele, if true "near twin" developments cannot be found.

Once the boundaries of the trade area have been defined through any of the above approaches, they can be adjusted to account for the effect of competing developments using the simple gravity approach introduced by Reilly (1929). This approach can be used to estimate how far the planned project's attraction extends, given its characteristics and distance from a specific competitor and the characteristics of the competitor. According to the modified gravity model presented by Converse (1949), the distance to the boundary of the trade area of center j (TAB$_j$) with respect to competing center k, depends on the distance between the two centers (d) and the relative attractiveness of the two centers, measured typically by their size (S_j, S_k) or other attributes (see Figure 8.1).

$$TAB_J = \frac{d}{1 + \sqrt{\left(\dfrac{S_k}{S_j}\right)}}$$

Figure 8.1   Estimating the trade area boundary with a competing center

Source: Converse (1949)

This technique can better account for multiple project characteristics in addition to size, if S_k and S_j are replaced by indices that are based on a comparative scoring system of the different characteristics (including size) of the planned project and its competitors. For example, the Total Amenity Indices that can be developed using the Competitive Differentials Technique (which is discussed in a subsequent section) can be used instead of S in order to better reflect the attractiveness of center *j* vis-à-vis center *k*.

The boundaries of the trade area should also be adjusted taking into account any socio-economic and cultural/ethnic barriers that may have an effect on visitation rates from particular population groups. For example, Hardin and Carr (2006) have found that higher income households may not patronize community centers in areas with high concentrations of households on public assistance even if they are the closest ones to their location. A final adjustment will be needed in order to align the boundaries of the development's trade area with the geographic boundaries of area definitions for which secondary population and income data is available, such as counties, census tracts, zip codes, or blocks. For example, if a census tract is used, Carn et al. (1988) suggest that it can be included in the trade area if more than 50% of its geographic area is contained within the line representing the maximum driving time from which customers can be drawn to the development under consideration. Carn et al. (1988) suggest that blocks may be used for neighborhood shopping centers, census tracts or zip codes for community shopping centers, and counties for regional shopping centers.

The most advanced and accurate approach in determining a retail development's trade area is the advanced probabilistic approach, which estimates econometrically the project's capture rates from the homogeneous residential zones that are within reasonable driving distances. These capture rates can then be used for more accurately defining the development's trade area by including all the residential zones with a minimum capture rate. However, its application requires a preliminary definition of the development's trade area using the techniques just discussed. This approach is discussed further in the section focusing on the estimation of the planned development's sales potential.

Stage 2: Analysis of the site/project and its competition

The analysis of the competition is very important in further refining the trade area boundaries using the gravity model discussed earlier. Furthermore, the analysis of all competing projects is crucial in accurately assessing the sales potential of the project under consideration in today's highly competitive environment. The new frontiers in retailing are quite relevant in analyzing the strength of the competition in terms of four aspects. These new-frontier aspects include the availability of the necessary technological infrastructure so as to allow wireless internet access at any point of the center; the extent to which the tenant mix includes strong entertainment, leisure, and service components that reinforce the center's destination character;

the variety of product assortment and the extent to which it appeals to a multicultural consumer base; and the resilience of the specific tenants of the center to internet penetration, which will depend on the types of goods they sell and the degree of integration of the physical and electronic shopping experience. Given the new trends in retailing, these factors can play a key role in determining the attractiveness of the project under consideration against the competing projects.

The analysis at this stage focuses both on location and project characteristics. In analyzing the location attributes of the competing projects, special attention needs to be given to the particular access to the site from the nearest major thoroughfare, as well as its broader access to consumers by car and public transportation; the attractiveness of the project's immediate environment; and the demographics and income characteristics of nearby residential neighborhoods. In analyzing the characteristics of competing projects the focus is on the main factors that affect their drawing power, including those relating to the new frontiers in retailing. These characteristics include, but are not limited to, the building-to-land area as a measure of the project's building density; the total Gross Leasing Area (GLA), which is an important center size indicator; GLA by major categories of retail goods, services, leisure, and entertainment, which provides the broader tenant mix of the development and can help assess its strength as a destination character; number of stores in different product lines as a measure of product mix, which was found to be one of the most important factors in affecting shopping center choice by millennials (Larsen, Selton, and Wright, 2015); technological infrastructure and access to the internet; landscaping, design, aesthetics, and appeal of internal layout; and parking availability.

Beyond the above, a detailed analysis of each competing project's tenant roster is required in order to better understand and evaluate the attractiveness of its detailed tenant mix and assess the variety and quality of the goods and services offered. Detailed tenant analysis identifies among others, anchor and non-anchor tenants, goods/services sold by each store, brand reputation, the particular providers of non-retail services, such as leisure, entertainment and other services, and the square footage occupied by each use and individual tenants. As indicated earlier, an important dimension of the tenant profile that has to be examined from a new-frontiers perspective is the extent to which they have integrated successfully in-store and offline shopping channels with online shopping channels.

Finally, information on the performance of competing centers such as rental rates and other tenant charges, visitation rates, sales per square foot, and occupancy rates can help not only better assess the strength of competing projects but also assess the potential performance of the project under consideration in terms of these indicators, taking into account its charac-teristics. However, performance measures and rental rate paid by each tenant or detailed categories of tenants, apply only to existing operating projects and it may be difficult to obtain, as they represent confidential information that the owners of competing projects are reluctant to share.

Once the detailed attributes of the planned project and competing developments are systematically and comprehensively listed they have to be analyzed in order to identify the project's strengths and weaknesses and assess its competitive position in attracting consumers. This comparative analysis can be carried out through the Competitive Differentials Technique, which is a non-econometric approach and does not require significant additional information beyond the information on the characteristics of the competing projects. This technique is discussed by Clapp (1987) and provides a reasonable methodology for quantifying the multi-dimensional differences between the planned development and the competing projects. The major issue in using this approach is that it cannot quantify scientifically the different weights assigned by consumers to the different center characteristics when making their shopping/patronage decisions. However, this may be a serious shortcoming in light of the rising importance of a handful of new-frontier related factors discussed earlier.

The differential effect of the various project characteristics on its attractiveness can be more accurately assessed using the econometric approach in which information on consumer shopping trips to and purchases at competing developments, in combination with information on the characteristics of these developments, is used to quantify the effect of the latter on the former (Weisbrod, Parcels, and Kern, 1984; DiPasquale and Wheaton, 1996). This methodology is discussed further in the estimation of the project's sales potential using the probabilistic approach. From a new-frontiers perspective, the use of this approach can provide a more accurate assessment of the attractiveness of the project under consideration against its competition.

Stage 3: Shopping pattern analysis

Given that the new frontiers of retailing have to do with how the consumers interface with electronic and in-store shopping, their basic demographic structures, and their cultural diversity, this stage is of critical importance in accurately assessing the project's potential sales and planned tenant mix. The third stage focuses on the consumer groups within the planned development's trade area and involves primary research that aims at better understanding their characteristics, their shopping choices and spending patterns. The information on consumer characteristics and preferences will help in determining the most suitable tenant mix for the particular type of development and improving the project's rating against the competition based on actual consumer preferences. The information on consumer spending patterns will provide the foundation for estimating the project's sales potential by analyzing how consumer patronage preferences are influenced by consumer characteristics and retail cluster characteristics.

The information collected in this stage focuses specifically on consumer profiling within the project's trade area in order to identify key characteristics such as age, income, household size, education, and ethnicity/culture (Sivitanidou, 2011; ICSC, 2005). This information will allow the segmentation of the population residing within the project's trade area into distinct age and ethnic groups, and better identify their shopping, leisure, and entertainment preferences, as well as their spending patterns in different product lines both in physical stores and on the internet. The information that it is typically collected at this stage via the consumer survey is described in more detail in Table 8.1. If the consumer survey is carried out systematically to include samples from all residential zones within the development's trade area, it can allow for the estimation of the project's capture rate through advanced probabilistic modeling, which is discussed in the next section.

Stage 4: Estimating the project's sales potential and supportable square footage

The main objective of the fourth stage is to derive estimates of the project's sales potential and supportable square footage as a means of evaluating the feasibility and viability of the development concept under consideration. The key figure that is needed in order to estimate the development's sales potential is its capture rate or market share, that is, the percentage of the trade area's sales that the project will capture. One approach suggested by Fanning et al. (1994) for estimating the project's capture rate is using the industry standard capture rate for the primary trade area, which is 70–80 percent, and adjusting it accordingly depending on the strength of the project's location vis-à-vis the location of the major citywide retail clusters. If the location is among the strongest ones then the industry standard capture rate can be applied

Table 8.1 Information collected for shopping pattern analysis

a	Detailed demographics of the population residing in the different homogeneous residential zones that compose the project's trade area including age, income, education level, ethnicity and household size
b	The different types of purchases made by each socioeconomic group and the specific retail clusters that are chosen for each type of purchase
c	The characteristics of all shopping trips made by the consumers, including travel mode, travel cost and time, as well as motivations for each trip, such as shopping, entertainment, leisure or other
d	The frequency of visits to the competing retail clusters by type of purchases
e	The amount spent on different types of purchases of goods and services in the different competing projects
f	Online purchases in terms of amounts and different product/service lines, as well as intentions in terms of future online purchases versus in-store visitations and purchases by line of trade

to the project, otherwise it should be adjusted downwards depending on how weak the project location is compared to the strongest ones. According to Sivitanidou (2011), the project's competitive position index, derived through the competitive differentials technique, is a more comprehensive and appropriate measure of the project's relevant attractiveness as it takes into account not only its location but many other project characteristics including those that would make it more competitive in the new retailing environment. As such, the competitive position index is a more appropriate weight for adjusting the industry standard capture rate, as opposed to a location rating.

In terms of coping with the new frontiers in retailing, the most accurate approach for estimating the project's capture rate and sales potential is the advanced probabilistic approach. This approach is more accurate in terms of assessing the implications of new frontiers in retailing because it can take into account the demographic and diversity/cultural characteristics of the particular consumer groups that reside within the project's trade area as well as their in-store and online shopping patterns and shopping preferences. Empirical evidence has shown that differences in certain consumer characteristics are associated with different in-store and online shopping patterns. For example, Li, Kuo and Rusell (2006) confirmed that higher income and educational levels are associated with higher online purchases while no statistically significant effect of age was found.

The information on consumer characteristics and their shopping patterns is collected in the third stage through consumer surveys in the different homogeneous residential zones that compose the project's trade area. This information, along with information on the major characteristics of competing developments, can be used to estimate econometrically the probability that a consumer residing in a specific residential zone k and belonging to a particular consumer group y, will choose to visit the planned development for satisfying its shopping needs for line of trade i. This probability can be used as the capture rate (CR_{iky}) of the project for the specific zone k, consumer group y and line of trade i and represents the percentage of sales that is expected to be captured from each residential zone in the development's trade area. For a detailed discussion of this methodology see Weisbrod, Parcels and Kern (1984), and DiPasquale and Wheaton (1996).

Once the capture rate for each line of trade i, residential zone k and demographic group y (CR_{iky}) is estimated, then the project's total potential sales by line of trade i (SP_i) can be estimated using the formula shown in Figure 8.2.

$$SP_i = \sum_y \sum_k HH_{yk}\, HI_{yk}\, e_{yi}(1 - o_{yi})CR_{iky}$$

where

SP_i : Potential sales of the project under consideration in line of trade i

HH_{yk} : Number of households in consumer group y, typically defined by income and age, in residential zone k

HI_{yk} : Average income of households in consumer group y and residential zone k

e_{yi} : Percent of average income of households in consumer group y spent on line of trade i for both online and in-store purchases

o_{yi} : Percent of average income spent on e-commerce purchases of line of trade i by households in consumer group y

Figure 8.2 Estimation of sales potential by line of trade

Sources: Carn et al. (1988); Sivitanidou (2011); author

Special caution is needed in applying the formula in order to ensure that the income measure used and the percentage of income spent on a particular line of trade e_{yi} are consistent. Such percentages can be calculated from information that is collected from the detailed consumer survey in the analysis of the shopping patterns in the development's trade area. In cases where a consumer survey is not carried out, such percentages can be obtained from detailed consumer expenditure reports that are usually provided by the national statistical service. In the USA, such information is provided by the Bureau of Labor Statistics (BLS). If this data is used for estimating e_{yi}, then the income measure that needs to be used in order to carry out the calculation is income before taxes because that is the income measure used by BLS in its cross-tabulations of expenditures in different product lines by age and income groups.

It should be noted that the consumer expenditure reports published by the BLS do not separate the spending for purchases in physical stores and purchases on the internet. Thus, if e_{yi} is derived using the BLS data it will refer to both in-store and online purchases by households. However, in terms of assessing the supportable square footage in a planned development, the analyst needs to estimate the area's potential sales in physical stores. That is the reason for taking into account in the formula the percentage of purchases carried out online (o_{yi}). In cases where a detailed consumer survey in the project's trade area has been carried out, this percentage should be available from the data collected through the survey. If a survey has not been carried out, an overall national percentage by line of trade can be used, based on the data provided by the Annual Retail Trade reports published by the Bureau of the Census. However, in such a case, the analyst needs to critically evaluate whether there is any strong rationale that would justify a significant deviation of the online purchases of the particular population within the development's trade area from the national average and accordingly adjust the figure used for o_{yi}.

Once potential sales in each line of trade are estimated, then the analyst can estimate the supportable square footage by dividing these figures by the sales per square foot for the respective line of trade. Data for sales per square foot by line of trade and type of shopping center can be obtained from industry organizations, such as the International Council of Shopping Centers.

In sum, in order to accurately assess the project's true sales potential and viability of proposed tenant mix, the estimation of the project's sales potential and supportable square footage needs to explicitly take into account the online and in-store spending patterns of the consumers within the project's trade area. This is the most important analytical implication of the new frontiers in retailing, which retail property developers and their analysts should by no means ignore.

Conclusion

In this chapter we have discussed the new frontiers in retailing as well as the four basic stages of retail market analysis and the relevance of these new frontiers in each stage. The new frontiers in retailing are highlighted by the consistent and rapid rise of e-commerce's share in total retail sales and the resultant competition with bricks-and-mortar retailing; the increasing use of multi-channel strategies by leading retailers; the changing role of the physical store so as to complement the online shopping experience; and the changing consumer base, with significant implications in terms of shopping preferences and spending patterns. These new frontiers in retailing have clear implications in terms of assessing the potential performance and viability of a retail development project in this new environment. A basic component of such an assessment is the market study.

The new-frontier aspects are not relevant in the first of the four basic stages of a retail market study but they are quite relevant in the other three. In particular, in the first stage, which is the preliminary definition of the project's trade area, the new-frontier aspects are not relevant as the preliminary borders of the trade area are defined mainly on the basis of a maximum driving time. However, in evaluating in the second stage – the project against the competition – project attributes pertaining to the new frontiers in retailing must be taken into account along with other location and project characteristics in order to correctly evaluate project attractiveness in this new retailing regime. The third stage focuses on detailed analysis of consumer character-istics and shopping patterns within the project's trade area. This stage is the most critical in understanding the implications of new-frontier aspects on the particular project by assessing the online and in-store spending patterns, as well as the shopping/leisure/entertainment preferences of its potential clientele. Within this context, the consumer survey of shopping preferences and spending patterns within the project's trade area becomes one of the most critical components of the retail market study in the new retailing environment. The consumer survey and analysis carried out in the third stage provide a solid basis for estimating in the fourth stage the development's true sales potential and supportable square footage by line of trade in a way that takes into account all crucial new-frontier aspects.

References

AT Kearney (2014). On solid ground: brick-and-mortar is the foundation of omnichannel retailing. Available at: www.atkearney.com/documents/10192/4683364/On+Solid+Ground.pdf/f96d82ce-e40c-450d-97bb-884b017f4cd7

Carn, N., Rabianksi, J., Racster, R., and Seldin, M. (1988). *Real Estate Market Analysis: Techniques and Applications*. Englewood Cliffs, NJ: Prentice-Hall.

Clapp, J. (1987). *Handbook for Real Estate Market Analysis*. Englewood Cliffs, NJ: Prentice-Hall.

Converse, P. D. (1949). New laws of retail gravitation. *Journal of Marketing*, 14, 379–384.

Damesick, P. (2001). E-commerce and UK retail property: Trends and issues. *Briefings in Real Estate Finance*, 1(1), 18–27.

DiPasquale, D. and Wheaton, W. (1996). *Urban Economics and Real Estate Markets*. Englewood Cliffs, NJ: Prentice-Hall.

Evans, P. (2014). Non-retail tenants gain in importance. *Retail Property Insights*, 21(2), 36–42.

Fanning, S., Grissom, T., and Pearson T. (1994). *Market Analysis for Valuation Appraisals*. Chicago, IL: The Appraisal Institute.

Hardin, W. G. and Carr, J. (2006). Disaggregating neighborhood and community center property types. *Journal of Real Estate Research*, 28(2), 167–192.

Henderson Research (2013). The impact of technology on real estate: Implications for retail and logistics. Available at: www.propertysales.com/Articles/Article/9036/file1.pdf

Hortacsu, A. and Syverson, C. (2015). The ongoing evolution of US retail: A format tug-of-war. *Journal of Economic Perspectives*, 29(4), 89–112.

International Council of Shopping Centers (ICSC) (2005). *Market Research for Shopping Centers*. New York, NY: International Council of Shopping Centers.

Larsen, V., Shelton, R., and Wright, N. D. (2015). Shopping center attitudes: An empirical test of predictive attitudes. *International Academy of Marketing Studies Journal*, 19(2), 93–102.

Li, H., Kuo, C., and Rusell, M. G. (2006). The impact of perceived channel utilities, shopping orientations, and demographics on the consumer's online buying behavior. *Journal of Computer Mediated Communication*, 5(2), 1–23. Available at: http://onlinelibrary.wiley.com/doi/10.1111/j.1083-6101.1999.tb00336.x/abstract

Piotrowitcz, W. and Cuthbertson, R. (2014). Introduction to the special issue information technology in retail: Toward omnichannel retailing. *International Journal of Electronic Commerce*, 18(4), 5–15.

Platt Retail Institute (2014). The Future of Retail: A Perspective on Emerging Technology and Store Formats. Available at: https://nrf.com/system/tdf/Documents/retail%20library/whitepapers/2014-0224-Platt_Retail_Institute-The_Future_of_Retail.pdf?file=1&title=The%20Future%20of%20Retail:%20A%20Perspective%20on%20Emerging%20Technology%20and%20Store%20Formats

Reilly, W. (1929). *Methods for the Study of Retail Relationships*. Austin, TX: Bureau of Business Research, University of Texas.

Schmitz, A. and Brett, D. (2001). *Real Estate Market Analysis: A Case Study Approach*. Washington, DC: Urban Land Institute.

Sivitanidou, R. (2011). *Market Analysis for Real Estate*. Unpublished manuscript

Verhoef, P. C., Kannan, P. K., and Inman, J. J. (2015). From multi-channel retailing to omni-channel retailing: Introduction to the special issue on multi-channel retailing. *Journal of Retailing*, 91(2), 174–181.

Weisbrod, G., Parcels, R., and Kern, C. (1984). A disaggregate model for predicting shopping area market attraction. *Journal of Retailing*, 60, 65–83.

Part III

Organization and management of real estate development

Private sector-led urban development

Characteristics, typologies, and practices

Erwin Heurkens

Abstract

The role of the private sector in shaping the built environment is historically evident and seemingly increasing in contemporary urban development practice. Traditionally, real estate developers and investors play a crucial role in realizing and financing urban development and real estate projects. In addition, other private sector actors such as owners, entrepreneurs, local citizens, and all sorts of corporations increasingly influence city-making, as redeveloping real estate and regenerating urban areas are becoming more common. At the same time, public organizations such as local planning authorities increasingly focus on facilitating such private development initiatives. Especially in developed (Anglo-Saxon) countries, a new phenomenon called private sector-led urban development is occurring, in which private actors perform a leading role in managing the delivery of urban areas and real estate. Nevertheless, this trend in urban development is not fully explored and understood in real estate and planning literature. Therefore, this chapter deals with private sector-led urban development by discussing various characteristics, typologies and practices. In particular, it discusses the contemporary and potential future role of private actors as a driving force behind developing urban areas and real estate. Thereby, it aims to contribute new concepts and empirical insights about the management and organization of urban and real estate development.

Introduction

Worldwide, a trend towards more private sector influences in real estate development, urban development, and spatial planning can be noticed (Andersson and Moroni, 2014; Glasze et al., 2011; Heurkens et al., 2015; van der Krabben and Heurkens, 2015). The trend of 'neoliberalization' of space, planning, and development (Hackworth, 2007; Lovering, 2009; Olesen, 2013; Peck and Tickell, 2002) in the built environment can be partly attributed to the decreasing hierarchical role of government in planning and subsequent need for governance arrangements. For instance, in many Western countries, public–private development partnerships came into being aimed at combining public with private interests in urban and real estate development processes.

As a result of these trends, state–market relationships are constantly being redefined and reshaped. More specifically, Heurkens (2012) illustrates that in the Netherlands the state to an increasing extent delegates planning and development powers to the market and civil society actors (Figure 9.1). This has created urban development practices that are characterized as 'private sector-led' (Heurkens, 2010; 2012; Heurkens and Hobma, 2014), whereby private actors take a leading role and public actors adopt a facilitating role in managing the delivery of urban development projects. This phenomenon, however, can be noticed worldwide, especially in developed countries (Squires and Heurkens, 2015; 2016), which means that local planning authorities increasingly share or delegate responsibilities and powers to private sector actors such as developers, investors, and communities in shaping and governing the built environment. In this regard, it is remarkable that only a few other authors in the field of urban studies (Adams et al., 2012; Coiacetto, 2007; Henderson, 2010) emphasize the need to understand the role of private sector actors in urban development. Moreover, it seems that the diversity of possible private sector actors that engage in urban development is worthwhile exploring as their interests and goals might vary.

Therefore, this chapter aims to shed a light on the role of the private sector in contemporary urban development, with special attention being paid to developed countries. The chapter aims to answer the main question: to what extent do and can various private actors play a leading role in urban development projects? It does so by discussing the main characteristics of private sector-led urban development as a form of public–private partnership. In addition, a description is given about various types of private actors that can potentially take responsibility in managing private sector-led urban development projects. This is followed by empirical research findings from studies carried out under supervision of the author, comprising the roles private actors play in delivering urban development projects. The conclusion elaborates on the relevance of the findings for and future direction of research on urban and real estate development.

Private sector-led urban development characteristics

This section elaborates on the institutional characteristics of private sector-led urban development. Private sector-led urban development can be considered as a form of public–private partnership (PPP). PPPs were first introduced in the 1980s (Osborne, 2000; Dubben and Williams, 2009) as an institutional instrument incorporating organizational, legal, and financial aspects for the cooperation between public and private organizations. The main reason for introducing PPPs was the need in the public sector for private sector investment to deliver public services. Within urban development PPPs also became an effective means for local planning authorities and real estate developers to deliver projects. In general, urban development PPPs with regard to the relationship between the public and private sector, share overarching principles including:

- working together, some sort of organizational collaboration (Holland, 1984);
- reaching mutual agreements through legal contracts (Harding, 1990);
- mobilizing a coalition of interests (Bailey, 1994);
- achieving mutual benefits, common objectives (Klijn and Teisman, 2003);
- defining institutionalized arrangements (Bult-Spiering and Dewulf, 2006);
- combining organizational form with management strategies (Kort and Klijn, 2011).

Despite these universal characteristics of PPPs, several models exist in practice that do justice to context-specific conditions for and characteristics of projects (Chan and Cheung, 2014). In

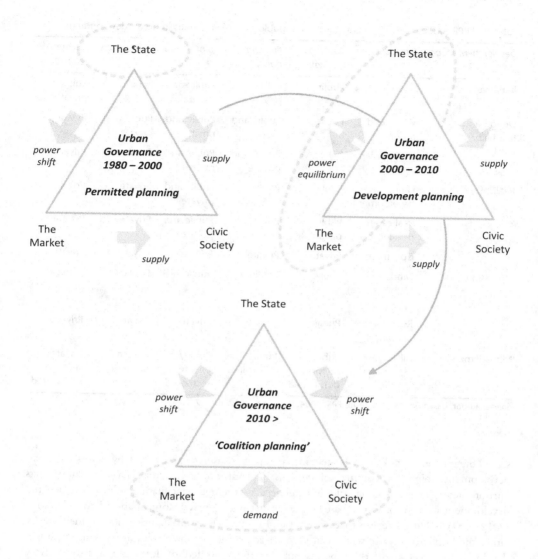

Figure 9.1   Dutch urban governance shifts since the 1980s

Source: Heurkens, 2012, p. 140

this regard, Dubben and Williams (2009, p. 228) argue that "the term [PPP] can cover a spectrum of models from relatively short-term management contracts involving limited capital expenditure, concession contracts involving the design and build of capital assets along with the provision of services and the financing of the project construction and operations, [as well as] joint ventures and partial privatisations with a sharing of ownership and responsibilities between the public and private sectors." In order to understand what is meant by private sector-led urban development, it is therefore useful to relate it to PPP model classifications.

Table 9.1 illustrates the various roles public and private actors perform in different PPP models for consecutive development stages. Heurkens (2012, p. 56) defines roles as "a coherent

Table 9.1 PPP models, development stages, and public and private roles in urban development

Development stage	Sub-stage	Public realization	Building rights	Joint ventures	Concessions	Private realization
Initiative	–	Public	Public or private	Public or private	Public	Public or private
Design and feasibility	Vision and program	Public	Public and private	Public and private	Public and private	Private
	Design plan	Public	Public and private	Public and private	Private	Private
Realization	Land development	Public	Public	Public and private	Private	Private
	Real estate development	Public or private	Private	Public and private	Private	Private
	Construction	Private	Private	Private	Private	Private
Operation	Land (public space) operation	Public	Public	Public	Public or private	Public or private
	Real estate operation	Public	Private	Private	Private	Private
Management	All stages	Public sector-dominated	Public sector-led	Public/ private sector-led	Private sector-led	Private sector-dominated

Source: author, based on Kenniscentrum PPS, 2004; Heurkens, 2012, p. 149)

set of organisational tasks and related management measures carried out by actors in urban development projects." According to Heurkens, in private sector-led urban development projects "private actors take a leading role and public actors adopt a facilitating role to manage the development of an urban area, based on a formal public–private organisational role division" (2012, p. 57). Table 9.1 reveals that within each PPP model, within different development stages, either the public or private sector can play a 'leading role' or share roles in working on urban development projects. Heurkens points out that the term 'led' or 'leading' refers to "actively steering [or managing] an urban development process into a preferred direction" (2012, p. 58).

What also becomes apparent from Table 9.1 is that private sector-led urban developments are legally typified as 'concessions'. In concession partnerships typically public authorities (e.g. local planning authorities, public land development bodies) commission private development companies (e.g. real estate developers, development investors, development consortia) to design, build, finance, (and sometimes) manage and operate urban development projects. Internationally, concessions are well known for their application in infrastructure projects and real estate development projects. In principle, according to Bult-Spiering and Dewulf (2006, p. 50) "concession contracting is known variously as private finance initiative (PFI), design-build-finance-maintain (DBFM), design-build-finance-operate (DBFO), build-operate-transfer (BOT) and by many other names."

Heurkens (2012) argues that development practices in Western countries witness an increased application of concessions at the urban development scale. This offers opportunities for public planning authorities to offset financial risks associated with large-scale land development while setting planning guidelines requirements. For private real estate developers and investors, urban development concessions create the possibility to lead projects thereby shaping and influencing project features such as financial feasibility and designs during various stages of the development process. The following legal definition of concessions in urban development is applicable:

> A concession in urban development is a contract form with clear preconditioned (financial) agreements between public and private parties, in which a conscious choice from public parties has been made to transfer risks, revenues, and responsibilities for plan development, land preparation, land and real estate development and possible operation for the entire development plan towards private parties, within a previously defined public brief in which the objective is to create an effective and efficient task division and a clear separation of public and private responsibilities.
>
> *(Gijzen, 2009, p. 19)*

Bregman and De Win (2005) argue that at the outset of concession developments, private actors predominantly own land, or in some cases acquire land from local authorities. Other key concession characteristics are: fixed financial agreements, limited public risks, predefined public conditions for development that function as a framework for developers to design plans, and a combined private land and real estate development (Heurkens and Peek, 2010). Wolting (2006) emphasizes that local authorities in concessions deliberately choose to limit their influence by solely predefining conditions for development. As a result of the land and real estate development undertaken by private actors the risks and revenues are also attributed to private developers. After project delivery, developers often transfer the land to public actors on the basis of agreed conditions, so public actors own and maintain the public space.

Figure 9.2 shows the public and private roles in land and real estate development and land and real estate operation for the concession model (Heurkens, 2012). In addition, this illustration also shows that projects based on the concession model are characterized as 'private sector-led,' as land and real estate development and real estate operation are in the hands of the private sector.

Private sector-led urban development typologies

This section elaborates on four typologies of private sector-led urban development, as a variety of private sector actors such as developers, investors, communities, and corporations can be active in urban development. Therefore, the typologies are developed as sub-categories of private sector actors that could perform a leading role in urban development, and does justice to different institutional origins of private actors involved.

Figure 9.3 illustrates the four typologies of private sector-led urban development in relation to the degree of involvement of actors (horizontal axis) and type of development strategy (vertical axis). For instance, 'short-term involvement' indicates a limited responsibility of private actors for the urban development project itself until delivery. 'Long-term involvement' means being responsible for urban real estate operation after completion of a project (e.g. land and real estate ownership, providing services). In addition, actors deploying 'integrated development strategies' focus on projects that are substantive in scale (e.g. urban development with various real estate functions coupled with public facilities and infrastructure development), and in risk (e.g. large share of debt finance and up-front investment). 'Incremental development strategies'

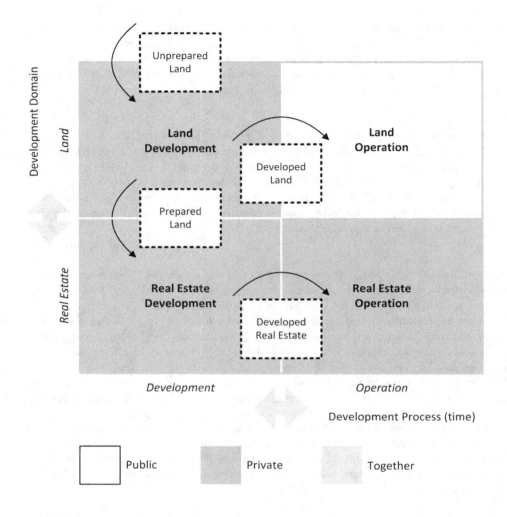

Figure 9.2 Public and private roles in the PPP concession model
Source: Heurkens, 2012, p. 154

are limited in scale (e.g. real estate development on a plot basis), and in risk (e.g. large share of equity finance, quicker investment returns).

Conceptually, the following descriptions indicate the basic characteristics of the various private actors and their (potential) role in developing urban projects.

Developer-led urban development

Developers are traditional real estate industry agencies. They typically buy land for real estate development with a combination of debt and private equity finance, and deliver projects for clients (either real estate investors or owners), upon which they normally complete and leave

Figure 9.3 Typologies of private sector-led urban development
Source: Author

development projects with a decent profit, typifying them as trader–developers. Developers can be financial–organizationally linked to banks, investors, construction companies, or be independent companies. Potentially, their leadership role in urban development projects could increase by becoming active in real estate operation (e.g. by offering building services), and incorporating social–sustainable measures throughout the entire development process.

Investor-led urban development

Investors are also traditional real estate industry agencies. They can either be institutional investors, investment banks, or development investors, and pursue real estate investment returns by purchasing real estate. Institutional investors typically buy real estate from developers, and often look for additional investors (e.g. investment banks, pension funds) to spread investment risk and share investment returns through cash flows during real estate operation. Development investors are actively involved in the development stages as well, as they specifically develop real estate for their own investment portfolio. Potentially, their leadership role in urban development projects could increase by becoming active in real estate (re)development by developing directly for their own investment portfolio.

Community-led urban development

Communities are non-traditional real estate industry agencies as their core expertise often is not real estate. Communities are locally rooted actors such as property owners, entrepreneurs, or local citizens that engage in urban development. Property owners search for ways to (re)develop or (re)invest in their real estate, entrepreneurs look for business opportunities in urban redevelopment, and citizens aim to improve or protect their immediate living environment. Often they operate on a building scale and adopt incremental strategies to optimize real estate operation and building and public space and infrastructure use. Potentially, their leadership role in urban development projects could increase by linking organization with investment capacity in neighborhood regeneration.

113

Corporation-led urban development

Corporations are considered as non-traditional real estate industry agencies, as their core business is not real estate. However, corporations such as technology and energy companies can be influential in urban development. Often such companies focus on the operation stage of real estate and urban areas by offering services such as energy management or data monitoring amongst others. In addition, multinational corporations can play an important role in shaping the built environment. Potentially, their leadership role in urban development projects could increase by extending their business models towards the real estate development stages.

Private sector-led urban development practices

The typology descriptions above indicate the variety of private actors that could take the lead in urban development projects. They offer conceptual insight into the potential role of various private actors in urban development practice. Hereafter, a collection of empirical findings from graduation projects conducted from 2010–2015 under supervision of the author are presented. The studies are set against the background of an increased role of private sector actors in Dutch urban development practice, and consequently investigate the phenomenon in Dutch practice and other foreign private sector-led urban development practices. The aim of these studies is to examine empirical characteristics of different private sector-led urban development typologies, and to identify to what extent the various private actors are able to play a leading role in contemporary urban development.

Developer-led urban development practices

Real estate developers in most developed countries around the globe face the challenge of becoming innovative businesses that develop sustainable real estate and urban areas (Squires and Heurkens, 2015). For example, since the start of the economic crisis in 2008, the Dutch development industry has been subject to some societal trends and market changes that have forced them to change roles and strategies. The decreased demand for office and retail space, limited debt funding opportunities, increased user involvement, stalling government investments, and the entrance of new competitive companies to the market, forced Dutch real estate developers to change. However, this change is obstructed by difficulties for developers to finance development and to make projects feasible in post-crisis times (Schiltmans, 2013; Veseli, 2014; Van der Bent, 2016). As a result their role as an indispensable player within the urban development process is being redefined and questioned (Putman, 2010; Heurkens, 2013).

Putman (2010) argues that for Dutch developers a strategic shift towards supplementing traditional real estate development tasks with investment activities could be a preferable path. This would involve a financial model similar to that of development investors, which focus on generating long-term yields instead of short-term profits upon project completion. More specifically, investing in innovations, applying new business models, and adopting sustainability principles are among the strategies that seem necessary for Dutch developers to remain a leading player within urban development. However, Panteia (2013) reveals that the Dutch real estate industry ranks among the least innovative sectors nationwide. At the same time, with the rise of private sector-led projects, developers are expected to take a leading role in various stages of the development process. This section highlights findings from various research projects conducted in the last two years specifically focused on the changing role of Dutch real estate developers in urban development in relation to innovation, business models, and sustainability.

A survey amongst the Dutch real estate development industry conducted by Haak (2015) reveals that 84% of developers currently adopt some form of innovation. Innovations can be found mainly in operations (57%) such as increased user involvement (cf. Geesing, 2015), products/services (45%) such as introducing sustainable real estate concepts, and business models (24%) such as vertical supply chain integration, but scarcely within Dutch developer business strategies. As such, innovations are rarely linked to management innovations, which raises the question whether innovations have become structurally embedded within organizations. On top of that, Haak and Heurkens (2015) argue that most innovations can hardly be considered innovative, as a number of developers list similar innovations. In addition, developers state that 88% of innovations are successful in terms of contributing to the company's business goals. This finding seems questionable, as other sectors typically show success rates of 6–35%. Overall, Haak (2015) concludes that Dutch real estate developers show some signs of becoming innovative albeit with some caution, which indicates a slightly changing role.

This observation resonates with survey findings from Bogers (2015) on Dutch real estate developer business model changes. Dutch developers have made changes to their business model component's networks such as collaboration with other stakeholders, value propositions such as development concepts, and client relationships such as increased participation of clients. Bogers argues that difficulties in securing development financing has been the main reason for accelerating these business model changes. Although Dutch developers state that sustainability has become increasingly included in concepts, products, and value propositions for clients, it remains unclear whether these changes are structural or temporal. In this regard, he argues that once the market demand for real estate picks up development could be 'business as usual,' as most indicated changes are hardly backed up by fundamental business strategy changes. Overall Bogers (2015) therefore concludes that Dutch real estate developers just gradually adjust their business models to appearing social–economic trends, but that a fundamentally changing role is not notable yet.

In addition, Buskens (2015) raises the question whether Dutch real estate developers are able to commit themselves to deliver more sustainable urban development projects. He indicates that the main reasons for commitment to sustainability are: company marketing (70%), social responsibility (70%), regulatory requirements (57%), competitive advantage (50%), and client support (45%). Buskens observes that Dutch developers, to a large extent, equalize sustainability to ecological environment-friendly aspects, leaving out economic, social, and spatial components in the equation. Moreover, sustainable solutions are often limited to the scale of buildings instead of urban areas, and are rather considered as an end product than being an integrated part of development processes. This has generated a widespread conception within the Dutch development industry that sustainability is something that is dealt with only when requested by governments or clients, rather than being the central focal point of real estate development and investment activity. In conclusion, Buskens and Heurkens (2016) argue that Dutch developers are hardly aware of the multifaceted aspects of sustainable urban development, and do not proactively implement sustainability in their company culture and development strategies.

Investor-led urban development practices

Another witnessed global phenomenon is the increased involvement of development investors in urban development. They are expected to recoup the costs of real estate development through their investment returns. Thus, they can be long-term committed private actors with an ability to deploy integrated development and investment strategies (Sturm et al., 2014).

Examples of such mostly Anglo-Saxon development investment companies include: Argent, Grosvenor (United Kingdom); Forest City, CIM Group (United States of America); Lendlease (Australia); Wheelock Properties (Hong Kong); and Amvest (The Netherlands). Contrary to these, institutional investors such as pension funds or insurance companies typically only buy real estate from developers or owners for their own investment portfolio and do not take on a leading role in development itself (Steigenga, 2015; Stumpel, 2014; Stumpel and Heurkens, 2014). Therefore, above all, it seems fruitful to explore to what extent development investors can play a leading role in urban projects.

Recent studies (Buskens, 2015; Huijbregts, 2017; Regales, 2017; Sturm, 2014; Sturm et al., 2014) indicate that real estate development investors show an increased attention towards community involvement, new development coalitions, corporate social responsible investments, place-making, and sustainable urban and real estate development. Such development investors are taking a leading role throughout each phase of the development process, from development initiative to land and real estate operation and management. Moreover, this seems to be profoundly incorporated in the development/investors' business and management principles. According to Sturm et al. (2014) development investors are deemed to take a leading role in development processes due to their multi-sectoral area-focused development and investment approach. The way investment and finance are arranged by real estate development investors can be through institutional capital (real estate funds or bank loans) and/or private equity. Carrying out urban development themselves then becomes a strategic means to secure investment returns for the company and its shareholders.

Community-led urban development practices

Communities are said to become more involved in urban development as the focus of developing has shifted towards transforming, redeveloping, and regenerating existing real estate and urban areas. In this typology of private sector-led urban development, (local) communities can consist of property owners, entrepreneurs, and citizen(s) groups. This category thus excludes investors who own real estate, but could include entrepreneurs who aim to redevelop their own property. It also involves local communities that collectively organize themselves to regenerate neighborhoods, for instance by using crowdsourcing and subsequent crowdfunding. In terms of their role in urban development not much is known. However, what can be noticed in practice is that real estate owners increasingly form private–private partnerships (van der Krabben, 2014) with developers and investors with knowledge about real estate development, aimed at bringing about change in their immediate environment.

For instance, Van den Berg identifies that real estate owners form collective 'development coalitions' with other owners in the area. "Development coalitions are a form of private–private collaboration aimed at bringing about change in the environment, by strategically steering on network relations and/or directly by signing formal contractual agreements" (Van den Berg, 2014, p. 7). Based on empirical research, Van den Berg (2014) concludes that private–private cooperation requires leaders that connect the interests of all actors involved, and builds external relations with public and private organizations. Also, he concludes that in addition to collective networking and vision-making, it is an absolute necessity to formalize private–private collaboration in order to be able to collectively plan, finance, and redevelop real estate and urban areas.

The co-creation of real estate and urban areas is dependent on close cooperation between local communities and established actors such as local planning authorities and real estate developers and investors (Straub, 2012). The relationships between these actors are changing constantly, and require traditional planning and real estate actors to search for new ways of collaboration. Labruyere (2015) argues that it is of upmost importance that public authorities consciously facilitate private–private partnerships in their initiatives to redevelop areas and real estate. For instance, when owners collectively aim to reduce vacancies, it can be an effective strategy to extend the scope of a project towards investment in public space, which results in an attractive urban area and not just redeveloped real estate.

Corporation-led urban development practices

Finally, corporations are seen as a sub-type of actors of private sector-led urban development. Vande Putte and De Jonge (2008) indicate that corporations have played and continue to play an important role in city-making. For example, throughout history, multinational corporations not only developed office work environments but also invested in developing housing for employees. More recently, it seems that technological corporations such as Siemens, IBM, and Cisco amongst others are becoming interested and engaged in urban development, which can be partly attributed to the movement for smart cities. Also, non-traditional real estate companies such as IKEA, Vulcan Inc. and Tata Steel are developing real estate and urban areas, which can be seen as an addition to or extension of corporate strategies. Moreover, energy companies seem to become more active in urban real estate projects by decentralizing their operations and activities. In a broader perspective, Potters (2015) argues that corporations can be seen as new international (land and real estate) market entrants. They finance development projects with private equity, which forms an answer to stalling real estate development and investment activity in the post-GFC period in many developed countries. Also, "several mega trends, growing competition and an increased emphasis on the societal responsibilities of corporations demand them to continuously reassess their strategic focus" (Potters, 2015, p. ix).

Based on international comparative research, Potters and Heurkens (2015) conclude that two different types of corporations emerge in real estate practice: the 'developing multinational,' and the 'technological partner.' The first category of corporations can be seen as developers that are part of a mother company that is a non-traditional real estate company (such as InterIkea, Vulcan Inc. or Tata Steel). They perform activities throughout all phases of urban development processes, adopt the role of holistic developer, have their activities organized in an independent subsidiary, and contribute to their corporate strategy by creating economic value and long-term returns on investment and improving communities (Potters and Heurkens, 2015). Contrary to this, technological partners (such as Siemens, Arcadis, and Cisco) have a technological core business and perform advisory activities in urban development, offer (technological) services, and are only marginally involved in shaping actual urban development processes. Van der Post (2011) found similar results for energy-service companies, which do not take on urban development themselves but offer advice and technological services, resulting in an active involvement in the operational phases of real estate and urban areas.

Discussion

The previous sections explored various typologies of private sector-led urban development by studying the contemporary phenomenon within practice. These studies reveal the diversity of private sector actors involved in urban development, and current trends and issues associated with their role in the built environment. This section discusses some of the findings, by answering the central question of this chapter: To what extent do and can various private actors play a leading role in urban development projects? Table 9.2 illustrates the main exploratory

Table 9.2 Empirical findings from private sector-led urban development practices

Private sector-led urban development	Integrated vs incremental development strategies	Short-term vs long-term actor involvement	Contemporary leading role	Potential leading role
Developer-led	Both	Short-term	Scarcely	Possibly
Investor-led	Both	Long-term	Yes	Certainly
Community-led	Incremental	Short-term	Scarcely	Possibly
Corporation-led	Both	Both	Yes	Certainly

empirical findings for the various private sector-led urban development typologies in terms of development strategies, actor involvement, and contemporary and potential leading roles.

In brief, the studies of developer-led urban development practices show that real estate developers seem rarely capable of adjusting their traditional development role with short-term focus towards a more leading role. There is no sign of a fundamental extension of the role towards land and real estate operation, at least in the Netherlands. Nonetheless, earlier studies (Heurkens, 2012; Heurkens and Peek, 2010; Gijzen, 2009) reveal that development concessions are becoming more commonly used in Dutch practice. At the same time, innovations, changes in business models, and approaches to sustainability, point towards a sector that is following trends rather than being a frontrunner.

Quite contrary to this, investor-led urban development practices show that development investors in particular already take a leading role in urban development projects. Their long-term focus on real estate operation and investment returns necessitates a close involvement in the earlier stages of development and managing design and place quality. However, development investors are not widely represented in the real estate development industry when compared to the share of traditional real estate developers. Despite this it seems to be interesting to monitor how development investors play out their role in urban real estate development and to see whether their values and objectives influence the real estate sector.

In addition, within community-led urban development practices, property owners and entrepreneurs seldom solely take on a leading role in redevelopment. The studies illustrate that they often use incremental development strategies to make improvements to their own properties, or seek business opportunities at building scale. Additionally, owners often are committed to short-term redeveloping of real estate, with interventions that don't require large amounts of investment. Nonetheless, there seems to be potential for a more leading role, once private–private partnerships are effectuated and collaboration is sought with stakeholders beyond project boundaries.

Finally, within corporation-led urban development practices it became clear that in essence only multinational corporations currently and potentially play a leading role in such projects. In terms of commitment of corporations, it can be said that developing multinationals show an interest in being involved in land and real estate development and its operation, while techno-logical partners only play a role in real estate operation. However, no conclusion can be drawn on the corporations' use of integrated or incremental development strategies. Moreover, for traditional private real estate companies it might be fruitful to partner with technological corporations in order to introduce (technological) innovations within real estate and urban projects.

Conclusions

This chapter dealt with the phenomenon of private sector-led urban development by discussing various characteristics, typologies and practices. In terms of characteristics, private sector-led urban developments are classified as PPP concessions. Such formalized partnerships are based on contracts between public and private actors, which results in private actors bearing responsibility for land development, real estate development and (often) real estate operation, and performing a leading role throughout various stages of the development process. A closer conceptual examination of the subject suggests that four typologies of private sector-led urban development exist, namely, developer-led, investor-led, community-led, and corporation-led urban development. Practices of each of these typologies indicate the wide variety of development strategies, actor commitment to urban development, and actor abilities to play a leading role in development processes. Thereby, this chapter attempted to show the great diversity of private sector actors currently involved in shaping and delivering urban areas and real estate.

Future research on private sector-led urban development envisages a focus on the relationship with sustainability. As neoliberalism and sustainability seem two structural trends influencing changes in the built environment (Crouch, 2012), it seems plausible and worthwhile to study if and how private actors can develop sustainable urban areas that are economically viable, socially responsible, and environmentally friendly (Heurkens, 2016). In addition, future research in this field foresees studying how public actors can facilitate and incentivize private actors to develop sustainably, and to understand which institutional conditions can overcome market barriers to do so. It would be interesting for both real estate development and planning practice and theory to move beyond general perceptions of the role of the private sector in sustainable urban development. In a time of government retrenchments in urban development and privatization of planning powers (Hobma and Heurkens, 2015), such a research focus might provide ground for a well-informed discussion about the balancing of private with public interests in contemporary sustainable urban and real estate development.

References

Adams, D., Croudace, R., and Tiesdell, S. 2012, "Exploring the 'notional property developer' as a policy construct," *Urban Studies*, vol. 49, no. 12, 2577–2596.

Andersson, D. A. and Moroni, S. (eds) 2014, *Cities and private planning: property rights, entrepreneurship and transaction costs*, Edward Elgar, Cheltenham.

Bailey, N. 1994, "Towards a research agenda for public–private partnerships in the 1990's," *Local Economy*, vol. 8, 292–306.

Bogers, B. A. 2015, "Post global financial crisis property development: An assessment of the impact of the post-GFC environment on Dutch property developer's business models," Master's thesis, Delft University of Technology.

Bregman, A. G. and Win, R. W. J. J. 2005, *Publiek–private samenwerking bij ruimtelijke inrichting en haar exploitatie*, Kluwer, Deventer.

Bult-Spiering, M. and Dewulf, G. 2006, *Strategic issues in public–private partnerships: An international perspective*, Blackwell, Oxford.

Buskens, B. 2015, "De duurzame ontwikkelaar: Hoe en waarom projectontwikkelaars zich kunnen committeren aan duurzame gebiedsontwikkeling," Master's thesis, Delft University of Technology.

Buskens, B. and Heurkens, E. 2016, "De duurzame private gebiedsontwikkelaar," *Real Estate Research Quarterly*, vol. 15, no. 3, 38–46.

Chan, A. P. C. and Cheung, E. (eds) 2014, *Public–private partnerships in international construction: Learning from case studies*, Routledge, London.

Coiacetto, E. 2007, "The role of the development industry in shaping urban social space: A conceptual model," *Geographical Research*, vol. 45, no. 4, 340–347.

Crouch, C. 2012, "Sustainability, neoliberalism, and the moral quality of capitalism," *Business and Professional Ethics Journal*, vol. 31, no. 2, 363–374.

Dubben, N. and Williams, B. 2009, *Partnerships in urban property development*. Chichester: Wiley-Blackwell.

Geesing, C. 2015, "Public engagement from a developer's perspective," Master's thesis, Delft University of Technology.

Gijzen, M. H. M. 2009, "Zonder loslaten geen concessie: Inzicht in de recente toepassing van deze publiek–private samenwerkingsvorm in de Nederlandse gebiedsontwikkelingspraktijk met 'evidence-based' verbetervoorstellen," Master's City Developer thesis, Erasmus University Rotterdam.

Glasze, G., Webster, C., and Frantz, K. 2011, *Private cities: global and local perspectives*, Routledge, London.

Haak, M. 2015, "De veranderende ontwikkelaar: Succes door innovatie," Master's thesis, Delft University of Technology.

Haak, M. and Heurkens, E. 2015, "Innovatie bij vastgoedontwikkelaars: typologieën en strategieën," *Real Estate Research Quarterly*, vol. 14, no. 2, 48–54.

Hackworth, J. 2007, *The neoliberal city: governance, ideology and development in American urbanism*, Cornell University Press, Ithaca, NY.

Harding, A. 1990, "Public-private partnerships in urban regeneration," in M. Campbell (ed.), *Local economic policy*, Cassell, London, pp. 108–127.

Henderson, S. 2010, "Developer collaboration in urban land development: Partnership working in Paddington, London," *Environment and Planning C: Government and Policy*, vol. 28, 165–185.

Heurkens, E. 2010, "Private sector-led urban development: Dutch case study," *Real Estate Research Quarterly*, vol. 9, no. 2, 29–34.

Heurkens, E. 2012, "Private sector-led urban development projects: Management, partnerships and effects in the Netherlands and the UK," *A+BE | Architecture and the Built Environment*, [S.l.], no. 4, 1–480.

Heurkens, E. 2013, "Een nieuwe rolverdeling: privaat 'in the lead,' publiek faciliteert," *VHV Bulletin*, vol. 40, no. 3, 15–16.

Heurkens, E. 2016, "Institutional conditions for sustainable private sector-led urban development projects: A conceptual model," in ZEBAU – Centre for Energy, Construction and the Environment (eds), *Proceedings of the International Conference on Sustainable Built Environment: Strategies – Stakeholders – Success Factors*, HafenCity University, Hamburg, pp. 726–735.

Heurkens, E. and Hobma, F. 2014, "Private sector-led urban development projects: Comparative insights from planning practices in the Netherlands and the UK," *Planning Practice and Research*, vol. 29, no. 4, 350–369.

Heurkens, E. and Peek, B. 2010, "Effecten van de toepassing van het concessiemodel bij gebiedsontwikkeling," *Real Estate Magazine*, vol. 71, 42–45.

Heurkens, E., Adams, D., and Hobma, F. 2015, "Planners as market actors: The role of local planning authorities in the UK's urban regeneration practice," *Town Planning Review*, vol. 86, no. 6, 625–650.

Hobma, F. and Heurkens, E. 2015, "Netherlands," in S. Mitschang (ed.) *Privatisation of planning powers and urban infrastructure*, Peter Lang Verlag, Frankfurt am Main, pp. 209–211.

Holland, R. C. 1984, "The new era in public–private partnerships," in P. R. Porter and D. C. Sweet (eds), *Rebuilding America's cities: roads to recovery*, Center for Urban Policy Research, New Brunswick, NJ, pp. 121–151.

Huijbregts, R. 2017, "The social responsible developing investor," Masters thesis, Delft University of Technology.

Kenniscentrum PPS 2004, *Samenwerkingsmodellen en de juridische vormgeving daarvan bij pps bij gebiedsontwikkeling*, Ministerie van Financiën, The Hague.

Klijn, E-H. and Teisman, G. R. 2003, "Institutional and strategic barriers to public–private partnership: an analysis of Dutch cases," *Public Money and Management*, vol. 23, no. 3, 137–146.

Kort, M. and Klijn, E.-H. 2011, "Public–private partnerships in urban regeneration projects: organizational form or managerial capacity?" *Public Administration Review*, vol. 71, no. 4, 618–626.

Labruyere, A. 2015, "Leegstandsaanpak doe je samen: Mogelijkheden voor een gezamenlijke aanpak van kantorenleegstand op gebiedsniveau," Master's thesis, Delft University of Technology.

Lovering, J. 2009, "The recession and the end of planning as we have known it," *International Planning Studies*, vol. 14, no. 1, 1–6.

Olesen, K. 2013, "The neoliberalisation of strategic spatial planning," *Planning Theory* [online] 1–16, available at: http://plt.sagepub.com/content/early/2013/08/28/1473095213499340 (accessed August 23, 2016).

Osborne, S. P. (ed.) 2000, *Public–private partnerships: theory and practice in international perspective*, Routledge, London.

Panteia 2013, *De innovativiteit van het MKB in 2013*, Panteia, Zoetemeer.

Peck, J. and Tickell, A. 2002, "Neoliberalising space," in N. Brenner and N. Theodore (eds), *Spaces of neoliberalism: urban restructuring in North America and Western Europe*, Blackwell, Oxford, pp. 37–57.

Potters, B. 2015, "Corporation-led urban development: Motives for multinational corporations to engage in urban development projects," Master's thesis, Delft University of Technology.

Potters, B. and Heurkens, E. W. T. M. 2015, "Corporation-led urban development," *BOSS Magazine*, vol. 53, 26–31.

Putman, M. 2010, "Een nieuwe ontwikkelaar?: Een toekomstperspectief voor de projectontwikkelaar in gebiedsontwikkeling," Master's City Developer thesis, Erasmus University Rotterdam.

Regales, C. 2017, "Stimulating sustainable urban developments," Master's thesis, Delft University of Technology.

Schiltmans, A. 2013, "Crisis-bestendig ontwikkelen," Master's City Developer thesis, Erasmus University Rotterdam.

Squires, G. and Heurkens, E. (eds) 2015, *International approaches to real estate development*, Routledge, London.

Squires, G. and Heurkens, E. 2016, "Methods and models for international comparative approaches to real estate development," *Land Use Policy*, vol. 50, 573–581.

Steigenga, P. 2015, "Beleggers in gebiedsontwikkeling: Incentives en belemmeringen voor een actieve rol van institutionele beleggers bij de herontwikkeling van binnenstedelijke gebieden," Master's thesis, Delft University of Technology.

Straub, S. 2012, "Co-creation in real-estate: A framework to steer upon value-creating firm–consumer relations in demand-driven development," Master's thesis, Delft University of Technology.

Stumpel, B. 2014, "Urban development trust: Possibilities and limitations for Dutch urban area development," Master's thesis, Delft University of Technology.

Stumpel, B. and Heurkens, E. 2014, "Gebiedsfondsen: lessen uit Engeland voor Nederlandse gebiedsontwikkeling," *Real Estate Research Quarterly*, vol. 13, no. 4, 32–39.

Sturm, C. 2014, "Investor-led urban development: Lessons from community-oriented investment strategies applied by developing investors in the United States," Master's thesis, Delft University of Technology.

Sturm, C., Heurkens, E., and Bol, N. 2014, "Belegger-gestuurde gebiedsontwikkeling: lessen binnenstedelijke herontwikkelingsprojecten in de V.S.," *Real Estate Research Quarterly*, vol. 13, no. 2, 23–32.

Van den Berg, H. 2014, "Gebiedscoalities: Sturen in binnenstedelijke gebiedsontwikkelingen," Master's City Developer thesis, Erasmus University Rotterdam.

Van der Bent, B. 2016, "Financiële haalbaarheid van gebiedsconcessies: De invloed van risico's en enkele ontwikkelmethoden op de financiële haalbaarheid van gebiedsconcessies," Master's thesis, Delft University of Technology.

Van der Krabben, E. 2014, "Private–private cooperation in urban redevelopment projects," *Proceedings of the International Federation of Surveyors Congress: Engaging the Challenges, Enhancing the Relevance*, Kuala Lumpur, viewed on 23 August 2016, www.fig.net/resources/proceedings/fig_proceedings/fig2014/papers/ts08f/TS08F_van_der_krabben_7056.pdf.

Van der Krabben, E. and Heurkens, E. 2015, "Netherlands: A search for alternative public–private development strategies from neighbouring countries," in G. Squires and E. W. T. M. Heurkens (eds), *International approaches to real estate development*, London, Routledge, pp. 66–81.

Van der Post, H. 2011, "New kid on the block: Lokaal energiebedrijf als nieuwe speller in gebiedsontwikkeling," Master's City Developer thesis, Erasmus University Rotterdam.

Vande Putte, H. and De Jonge, H. (eds) 2008, *Corporations and cities: envisioning corporate real estate in the urban future*, Publikatiebureau Bouwkunde, Delft.

Veseli, B. 2014, "De business case van gebiedsontwikkelingsprojecten: Mechanismen die kunnen bijdragen aan het sluitend krijgen van de business case," Master's thesis, Delft University of Technology.

Wolting, B. 2006, *PPS en gebiedsontwikkeling*, SDU Uitgevers, The Hague.

10

Building and leading high performance real estate companies

Frank Apeseche

Abstract

There is plenty of literature published on conceiving, designing, building, financing, and acquiring real estate, yet not much has been written on how to build high performing real estate organizations. To be an enduring company that can effectively construct, acquire, finance, and manage the hard assets of the built environment, the organization responsible for making it happen must be hired, trained, motivated, and appropriately led. In fact, it can be argued that the true assets of an enterprise are the ones that "go down the elevator" each night. It can also be said that the job of high-quality leadership is to inspire these employees to return day after day and continue to perform great work for their companies. This paper examines many of the critical factors that are essential for developing high performing real estate organizations. The chapter addresses how to bring together four factors that form the competitive fabric of superior real estate companies: strategy, talent, culture, and capital. The chapter explores how to construct a real estate industry value chain and position the core competencies of a company on the value chain so it can successfully compete. It then examines how executives can build their own personal leadership skills necessary to oversee high performing companies. The important steps for hiring and training employees, and amalgamating talent into a productive company culture are then discussed. Finally, the chapter examines how strategy, talent, and culture can be effectively utilized to attract desirable, low-cost capital for future investment and development activities.

Introduction

Real estate companies, by virtue of the industry in which they compete, are predominantly developed and managed by their founding entrepreneurs. A few ultimately grow to become institutional "category killers"; others survive to successfully compete in a crowded and competitive landscape. Many, however, eventually close down. In fact, according to a report by McKinsey and Company, "less than 30% of family businesses survive into the third generation of family ownership" (Caspar et al., 2010). Other references have separately documented the increase in delistings and privatizations of public REITS that were not able to maintain their public competitiveness (Chapman, 2005; Filisko, 2007; Jacobius, 2015).

Sometimes companies cease to exist due to material economic events that stop them from being competitive, sometimes due to poorly executed strategies, and other times due to the founder neglecting to institutionalize a lasting organization to succeed him or her.

This chapter identifies and discusses some of the critical ingredients required to start, grow and lead high performing, sustainable real estate enterprises. It is comprised of four sections:

1   Identifying and acquiring the necessary personal skills to effectively lead
2   Building a competitive strategy that can access capital
3   Hiring, developing, and managing talent
4   Creating a sustainable, high performing culture.

In each section the critical items that drive success are noted and the techniques for developing these are highlighted.

There are an abundant number of real estate companies in existence today, but only a select few are considered "the highest performing" enterprises. Some noteworthy US-founded examples include Avalon Bay Communities, Boston Properties, Taubman, the Blackstone Group, CBRE, Morgan Stanley, and Marriott International. What attributes do these "high performing" real estate companies have in common? First, they execute their investment, development, or other services in a consistent, high-quality manner; they also have sustained their superior quality over multiple real estate cycles. Second, these companies have refined their competitive strategy to provide customers with significant perceived value. They have done this by either outperforming their competition or by perfecting a business model that is both unique and in high demand. Third, these elite enterprises continuously access low-cost capital to fund their business requirements. They do this by raising money around superior strategies, and then proactively managing investors' expectations to maintain their long-term loyalty. Finally, these companies build organizations that are talented, passionate, and able to thrive through leadership transitions from the founder or former executives to new corporate chiefs. They accomplish this, in part, by conscientiously developing younger talent into future leaders so long-term prosperity can be maintained. So how does one begin the process of building a high performing real estate company? It starts with identifying the key qualities required to be a successful company leader.

Identifying and acquiring the necessary personal skills to successfully lead

This process begins with assessing the leadership qualifications of the founder or highest managing executive and identifying those skills that may be underdeveloped yet are necessary to build and lead the entire enterprise. What a managing executive may lack inherently, he/she can either develop or access by partnering with other senior managers in the organization who are strong in those areas where the executive may not be. Most of the greatest organizations are actually led by "executive teams" that have individual members with complementary skills, and yet as a collective group exhibit the full gamut of managerial and leadership capabilities that are essential to drive the company forward. Being able to utilize a full range of such leadership skills is necessary for building long-lasting, competitive, sustainable enterprises.

What are the fundamental skills that are required to manage vibrant real estate companies? Seven skills, which progressively build on each other and help form the basis of high-functioning corporate leadership, are listed below. As with many normal progressions, less tenured employees tend to start at the top of the list perfecting the first two sets of skills initially, while qualified executives, as a collective group, have had to master all seven (Armstrong, 2011; Collins, 2001).

1 Technical skills: These are the foundational skills good practitioners develop early in their careers. They include financial and quantitative analysis, underwriting, project management, written and oral communication, transaction sourcing, legal document review, and fundamentals of negotiations.

2 People skills: These include the ability to get along with a broad constituency of professionals, as well as the ability to create and communicate win–win opportunities with important established relationships.

3 Managerial skills: These encompass the ability to effectively delegate and still ensure that progress of key initiatives is accomplished. These also include the ability to train and mentor employees, build effective teams with complementary skills, and establish high-functioning controls that enable the teams to successfully focus and execute. Great managers are great motivators and organizers of people.

4 Industry knowledge and vital industry relationships: These incorporate a deep understanding of the industry and its trends, insightful knowledge of the competitors and what strategic initiatives they are developing, and sound relationships with those who influence the dynamics of the industry.

5 Financial stewardship: Financial stewardship involves the ability to understand the vital financial metrics that must be implemented to compete and then thrive. It also incorporates the ability to direct company activities that take advantage of financially beneficial opportunities and avoid initiatives that will create undue risk. Finally, financial stewardship includes the ability to develop and successfully execute a financial plan.

6 Strategy and marketing expertise. These require a very sophisticated level of skill because they are the most nuanced, yet are key to driving a company's competitive superiority. Strategy development is the ability to comprehend where the company can best compete in its industry segment and then put in place the necessary items to enable the company to effectively execute the strategy. Marketing deals with the communication of the company's vision in a way that inspires its stakeholders (employees, investors, customers, agents) to engage and help achieve its potential. When the dynamics of an industry or its competitive forces change, these skills must also be employed to alter the company's strategic direction.

7 Strongly Grounded Ethics. The best corporate leaders understand that maintaining a long-lasting, prestigious industry position requires business behaviors that all stakeholders can be proud of. Being able to personally embrace and then permeate strong ethical competency through the organization is essential for maintaining effective corporate values.

As one progresses from mastering technical skills to having strong strategic marketing and ethical stewardship capabilities, emotional intelligence overrides analytical intelligence. Since the real estate industry requires a high amount of human interaction, it is essential that executive leaders are adept at interpersonal interactions. Assessing one's skills before taking on the challenge of building and leading a real estate company is critical. A good assessment may lead one to work on developing personal skills that are not yet mastered, partner with others who have complementary skills, or hire key managers into the organization who can bolster any underdeveloped leadership capabilities needed to run and govern a robust enterprise.

Building a competitive strategy that can access capital

The first part of constructing a successful competitive strategy is to truly understand the company's unique core competencies and then to translate these into service offerings that customers and capital providers truly desire. It is important to determine early on if there is a large enough demand for the company's service offerings in the markets where it plans to compete. To help gauge this, one should ascertain if the company's customers will pay enough for the services purchased in order for the company to thrive. This becomes more probable if the company can provide its customers with something that the competition cannot, or alternatively, if it can do something significantly better than the competitors.

The second part of this process involves "mapping out the value chain" for the industry segment in which the company will compete, and then analyzing what part of that "value chain" the company plans to focus on. The "value chain" describes the necessary activities that firms competing in a particular real estate industry segment must perform for their customers. These activities are the key items that customers desire and will pay to have produced (Porter, 1985).

An example value chain for the real estate investment/development market segment is highlighted in Figure 10.1. It is important to note that industry segment value chains can be developed for all parts of the consolidated real estate industry. In addition, more detailed subsidiary value chains can be developed for each activity listed in a segment value chain. Figure 10.2 portrays an example of subsidiary value chains. Subsidiary value chains of a particular real estate market segment may highlight opportunities where very small, ultra-focused firms can successfully compete.

Exhibit 1:   Value Chain for RE Investment/Development Market Segment

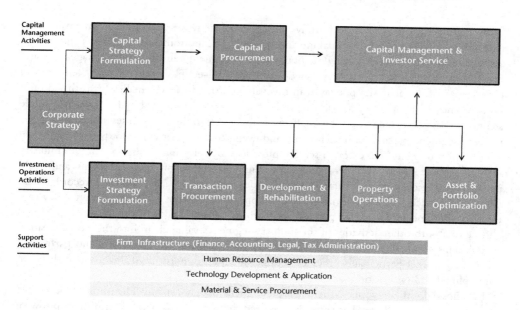

Figure 10.1   Value chain for real estate investment/development market segment

Source: Apeseche, 2012

Exhibit 2.   <u>Example of Subsidiary Value Chains</u> (where specialized firms might focus)

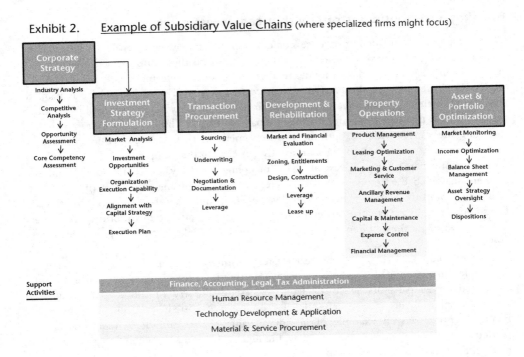

Figure 10.2 Example of subsidiary value chains (where specialized firms might focus)

Source: Apeseche, 2012

In Figure 10.1 the top row portrays the essential activities associated with raising and managing capital necessary for making investments. The bottom row represents the critical activities associated with executing the purchase, development, rehabilitation, management, and disposition of land and constructed assets. Very large, well-established firms have the luxury of pursuing all of the activities portrayed in this value chain. However, smaller firms that do not have the means or capability to engage in all activities can also succeed by focusing on a single activity or subset of activities. These firms may service other companies operating in the industry segment, partner with such firms, and even outsource activities not part of their own core mission to additional external service providers. By selecting where to enter an industry segment's value chain, younger companies can concentrate on their strengths and build necessary experience to gain increased market penetration and a healthy track record. These firms, when successful, will ultimately have the option to grow in a variety of ways. They may elect to continue to focus and partner with others or they may, with increased experience, choose to absorb other activities of the industry segment's value chain into their own business model. This approach of initially concentrating on one's core competencies and partnering with other market participants provides younger firms an opportunity to evolve and thrive in a disciplined and systematic way.

But how do these young "sharp shooter" firms access capital? The most experienced competitors in the industry have already mastered the activities associated with the top row of the value chain. They can repeatedly raise low-cost capital from well-endowed institutions and/or the public marketplace. Younger firms, however, can still participate in this industry segment by partnering with established experts who actively raise and manage low-cost capital. Many large capital managers frequently like to partner or "allocate" their capital to outstanding operating "sharp shooters." The large capital management companies such as Apollo, Blackstone, Carlyle, etc., have chosen to specialize on what they do best. They develop robust real estate investment strategies and successfully market these to national and international capital sources. Then they proactively manage investor return and risk expectations to achieve long-lasting relationships. "Allocators," such as these, choose to optimize their investment flexibility by focusing on the top row of Figure 10.1. They then partner with local developers or value-add acquirers who can successfully outperform the activities on the bottom row. Many times the local operating partners, in turn, outsource their non-core competencies to other providers. For example, property management companies (those who specialize in providing superior leasing and building services per Figure 10.2) are frequently hired by local development and acquisition operating partners.

As such, one can see how successful, focused companies can compete by concentrating on specific activities of a real estate segment's value chain and then accessing capital by tapping into the balance sheets of expert fund managers. Most practitioners who grow real estate companies from start-up first begin by partnering with firms who have already perfected their institutional capital accumulation and investor management capability. Once these younger firms gain meaningful experience and expertise, they too can consider pursuing lower cost, direct institutional funding relationships. Generally "capital allocator" firms (those who provide investment funds to younger, less mature sharp shooters) will require strict project underwriting standards and significant project oversight. They will also demand control and discretion over material events and charge a higher cost of capital on their investment. However, the lessons and experience learned from working with these expert allocators can be very powerful for companies who want to eventually raise low cost institutional capital on their own.

When deciding where on the value chain to initially concentrate, some important strategic questions should be addressed:

1   Are you planning to absorb a piece of the value chain that is large enough for you to continue to grow and scale?
2   Are you absorbing a part of the value chain that is profitable enough?
3   Do your perceived "core competencies" truly match your focus?
4   Are you going to be competitive? Or are there a number of direct competitors that are so large and powerful in the places you want to concentrate that they will drive you out of business, not let you grow, or not let you generate appropriate profits?

In summary, thinking through a differentiated, implementable, and scalable competitive strategy and aligning it with the desired demand of industry investors and customers is a critical first step to establishing a sustainable profitable position. Starting with a focused concentration that plays to one's core competencies is a pragmatic and viable way to begin a company's successful evolution.

Hiring, developing, and managing talent

Many young real estate entrepreneurs believe their most important assets are the projects with which they are currently engaged. In actuality, when one oversees a real estate company, the most important assets are the employees who go down the elevator each night. And the job of a true leader is to ensure the same employees are motivated to take the elevator back up the following morning and continue to do great things for their company. How does one begin to think through the process of building a talented and effective organization?

First, the company's leadership must determine the skills and resources that are needed to accomplish mission-critical activities, those that are part of its own "personal value chain." The personal value chain is that part of the industry segment's value chain where the company will compete. Non-critical items can potentially be outsourced to other service providers. Once the skills required to address these activities are identified, compensation structures need to be developed before beginning the employee recruitment process.

Compensation initially tends to be above market rates to attract the most talented professionals. Compensation comes in three components: base salary, bonus incentives, and long-term equity compensation. To some degree, leaders can trade one of these components for another. For instance, a firm with a great strategic vision but limited current cash flow may be able to offer a lower base salary in exchange for an attractive long-term equity plan. Of the three components, the equity piece is by far the most complex and potentially costly to the company. It is, however, the piece that will most likely attract and retain top talent. Although there is no single formula that defines when, how, or how much equity to grant, there are a number of important items to consider. When giving equity to employees, make sure to address the following:

1   Will the employee contribute to the overall success of the firm over the long term? If not, then incentives should be structured as shorter-term bonus awards.
2   What type of equity units will be issued? Many companies grant stock options or carried interest to employees. These instruments are linked to the future appreciation of assets. Others provide employees with loans so they can directly purchase common stock, restricted stock, or limited partnership interests. By doing so, these employees participate in the company's value that has been previously created as well.
3   What entity will equity interest be given in? Will equity be granted in a particular project (something the employee might be directly responsible for), a portfolio of projects or the entire company? The most senior executives usually receive equity in the entire company. Additionally, most executives of publicly traded companies receive stock options tied to the appreciation of the entire enterprise. Many private development firms, however, grant employees equity only in the specific projects for which they are responsible.
4   What percentage of equity should an employee receive, given that the entire equity available only adds up to 100%? A desirable approach for most companies is to grant allotments of equity to an employee over time, rather than bestow everything up front (prior to knowing if the employee can perform successfully over the longer term).
5   What type of vesting schedule should employees receive? Vesting is how a specific tranche of equity is earned. Employers can structure both performance-based and time-based vesting schedules. Time-based schedules tend to be 3–6 years in length, while performance-based vesting is usually tied to the completion of a meaningful project or event.
6   How much equity should be held back in reserve for future hires?

Hiring the first round of employees smartly can reduce future compensation expenses. This works well when early senior hires are prescreened to have both the requisite technical skills as well as high-quality interpersonal and managerial skills that will be needed to build future teams and train employees. For example, a long-term veteran executive with special talent in land development and asset construction should be able to help build a future team of subordinates who can be trained internally and therefore reduce the need to continue overpaying for pre-established experience down the road.

Organizational reporting also needs to be thought through up front. Individual and department reporting hierarchies should directly link to clear accountability for the key activities in a company's personal value chain. These activities are the imperative items that investors and customers desire and will pay for. Efficient and effective organizational reporting will establish clear lines of responsibility, minimize department overlaps, and alleviate future expense burdens by avoiding redundancy (Bossidy et al., 2011).

After compensation models are established and the organizational structure is developed, candidates can be screened for skills that are important to the company. The employee hiring process should be fully vetted throughout the organization, because hiring those who can truly make a contribution to the enterprise is paramount to the development of high functioning firms. After new hires are on board, they must be trained, mentored, and later evaluated based on their actual performance. Training, mentoring, and evaluating employees are probably the most important building blocks to developing lasting talent in a company, but too frequently are partially or entirely ignored. Making the workforce as effective as possible requires constant training, timely and consistent reinforcement, and a transparent road map, with active mentorship, that enables employees to see their own career advancement opportunities.

Ironically the most frequent cause of voluntary employee resignation is when a direct boss does not honor his or her commitment to accomplish these items. High performing organizations, on the other hand, hire candidates that offer the potential for significant future company contribution, and then train and motivate these individuals to succeed. These organizations proactively establish formal career advancement and succession structures in their policies and practices creating effective advancement and succession at all company levels. There are numerous techniques for training and motivating employees. Some of the most important elements to consider are:

1 Insure a consistent and rigorous individual performance system is in place. Key skills for superior job performance should be mapped out with each skill given a measure of importance relative to the overall goals of the company. Specific observations of individual performance against plan should be clearly delineated and then discussed in detail with the employee. Finally, a personal growth plan should be developed and reviewed with every employee at least once per year.

2 Emphasize honesty, integrity and transparency when discussing an employee's compensation, performance review, or career trajectory status.

3 Since most employees change jobs because they believe their boss does not treat them fairly, the direct boss-to-employee relationship is pivotal to long-term talent development and employee retention. Therefore, the highest priority should be given to routine, high frequency, informal employer/employee contact points, along with consistent, proactive, and balanced formal reviews. To be effective this activity must be endorsed and absolutely required by the most senior leaders of the company.

4 Acknowledging an employee at the moment when he or she produces extraordinary results is the most impactful way a boss can reinforce future desired behavior.

5 Bosses who are disappointed by employee behavior often disengage when just the opposite needs to happen. The employees who falter should be called out quickly and without emotion. The expected behavior should be discussed, without making it a personal attack, and the impact of poor behavior on both the organization and the employee should be clearly explained. No employee should be surprised during a formal

review, and no employee should be unaware of how his or her individual performance links to compensation and ultimate career advancement.

6   On the other hand, when bosses see that an underperforming employee truly lacks the capacity or desire to improve his or her future contribution, they should move swiftly to counsel this person out of the company. This is because allowing underperformers to "hang on" sends a clear and demotivating message to the rest of the contributing organization.

7   For exceptional employees who show the potential for future executive leadership, a formal mentoring leadership program should be established. Many companies do an adequate job supporting employees through the first three personal leadership skills mentioned earlier in this chapter (technical, interpersonal, and managerial), but most fall short on the later four (industry influence, financial stewardship, strategy and marketing, and ethics). If new leaders are to be successfully cultivated, the infrastructure for supporting these individuals should be established well in advance.

8   Chief executives or founders must set a clear example for the company by adhering to these practices with their own direct reports. If leaders are allowed to underperform or exhibit poor supervisory behaviors with their employees, the rest of the organization will inevitably suffer.

In summary, hiring and managing talent starts with identifying the important skills needed to accomplish mission-critical tasks within a company's personal value chain. Compensation models and organizational reporting must be clearly established before the hiring process begins. After new employees are hired, they need to be actively trained, mentored, and evaluated for performance. Their personal performance and contribution should directly tie to their own compensation and career advancement opportunities.

Creating a sustainable, high performing culture

Superior performing real estate companies have high performing corporate cultures. These companies are able to garner market information, see changing trends before others do, and most importantly, take calculated decisive action to adjust current plans, make new plans, or conscientiously abort and avoid others. High performing companies optimize their individual departments' technical and managerial talents by working collaboratively across functions to effectuate impactful change. How do leaders move beyond developing individual employee and departmental contributions to building a high functioning corporate culture? Work begins at the top with the chief executive or founder. These leaders must set the expectation of high performance for their companies and consistently model these behaviors themselves for others to see. Then the various departments and divisions of a company must be trained to behave the same way.

Just as employees have to be motivated and managed to over-perform, so do departments and divisions that execute critical company activities. Linking the organizational elements of a company to corporate goals that inspire department collaboration is a first step in developing a high functioning culture. Multifaceted organizations that perform a variety of critical business activities need to proactively connect the expertise of their departments in order to advance their most sophisticated initiatives. This becomes even more important when a real estate company operates simultaneously in several disparate geographical locations (a distributed organization). Here decisions have to be made efficiently and at all organizational levels in order to react nimbly to changing market conditions. Information in high performance companies flows rapidly and efficiently throughout the organization. Resources that need to be quickly assembled or deployed, due to "an event," are capable and willing to help in a manner that is both actionable and impactful. It is not a coincidence that such high-quality teamwork tends to abound in high-functioning organizations.

Motivating departments to cross-collaborate on critical activities with a bias for action begins by visually showing the impact of how each department contributes to the other's success as well as to the overall major goals of the company. To exemplify, Figure 10.3 describes a technique for linking the contributions of various departments to the consolidated investment goals of a multifaceted real estate investment company.

1   Here the company's goal is to achieve a 15.5% levered return on its investment portfolios. Assume each portfolio is comprised of a high volume of individual projects.
2   The driving components of a particular project's total return have been disaggregated, so each department that participates in the value creation process can see how it specifically contributes to the overall return. If multiple projects consolidate to the company's corporate return objective, then the same graphic can be used to show how critical departments must come together to generate the company's consolidated investment goal.
3   Examining the property management division in this example, one can clearly see that if it increases rents, maximizes occupancy, and holds down expense growth, its contribution to the company's overall return goal can be achieved. Alternatively, one can surmise how the acquisition division can help the company achieve its total return goal by negotiating the acquisition price of a new investment very well, identifying projects that have the

Exhibit 3.   <u>Linking the Organization to a Deal's Total Return Goals</u>

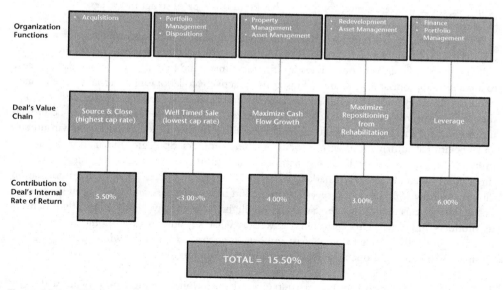

Figure 10.3   Linking the organization to a deal's total return goals

Source: Apeseche, 2012

potential to generate greater future rent growth, or projects that will likely receive the greatest interest from buyers at their time of disposition. The other departments such as rehabilitation/redevelopment, finance, etc., can apply the same principals to analyze their contribution as well. If one department cannot achieve its return objectives, it becomes evident how other departments can cross-collaborate to make up the difference, and ultimately still accomplish the corporate objective.

4   Because everyone has an important and understandable role in achieving the company's overall investment goal, it is easy to see why the culture of this company is one that consists of superior cross-collaboration between various departments and, equally, great respect for fellow workers.

It is important for the chief executive or founder to take a proactive role in establishing the connectivity that links each department and division to an overall collaborative corporate agenda.

Teaching the organization to act as an owner is the next challenge. Savvy business owner/entrepreneurs have a unique ability to quickly garner critical industry and competitor information, analyze what it means, and take swift, impactful action. These leaders are calculated risk takers; they understand that acting is about taking risks, and are constantly weighing, adjusting, and trying to reduce these risks as progress is made. They tend to always favor action over no action. In other words "execution" trumps "analysis." So how can a leader teach the organization and all of its divisions to act the same way?

This begins by developing a high-quality information infrastructure and then showing the organization how to translate good information into actionable decisions. Defining what is the most imperative actionable information that a company needs to have is difficult to do. This requires that senior leadership have a stake in defining it, as well as in building an environment that will promote sharp analysis and decisive execution. Because market cycle timing (and speed of implementation when one sees market changes occurring) is so important to achieving success in real estate, companies that build this type of structure successfully are likely to outperform less able competitors.

The first step in this process is to define what competitive and operational information is critical for the company to focus on. This job needs to have the engagement of the owner or chief executive officer. Without this leadership, companies tend to collect too much data, neglect to analyze it appropriately, and are lackluster in proactively collaborating internally to effectuate good results. They also are unable to leverage this information in a way to strategically catapult themselves forward (see Armstrong, 2011; Collins, 2001). Actionable information can be broken into two major categories: strategic information, with a focus on the competitive environment; and operational information, with a focus on the customers. Strategic information highlights shifts in the industry, what local and national competitors are up to and ongoing changes to supply and demand factors in markets where a company competes. Operational information illuminates how well the firm is serving current and new customers, and what it can specifically do to enhance this in a profitable way. It also highlights how effectively and efficiently it delivers its products and services to customers, as well as what key initiatives it should undertake to improve productivity. There are a few important factors to consider when developing the company's strategic and operational information infrastructure:

1   The specific data to capture for strategic and operational decision making must directly link to the strategic goals of the company, those that drive critical activities associated with the company's personal market segment value chain.

2   Informational metrics that are most beneficial help to determine what to focus on, and then measure how effective the initiatives taken have been.

3   When trying to distinguish between good data and vital company information consider the following: If the information being obtained and analyzed cannot lead to "actionable decisions," then it probably is not essential.

4   Valuable information today may be irrelevant in the future. Therefore it is imperative to consistently reassess and update what is most important at any given point in time.

Training the organization how to analyze and act upon information is equally, if not more, critical than the information itself. Here "less is more." In other words, companies that outperform tend to generate fewer reports but with meaningful actionable insights. They then utilize the information to successfully diagnose impacts from key trends and collaborate on execution. Executives can model their own personal behavior of good analysis, collaboration, and decision making publicly during monthly, quarterly, and annual meetings. They can ensure that a smaller number of very insightful reports are understood by all departments and actively utilized. In a group setting the executives can highlight issues, raise probing questions, and brainstorm opportunities. While the executives do not have to solve problems directly, they should emphasize to others how important it is that these issues are resolved well by the group. These behaviors, once understood by other senior leaders, can cascade through the organization with the complete endorsement and support of senior management. For a particular multifamily company, Figures 10.4 and 10.5 highlight how seven key reports are utilized throughout its organization during project site visits as well as for monthly, quarterly, and annual meetings. This company diagnoses problems and changes to market conditions quickly, because its employees are well acquainted with embracing valuable information, analyzing its impact, and cross-collaborating immediately to make impactful decisions faster than its competitors.

Exhibit 4.   Organizational Utilization of Seven Key Reports

Report Name	Purpose
1. Property Matrix	Property Behavior
2. Resident Survey	Property Behavior
3. History & Rolling Forecasts	Property Financials
4. Trend Metrics • Occupancy • Net Rental Income	Portfolio Insight
5. Yield Analysis	Portfolio Insight
6. Key Initiatives	Strategic Priorities
7. Market Reporting System	Market and Competitor Analysis

Figure 10.4   Organizational utilization of seven key reports

Source: Apeseche, 2012

Exhibit 5. 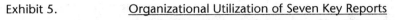 Organizational Utilization of Seven Key Reports

Key Meetings	Reports Used	Attendees
Monthly Property Manager Call	– Market Reporting System – History & Rolling Forecasts – Property Matrix – Resident Surveys	• All Property Managers • Property Management Executives • Asset Management
"Spot" Property Site Visits	– Market Reporting System – History & Rolling Forecasts – Property Matrix – Resident Surveys	• Asset Management • Executives
Quarterly Portfolio Reviews	– Market Reporting System – Trend Metrics – Yield Analysis – History & Rolling Forecasts – Key Initiatives	• Executives • Asset Management • Property Management VPs • Acquisition VPs • Human Resource SVP • CFO & Senior Controllers
Budget and Planning	– Market Reporting System – History/Rolling Forecasts – Property Matrix	• Same as Quarterly Portfolio Review
Quarterly Investor Meetings	– Market Reporting System – Trend Metrics – History & Rolling Forecasts	• Executives
Annual Investor Meetings	– Market Reporting System – Yield Analysis – History & Rolling Forecasts	• Executives

(Internal: Monthly Property Manager Call, "Spot" Property Site Visits, Quarterly Portfolio Reviews, Budget and Planning, Quarterly Investor Meetings; External: Annual Investor Meetings)

Figure 10.5 Organizational utilization of seven key reports extended

Source: Apeseche, 2012

When the organization is trained to diagnose and execute well, the information that flows through the enterprise is fast, efficient, and with purpose. As managers begin to diagnose problems as well as think and act as owners do, important strategic actions can be taken without the need to be bogged down by slow corporate approvals. Instead managers are free to execute. Because poor behaviors, such as blaming others for disappointing results, are replaced with good analytical and action-oriented behaviors, it becomes easy to debrief on executed initiatives with the goal of making even better decisions in the future. This is critical for decentralized real estate companies, working at multiple sites and in multiple markets. Figures 10.6 and 10.7 highlight the result of applying these concepts by comparing the information flow of a low performing real estate investment manager to a high performer.

High-performing companies reinforce high-quality "execution oriented" behavior by also linking it back to their compensation systems. Bonuses can be devised so part of an employee's incentive is tied to a subjective review of how he or she communicates, collaborates, and reinforces good team dynamics. Long-term equity grants in the consolidated enterprise will reinforce the concept of making decisions that maximize the strategic competitiveness of the overall company, rather than just individual departments.

Indeed, high performing cultures allow companies to become smarter, develop better strategies, operate tactically, and most important, execute with greater impact than their competitors. It is no wonder that companies with high performing cultures can accomplish much more. In summary:

Exhibit 6. Communication Flow of a Low Performer

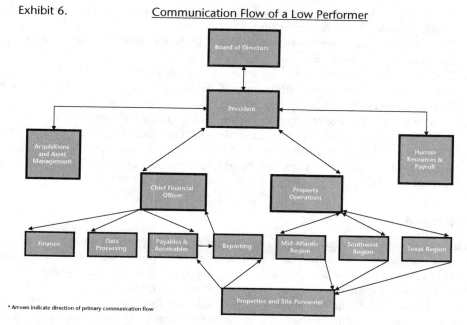

Figure 10.6 Communication flow of a low performer

Source: Apeseche, 2014

Exhibit 7. Communications Flow of a High Performer

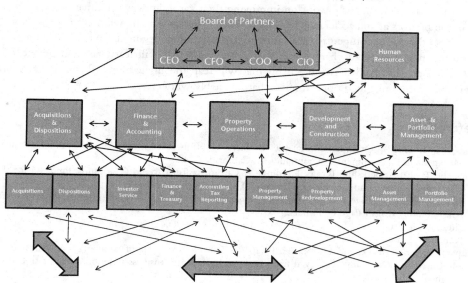

Figure 10.7 Communication flow of a high performer

Source: Apeseche, 2014

1   Companies with high performing cultures see opportunities and changes in market conditions faster than their competitors do.

2   This allows the high performers to take advantage of first-move arbitrage opportunities that may present themselves.

3   These companies have the option to actively engage offensively to enhance returns and financial results, or alternatively, defend against oncoming threats and economic downturns faster than the competition.

4   A company can optimize opportunities by using high-quality information that converts to actionable decisions, backed by an organization that can rapidly diagnose and implement.

5   These desired behaviors can be reinforced by ensuring that a company's compensation system rewards those that exhibit meaningful team collaboration followed by impactful execution.

Conclusion

Every real estate company has characteristics that make it unique. And as a result, it is difficult to define a single set of parameters that will optimize the performance of all companies participating in the consolidated industry. However, great companies do seem to have an ample supply of the following ingredients that help propel them to the head of the pack: STRATEGY, TALENT, CULTURE, and CAPITAL. Founders and chief executives who successfully build these items into their companies will develop firms that become superior competitors. In this chapter, some of the important considerations for creating high performing real estate companies, those with the best strategies, talent, culture, and capital, have been discussed. Once initially established, the goal of enriching an enterprise's performance should never cease, and it is one of the most vital roles an executive leader can perform for his or her company. Chief executives who also embrace the principal of "succession planning" will have even greater impact on bringing their firms to these highest levels of performance. And the best performing companies that continuously work on improving their capabilities will ultimately outdistance their competitors at an accelerated rate. These enterprises, as a result, will have the capacity to dominate the most competitive positions in their industry segments. As the markets and environment continue to change, it is these companies that will learn fast and be the first movers to readapt. And it is these firms that will reap the highest rewards from the future profitable and unimagined opportunities that inevitably will emerge.

References

Apeseche, F. (2012), Optimizing Strategic Business Controls, Presentation to Harvard's Advanced Management Development Program, July 14, Cambridge, MA.

Apeseche, F. (2014), Berkshire Realty Holdings: A Corporate Turnaround Case Study, Prepared for the Harvard Graduate School of Design, January 20, Cambridge, MA.

Armstrong, M. (2011), *How to Be an Even Better Manager: A complete A to Z of proven techniques and essential skills*, London, Kogan Page Publishers.

Bossidy, L., Charan, R., and Burck, C. (2011), *Execution: The discipline of getting things done*, New York, NY, Random House.

Caspar, C., Karina, A. D., and Elstrodt, H. P. (2010), The five attributes of enduring family businesses, McKinsey and Company. Retrieved from www.mckinsey.com/business-functions/organization/our-insights).

Chapman, P. (2005), Why public REITS are going private, *National Real Estate Investor*, November 2.

Collins, J. C. (2001), *Good to Great: Why some companies make the leap and others don't*, New York, NY, Random House.

Filisko, G. M. (2007), Privatization wave hits public REITS, *National Real Estate Investor*, February 1.
Jacobius, A. (2015), Managers snap up market-battered REITS, *Pensions and Investments*, September 7.
Porter, M. E. (1985), *Competitive Advantage: Creating and sustaining superior performance*, New York, NY, Free Press.

11

Strategic management systems for real estate development

Paul H. K. Ho

Abstract

The real estate development model can be seen as an input–transformation–output system in which developers must periodically respond to internal and external strategic issues over the course of the development process. Today, the most decisive issues are those arising from dynamic and unpredictable environments. This study proposes four main types of strategic management systems: proactive, reactive, planned and ad hoc. Proactive strategic management systems are adopted by large, powerful, and politically well-connected developers with future-oriented management systems. Planned strategic management systems are adopted by large developers with sophisticated strategic planning practices. Reactive strategic management systems are adopted by large- to medium-sized developers with formal structures of management. Ad hoc strategic management systems are adopted by medium- to small-sized developers that lack a systematic management process. Data were collected by means of a web-based questionnaire survey in Hong Kong. 160 questionnaires were sent out between July 2015 and September 2015, and 45 questionnaires were completed representing a response rate of 28 percent. Descriptive statistics, hierarchical and K-means cluster analyses and discriminant analysis were used to test the hypothesis that "there exist four significantly different kinds of strategic management systems adopted by developers, with each system having different capabilities for responding to strategic issues." The result generated from the cluster analysis was validated by the between-method triangulation. This study indicates that understanding emerging strategic issues is critical for the success of an organisation. Although not all of the potential threats and opportunities can be avoided or captured, developers would certainly increase their chances of success by improving their abilities to anticipate and deal with threats or opportunities.

Introduction

Before the early 1980s, when external organizational environments were relatively stable, developers could reasonably assume that what worked at that time would work equally well in the future. As the change in organizational environments was slow, a reactive-style management system was still workable, and business was business. Most critical decisions arose in response to internal rather than external issues. Developers could concentrate their efforts on improving the real estate development process itself. As a whole, the development process was not greatly vulnerable to interference from the environment (Ho, 2014).

However, since the early 1980s, things have become increasingly connected in a global world. The real estate development sector has become more dynamic and unpredictable. Globalization, urbanization, demographic change and technological advancement are constantly interacting with each other. Some of the resulting events, threats and issues have had an explosive effect on both local and global economies. A review of financial crises in the past 30 years shows that there have been a number of crisis events such as the Latin American sovereign debt crisis in 1982, the stock market crash in 1987, the junk bond crash in 1989, the Asian financial crisis in 1997–98, the dotcom bubble in 1999–2000, and the global financial crisis in 2007–8 (IFR, 2013). In these times, a major geopolitical event may push the global economy into a recession. Developers cannot afford to ignore changes happening beyond the real estate development sector. A lapse of attention in a volatile environment can result in being blindsided so that even a well-planned project may be foiled.

As organizational environments rapidly change, the management style must also change. Traditional management systems are no longer sufficiently meticulous to deal with today's dynamic environments (Ansoff, 1980). For any major management decision, it is necessary to foresee the timing, extent and severity of any potential crisis emerging from the environment. Indeed, some of these threats are quite predictable. However, so far, there has been limited research on the strategic management systems adopted by developers. Therefore, this study aims to develop a typology of strategic management systems for real estate development organizations based on their relative capabilities for responding to their changing environments.

Proposed management model for real estate development

In strategic management theory and practice, one of the most fundamental principles is that the short-term effectiveness and long-term survival of an organization depend on the actions it takes in responding to its environment. Developers are often bombarded by a continuous stream of diverse events and trends emerging from their environments. Some of these events and trends present possible threats or opportunities for an organization. Threats represent potential losses, whereas opportunities represent potential gains. Both threats and opportunities must be taken into consideration, as they may significantly affect a developer's ability to achieve its organizational and project-specific goals.

Based on the "open system" theory, the proposed management model for the real estate development can be seen as an "input–transformation–output system" as shown in Figure 11.1. In this system, an organization must from time to time respond to internal and external strategic issues over the course of the real estate development process. The whole development process consists of a series of stages from the beginning to the completion of a project. Typically, these stages include the feasibility study, land acquisition, funding arrangement, planning application, preliminary design, building and other approvals, detailed building design, on-site construction, and the marketing or letting/disposal of the completed properties. This process is similar to an industrial production process, where various resources (such as land, finances, designs and building materials) are input from the environment. These resources are then transformed into a final product (in this case, a property) which is ultimately output to the real estate market. During the development process, developers are required to deal with a number of strategic issues that are generated either within their internal organizations or from the external environment. The real estate development process can be managed on an "issue by issue" basis.

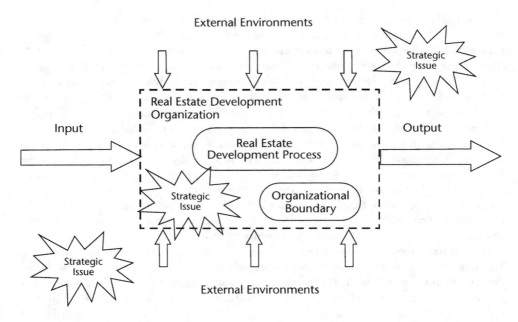

Figure 11.1   Basic management model for real estate development

Based on the recent study conducted by Ho (2014), 40 potential strategic issues are identified and listed in Table 11.1.

Proposed typologies of strategic management systems

The success of developers depends on their ability to manage various strategic issues during the real estate development process. More precisely, the success depends on the strategic management system adopted by individual developers. The strategic management system is that part of an organizational unit that forecasts, addresses and responds to emerging strategic issues that are potentially important for the organization. Based on previous studies on strategic management, this chapter proposes four main types of strategic management systems: proactive, reactive, planned and ad hoc. These four types of management systems seldom appear in their pure forms. Most systems are mixtures of two or more management styles. However, one of the four management systems is usually dominant.

Proactive system

If an organization is able to learn more about a strategic issue earlier, they will have more time to respond. Hence, the most important time to detect a strategic issue is when it is still emerging. Some active developers may be capable of taking appropriate actions within their organization or environment to alter the course of an issue's development. They may do so even before they have enough information to accurately estimate an issue's future effect. This organization deals with a strategic issue before it becomes established reality (i.e., before a threat strikes or before an opportunity is missed). This kind of capability is called a proactive system.

Table 11.1 Potential strategic issues

Economic and financial components	*Cultural and sociological components*
• GNP growth rate	• Population growth
• Money supply	• Marriage rate
• Local currency	• Quality of real estate property
• Unemployment	*Customer components*
• General inflation	• Total demand for real estate property
• Household wealth	• Elasticity of demand for real estate property
• Inflation in materials and labor costs	• Expectation on after-sale services
• Interest rate	• Real estate prices
• Sources of finance for development	• Speculative atmosphere
• Amounts of equity required for mortgages	*Industry-structure component*
• Periods of loan repayment for mortgages	• Barriers to entry into the industry
Political and legal components	• Vacancy rates
• Political climate	• Existing stocks in the public sector
• Taxation	• Existing stock in the private sector
• Supply of government land	• Supply trends from the public sector
• Provisions of public housing	• Supply trends from the private sector
• Provisions of infrastructure	• Technologies in design and construction
• Controls on pre-sale of real estate property	*Competitor component*
• Controls on land-use planning	• Competition in acquiring sites
• Controls on design and construction	• Competition in setting sale prices
• Controls on environments	• Competition in bidding for consultancy work
• Controls on health and safety in the workplace	• Competition in bidding for construction work

Proactive-style organizations believe that the future environment is subject to being created. This view is based on Child's (1972) strategic choice perspective or powerful organization theory which is grounded on two fundamental arguments. First, while there are always certain environmental constraints (be it social, economic, financial, legal, institutional or technological) and internal constraints (involving power structures, limited resources, defined information flows or structured technological processes), the decision makers can still retain a certain amount of discretion over their own actions. Second, large and powerful organizations are not always passive recipients of environmental influences. They often have enough power to manipulate and even control their environments (Child, 1972; Perrow, 1972). Organizations that adopt a proactive system imply that they have a high capability for responding to strategic issues.

Large and powerful developers always attempt to go beyond passively accepting whatever happens. They not only anticipate the threat and opportunity that might occur in the future, but also take an active role in molding the environment to their advantage. As Schwartz (2003) argues, it is inevitable that there will be many surprises in the future, but the driving forces underlying these surprises can be analyzed now. With better understanding of these driving forces, developers can both increase their capacity for responding to ultimate threats and improve their ability to grasp opportunities that might otherwise be missed. Therefore, it is definitely better to start preparing for future possibilities earlier rather than later. To do that, developers must establish and rely on some sort of active environmental scanning system that can warn them of any strategic issues that might pose potential threats, and also inform them of any opportunities that might be exploited through timely action. Proactive developers use such systems to continuously detect any prospective threats and to take real-time action in response.

As proactive developers are involved in monitoring issues at the time they begin to emerge, it is easier for them to make informed choices. They can choose to take immediate initiatives before the issue can unfold, or simply wait and implement defensive actions. The course of action would depend on the nature of the strategic issue, its probability of occurrence, potential effects and the management's attitude (Oomens and van den Bosch, 1999).

Reactive system

Some developers do not engage in environmental scanning and are thus not able to collect sufficient advance information about strategic issues. They can only respond to a strategic issue after it has fully materialized and has already had a definite effect on the organization. In effect, they treat strategic issues only after they become established realities. These management systems are called reactive systems.

Reactive-style organizations are satisfied with neither their past nor their future environments. They are focused only on their present environment (Ackoff, 1981). Unlike the proactive systems, they believe that their environments are so uncontrollable that they have limited capacity to change the situation. This argument is based on Aldrich's (1979) natural selection model or weak organization theory. Such powerlessness is partly due to weaknesses within the organization such as constraints on decision making due to inter-organizational dependence. The managers of reactive systems feel that only very large and politically well-connected organizations can have sufficient power to influence their environments. The large numbers of smaller or weaker organizations are unlikely to succeed in altering the environments in which they operate. Natural selection theorists believe that constraints such as organizational size, limited market share, short-term restrictions on production processes or inadequate abilities for information gathering largely determine each organization's character-istics and its outcomes. The scope for discretion by individual organizations is relatively limited (Hannan and Freeman, 1977; Aldrich, 1979). Therefore, "powerless" developers seldom try to alter the course of an issue's development. Organizations that adopt a reactive system imply that they have a relatively low capability for response to strategic issues.

After a strategic issue has already started to impact upon developers, they are unable to either wholly avoid a threat or to capture an opportunity. They would probably start their response as soon as practicable so as to limit the effect of an issue or catch an opportunity.

Planned system

Under normal circumstances, most developers can start to respond to a strategic issue after that issue has materialized. Many large organizations adopt periodic management systems such as long-range planning, strategic planning or strategic management to deal with issues soon after they become apparent. These systems are the dominant forms of management in the United States (Ackoff, 1981). Developers that operate such management systems may find themselves unable to respond to a strategic issue even after it has reached the materialization stage. This difficulty typically arises because certain strategic issues unfold well before the next planning cycle. In such a management system, strategic issues are not treated on an ongoing basis, but only in a periodic, planned manner. This approach is therefore called a planned system (Ansoff, 1975).

Planned-style organizations commonly believe that their future environments will be better than those of the present or the past. During their planning process, they normally try to predict their future environment and to make preparations for the changes they expect. To a certain extent, they take steps to avoid future threats and to exploit future opportunities (Ackoff, 1981). In this sense, they treat strategic issues in a proactive manner. Organizations that adopt such planned systems imply that they have relatively high capability for responding to strategic issues.

As planned systems tend to treat strategic issues in a proactive manner, one may argue that these systems are more effective than reactive systems (Ansoff and McDonnell, 1990). However, in reality, the future is often impossible to forecast accurately, particularly when the environment is complex and dynamic. The further into the future the managers try to forecast, the greater their error is likely to be. Therefore, the benefits of proactive efforts are restricted, at best, to plans made for the relatively near future.

For planned-style organizations, the most critical factor is their planning horizon, which determines when they will respond to a strategic issue. If the planning horizon is too long, it will be too late for the organization to take any action if the strategic issue has reached its mature stage because by then the threat can no longer be avoided or the opportunity has already passed. The main solution for planned-style organizations is to reduce the period of their planning horizon so that hopefully the planning activity will happen closer to the time when emerging strategic issues can be dealt with appropriately.

Ad hoc system

Many developers do not have any formal strategic planning or management system at all. The managers of such organizations typically respond to strategic issues one at a time, as and when they perceive them. Some may confront a strategic issue when its consequences are imminent, but most would prefer to delay their response until after the issue has had a significant effect on them. They treat strategic issues in an ad hoc manner. This approach is therefore called an ad hoc system (Ansoff and McDonnell, 1990).

Ad hoc-style organizations believe that their current situation is either good enough or else as good as can be reasonably expected. Hence, they seldom react to a strategic issue until their organizational performance is threatened. However, as the processing of strategic issues typically involves a small number of top management staff, the time taken in decision making may be short and the initiation of action may be prompt (Ansoff and McDonnell, 1990). While decisions are often taken in haste, this does not necessarily mean that the responses are poorly prepared. Some ad hoc organizations may consult internal or external experts. This practice is particularly apparent in small developers. In general, organizations that adopt an ad hoc system imply that they have a low capability for responding to strategic issues.

Proposed hypothesis

As discussed above, the strategic management system adopted by an organization represents not only its managerial behavior, but also its capability for responding to strategic issues. Based on the proposed typologies of strategic management systems, the following hypothesis is formulated:

> There exist four significantly different kinds of strategic management systems adopted by developers, with each system having different capabilities for responding to strategic issues.

This hypothesis is shown diagrammatically in Figure 11.2.

143

Figure 11.2 Relationships between strategic management systems and capabilities for response

Research methodology

Data collection

This study was conducted in Hong Kong. An invitation letter was sent to each respondent by an email, with a link to a web-based questionnaire. In total, 160 questionnaires were sent out between July 2015 and September 2015. Eventually, 45 questionnaires were completed, representing a response rate of 28 percent.

Questionnaire design

The data were collected by means of a web-based questionnaire survey which consisted of two sections. Section 1 of the questionnaire collected data on how developers classified their management systems. The respondents were requested to compare their organization as a whole with these four strategic management systems in terms of their own management styles and characteristics, and then to choose the management system that most accurately described their own organization.

Section 2 collected data about developers' capabilities for responding to the above-listed 40 strategic issues (Table 11.1). The respondents were requested to make judgements on their capabilities for responding to each of these strategic issues based on the 7-point Likert-type scale. The evaluation period was between mid-2014 and mid-2015.

Statistical analyses

SPSS was used to perform all statistical analyses. The data collected from section 1 of the questionnaire were analyzed by simple descriptive statistics, which were used to verify whether there were four different types of strategic management systems. The results were also used to validate the statistical results from section 2.

The data collected from section 2 of the questionnaire were analyzed by cluster analysis, which identified groups of developers with similar capabilities for response to strategic issues.

Cluster analysis was carried out in two stages: the first stage was to find the number of clusters, and the second stage was to generate the final solution. Between-method triangulation was pursued by comparing the results drawn from sections 1 and 2 of the questionnaire. As the F-test generated from the cluster analysis was conducted for descriptive purposes only, discriminant analysis was used to test the proposed hypothesis.

Data analyses and results

Step 1: Descriptive statistics

The descriptive statistical results are shown in Table 11.2. All of the participating organizations fit within the proposed four theoretical typologies of strategic management systems. This finding confirmed the hypothesis, albeit on a non-statistical basis, that there were four different clusters of strategic management systems.

Step 2: Hierarchical cluster analysis

Hierarchical cluster analysis was conducted to determine the number of clusters arising from the data collected in section 2 of the questionnaire. Table 11.3 is an agglomeration schedule, which shows how the clusters were combined. The process of cluster formation could be stopped when the increase in coefficients between two adjacent steps became "large" as this indicated that fairly dissimilar clusters were being combined.

Table 11.4 shows the changes in coefficients during the last few stages. It can be observed that the cluster formation could be stopped at stage 42, with a four-cluster solution.

Figure 11.3 shows the dendrogram that provides a visual representation of the distance between the clusters. A four-cluster solution could be found by excluding the large distances between the vertical lines connecting dissimilar clusters.

Step 3: K-means cluster analysis

Based on the four-cluster solution found in step 2, K-means clustering was then used to find the K-centers through an iterative process. After the iteration stopped, the cases were re-assigned to the final cluster centers as shown in Table 11.5.

Table 11.6 shows the cluster membership of each case or real estate development organization. The empirical results generated from the cluster analysis were in line with the results obtained from the top-level executives' subjective judgements, except for cases 19 and 34. Therefore, the overall rate of correctly cross-validated cases was high (96 percent).

Table 11.2 Descriptive statistics of strategic management systems

Strategic management systems	Case numbers	Number of cases	Percentages
Ad hoc	1–20	20	44.4%
Reactive	21–34	14	31.1%
Planned	35–41	7	15.6%
Proactive	42–45	4	8.9%

Table 11.3 Agglomeration schedule

Stage	Clusters combined		Coefficients	Stage cluster first appears		Next stage
	Cluster 1	Cluster 2		Cluster 1	Cluster 2	
1	12	16	9.500	0	0	6
2	3	15	19.000	0	0	13
3	43	45	29.500	0	0	20
4	34	37	40.000	0	0	15
5	6	7	51.000	0	0	21
.	.	.	.	.	.	.
.	.	.	.	.	.	.
40	21	24	846.623	36	37	43
41	1	4	942.690	39	12	43
42	34	42	1132.565	34	25	44
43	1	21	1454.788	41	40	44
44	1	34	2893.378	43	42	0

Note: The results in stages 6 to 39 are not shown, as these do not affect the overall interpretation and conclusion.

Table 11.4 Re-formed agglomeration table

No. of clusters	Stages	Coefficients between two stages	Change in coefficients
2	44–43	2893–1455	1438
3	43–42	1455–1133	322
4	42–41	1133–943	190
5	41–40	943–847	96

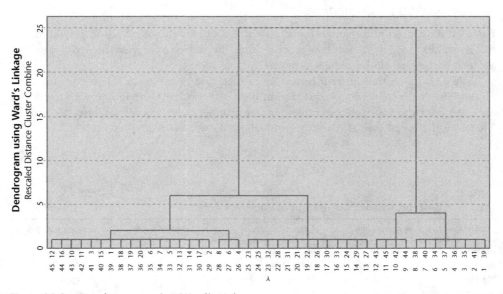

Figure 11.3   Dendrogram using Ward's Linkage

Table 11.5 Final cluster centers

Developers' capabilities for response to each strategic issue variable	Cluster 1 proactive	Cluster 2 reactive	Cluster 3 planned	Cluster 4 ad hoc
1 GNP growth rate	6.00	2.64	4.25	1.89
2 Money supply	6.00	2.43	3.88	1.95
3 Local currency	5.50	2.71	4.63	2.00
4 Unemployment	5.50	2.64	4.50	1.53
5 General inflation	5.25	2.71	4.25	1.74
6 Household wealth	5.50	2.79	4.13	2.00
7 Inflation in materials and labor costs	5.50	2.86	4.38	2.32
8 Interest rate	6.50	3.14	4.63	2.32
9 Finance sources for development	6.25	3.57	4.38	2.63
10 Amounts of equity for mortgages	6.25	3.29	4.63	2.68
11 Periods of loan repayment for mortgages	6.25	3.14	4.50	2.89
12 Political climate	6.00	2.93	4.25	2.26
13 Taxation	5.75	2.86	4.38	1.95
14 Supply of government land	6.00	2.93	4.13	2.68
15 Provisions of public housing	6.25	3.07	4.63	2.21
16 Provisions of infrastructure	5.75	3.07	4.13	2.47
17 Controls on pre-sale of real estate property	5.75	3.86	4.50	2.63
18 Controls on land-use planning	5.75	3.21	4.25	2.32
19 Controls on building design and construction	5.25	3.21	4.13	2.05
20 Controls on environmental protection	4.75	3.29	3.88	2.00
21 Controls on health and safety	5.50	3.21	4.38	2.32
22 Population growth	5.25	3.64	4.63	2.21
23 Marriage rate	5.25	3.07	4.25	1.84
24 Quality of real estate property	6.00	3.57	4.63	3.42
25 Total demand for real estate property	5.75	4.36	4.88	2.95
26 Elasticity of demand for real estate property	5.50	4.21	4.25	2.89
27 Expectations for after-sale services	6.50	4.07	4.63	3.37
28 Real estate prices	6.75	5.29	5.50	4.21
29 Speculative atmosphere	6.50	4.86	5.13	4.32
30 Barriers to entry in the industry	4.50	3.36	4.25	1.84
31 Vacancy rate	5.25	4.43	4.38	2.89
32 Existing stock in the public sector	5.25	3.71	4.25	2.42
33 Existing stock in the private sector	5.50	3.71	4.63	2.68
34 Supply trends from the public sector	5.00	3.71	4.50	2.63
35 Supply trends from the private sector	5.25	4.29	4.38	2.95
36 Technologies for design and construction	5.50	3.50	4.38	2.00
37 Competition in acquiring sites	6.25	3.43	4.63	3.21
38 Competition in setting sale prices	6.75	4.79	5.50	3.79
39 Competition in bidding for consultancy work	5.50	3.50	4.75	2.74
40 Competition in bidding for construction work	6.25	3.57	4.88	2.16
Mean	5.74	3.47	4.48	2.53

Step 4: Discriminant analysis

While the F-tests for all of the variables were significant at the 1 percent level, the F-test was used for descriptive purpose only. Therefore, discriminant analysis was used to verify whether these four clusters were indeed different. The Wilks' lambda is shown in Table 11.7.

Table 11.6 Cluster membership according to the developers' capabilities for response to strategic issues

Case number	Strategic management systems*	Cluster	Distance	Case number	Strategic management systems*	Cluster	Distance
1	Ad Hoc	4	3.674	24	Reactive	2	5.303
2	Ad Hoc	4	4.648	25	Reactive	2	3.738
3	Ad Hoc	4	4.337	26	Reactive	2	5.262
4	Ad Hoc	4	4.765	27	Reactive	2	5.343
5	Ad Hoc	4	4.831	28	Reactive	2	5.082
6	Ad Hoc	4	3.821	29	Reactive	2	5.449
7	Ad Hoc	4	3.996	30	Reactive	2	5.914
8	Ad Hoc	4	6.115	31	Reactive	2	5.276
9	Ad Hoc	4	6.462	32	Reactive	2	4.356
10	Ad Hoc	4	4.451	33	Reactive	2	5.462
11	Ad Hoc	4	4.100	34	Reactive	3**	3.194
12	Ad Hoc	4	3.475	35	Planned	3	3.702
13	Ad Hoc	4	4.693	36	Planned	3	3.457
14	Ad Hoc	4	5.117	37	Planned	3	3.899
15	Ad Hoc	4	4.132	38	Planned	3	3.899
16	Ad Hoc	4	4.331	39	Planned	3	3.702
17	Ad Hoc	4	4.239	40	Planned	3	3.194
18	Ad Hoc	4	5.122	41	Planned	3	4.382
19	Ad Hoc	2**	5.841	42	Proactive	1	3.473
20	Ad Hoc	4	5.914	43	Proactive	1	3.250
21	Reactive	2	4.469	44	Proactive	1	3.816
22	Reactive	2	4.823	45	Proactive	1	3.092
23	Reactive	2	4.306				

Notes: * Based on the data collected in section 1 of the questionnaire. ** Indicates inconsistent results.

Table 11.7 Wilks' lambda

Test of function(s)	Wilks' lambda	Chi-square	df	Sig.
1 through 3	.000	258.968	120	.000
2 through 3	.005	118.650	78	.002
3	.262	29.445	38	.839

The Wilks' lambda was very small (less than 0.0005). A chi-square transformation of the Wilks' lambda was used to determine significance, and the observed significance level was less than 0.0005. Therefore, the alternative hypothesis was accepted, confirming that there were four groups of strategic management systems, with strong differences between them.

The discriminant analysis also yielded a territorial map (see Figure 11.4), which shows the distribution of canonical discriminant functions. This figure clearly indicates the degree of separation between the four clusters. Therefore, it can be confirmed graphically that these four clusters do indeed differ.

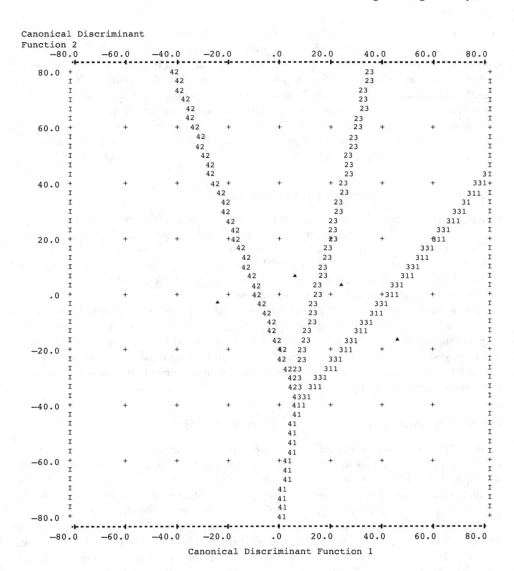

Figure 11.4 Territorial map

Discussion of findings

Having tested the hypothesis, it is necessary to analyze the characteristics of these four clusters of developers, based on their final clustering centers as shown in Table 11.5.

Ad hoc strategic management system – cluster no. 4

Developers in the ad hoc-style cluster generally have a low capability for response to strategic issues. However, they are still capable of responding to certain strategic issues. The first three strategic issues that exhibit relatively high values within this cluster are speculative atmosphere (4.32), real estate prices (4.21) and competition in setting sale prices (3.79). This finding suggested that developers in this cluster were not strongly affected by changes in the real estate market. This result could be explained by the exceptionally favorable conditions that prevailed in the local real estate market during the study period. This situation presented an opportunity rather than a threat, so that the environmental conditions had a positive effect on the performance of ad hoc developers. As the real estate market had a similar effect on all developers, it was found that the developers in the other three clusters also exhibited relatively high values on these variables.

Reactive strategic management system – cluster no. 2

Developers in the reactive-style cluster had their highest rankings in capability for response to real estate prices (5.29), speculative atmosphere (4.86) and competition in setting sale prices (4.79). As the real estate market was very favorable during the study period, the developers in all four clusters had relatively high values for these same three variables. Again, this indicated a consistent result that developers with relatively low management capability were still able to capture opportunities arising from the environment.

Reactive-style developers also showed relatively high values in their capacity to deal with vacancy rates (4.43), total demand for real estate property (4.36), new private real estate property supply trends (4.29) and elasticity of demand for real estate property (4.21). These results suggested that they had systematically monitored the general real estate market and planned for their developments accordingly. This finding implied that they had at least rudimentary formal planning systems.

Planned strategic management system – cluster no. 3

Small- to medium-sized developers with relatively low strategic management capability were able to capture opportunities arising from a favorable real estate market. The comparatively large developers in the planned-style cluster were also able to do so. This capability was reflected in the high values reported by planned-style developers for dealing with trends in real estate prices (5.51), setting real estate sale prices (5.51) and coping with a speculative atmosphere (5.13).

Planned-style developers also had relatively high values (above 4.00) for a number of other variables. These findings suggested that they used three different but related strategies. First, they placed a considerable emphasis on the selection and control of the design consultants and contractors in their projects, while ensuring maximum competition and value for money. This strategy was particularly apparent in government or public development organizations, in which accountability was one of the major concerns. Second, they actively monitored the real estate market, which suggested that they utilized sophisticated planning systems. Third, they strove for maximum competition in acquiring development sites, which reflected their strength in financial resources. All of these strategies suggested that this cluster of developers was large organizations with adequate resources, which used sophisticated planning systems for dealing with critical strategic issues in their real estate development process.

Proactive strategic management system – cluster no. 1

Proactive-style developers were characterized by exceptionally high capability for responding to a large number of significant variables. First, they ranked highest in ability to deal with real estate prices (6.75), to set real estate sale prices (6.75) and cope with a speculative atmosphere (6.50). These results suggested that they were able to gain significantly from a rising real estate market. Second, they were well able to deal with changes in expectations for after-sale services (6.50). These results were consistent with the previous findings. Third, they were little affected by issues related to interest rates (6.50), funding sources (6.25), the amounts of equity required for mortgages (6.25) or the periods of loan repayment for mortgages (6.25). All of these findings suggested that the financial resources of these developers were very strong. Indeed, large developers can easily raise funding from the stock and debt markets, and some developers have even established their own financial companies. They are able to convert adverse financial environments into opportunities.

Fourth, they had high values for their capacity in responding to a range of additional issues such as the provision of public housing by government (6.25), the acquisition of development sites (6.25) and the supply of government land (6.00). These results suggested that they have exerted a certain influence on government policies relating to the provision of public housing and private land. They were able not only to accurately forecast the future supply and demand of real estate, but also, to a certain extent, to manipulate the supply and demand. Fifth, they ranked high with respect to general environmental factors such as the GNP growth rate (6.00), money supply (6.00) and political climate (6.00). These findings suggested that they had established and applied various sorts of futuristic management systems to scan and monitor the general economic, political and money supply issues that could pose threats or create opportunities.

Conclusions

During the real estate development process, developers face numerous strategic issues emerging from turbulent environments such as major discontinuities in the economic, political, and social sectors. Some of these strategic issues can have significant effects on the performance of the developers' organizations and projects. Understanding such emerging strategic issues is critical for major decision making. Although not all of the potential threats and opportunities that emerge from the environment can be avoided or captured, developers would certainly increase their chances of success by improving their abilities to anticipate and deal with such threats or opportunities.

Based on different theoretical perspectives, this study has identified four major types of strategic management systems: the proactive, planned, reactive and ad hoc systems, which were confirmed through an empirical study. These strategic management systems are applicable not only to real estate development organizations, but also to organizations of other types. By building theoretical strategic management systems and investigating various strategic management practices, this study contributes to the better understanding of strategic management theory and practice, particularly in real estate development organizations.

As this study was conducted in Hong Kong, it was subject to two site-specific limitations. First, Hong Kong is relatively small. Although the research questionnaire was sent to most (if not all) developers in Hong Kong, the sample size was not very large, which is an unavoidable consequence of undertaking such research in a relatively small city. As a result, the number of cases in some particular clusters of developers might not be large enough to provide strong evidence. Second, the findings were specifically related to the local real estate development environment, although the proposed theoretical framework should also be applicable to other countries.

This study used a quantitative approach. It is recommended that a qualitative approach such as case studies can be utilized in future studies as a means to test the results obtained from quantitative data. In addition, this study was based on a perceptual approach to evaluating environments, as the data collected were based on the developers' own perceptions of their surroundings. In future studies, objective data can be collected and compared with the subjective data to yield a better understanding of the relevant theory.

References

Ackoff, R. L. (1981), *Creating the Corporate Planning*, John Wiley & Sons, New York.

Aldrich, H. E. (1979), *Organisations and Environments*, Prentice-Hall, Englewood Cliffs, NJ.

Ansoff, H. I. (1975), "Managing Strategic Surprise by Response to Weak Signals," *California Management Review*, Vol. 18, No. 2, 21–33.

Ansoff, H. I. (1980), "Strategy Issue Management," *Strategic Management Journal*, 1, 131–148.

Ansoff, H. I. and McDonnell, E. J. (1990), *Implanting Strategic Management*, Prentice-Hall, London.

Child, J. (1972), "Organisation Structure, Environment and Performance: The Role of Strategic Choice," *Sociology*, 6, 1–22.

Hannan, M. T. and Freeman, J. (1977), "The Population Ecology of Organisations," *American Journal of Sociology*, 82, 929–964.

Ho, P. H. K. (2014), "An Exploratory Study of the Impact of Organisational Environments on Property Development Firms," *Journal of Financial Management of Property and Construction*, Vol. 19, Issue 3, 226–245.

International Financing Review (IFR) (2013), "A History of the Past 40 Years in Financial Crises," IFR, 2000th Issue, retrieved 30 December 2015, www.ifre.com/a-history-of-the-past-40-years-in-financial-crises/21102949.fullarticle

Oomens, M. J. H. and van den Bosch, F. A. J. (1999), "Strategic Issue Management in Major European-Based Companies," *Long Range Planning*, Vol. 32, No. 1, 49–57.

Perrow, C. (1972), *Organisational Analysis: A Sociological View*, Wadsworth Publishing, Belmont.

Schwartz, P. (2003), *Inevitable Surprises: Thinking Ahead in a Time of Turbulence*, Gotham Books, New York.

12

Organizing public–private partnerships for real estate development through urban finance innovation

Michael Nadler

Abstract

By realizing large-scale real estate projects, stakeholders try to achieve sustainable urban developments. Public authorities and the private development sector need each other, but misunderstandings, delays, controversies, and failures are typical for many public–private partnerships (PPP). Potential reasons are the absence of a common target system, not clearly defined responsibilities, and wrong incentives in combination with organizational forms for a PPP. The following study presents an alternative PPP approach by applying urban finance innovations, which have become popular in the last ten years in European urban policy. To evaluate these financial innovations, we will first derive a common understanding of sustainable urban development through real estate projects and define the roles and interests of the involved stakeholders. Second, we will introduce an ideal type of real estate development project. In the third step, we apply potential financial innovations of a PPP solution. After evaluating the effects of these innovations, we will come to a final recommendation for organizing public–private partnerships for real estate development.

Introduction: Real estate projects as part of sustainable urban development strategies

Sustainable urban development has a prominent position for public decision makers and is one core aim of European Cohesion Policy (European Union, 2009; European Commission, 2014; European Parliament, 2014). However, with the introduction of the triple-bottom-line approach of Elkington (1997), also private companies aiming for sustainability need to perform not against a single, financial bottom line, but against the triple bottom line. In this context, large-scale real estate development projects are able to create positive externalities according to the triple-bottom line objectives. Mankiw (2014: 196) defines them as an impact of economic activities on the welfare of an uninvolved third party, for which no one pays or receives compensation. Thiel and Nadler (2015) have created a systematization of positive externalities from sustainable development projects by combing the triple-bottom-line approach with the concept of capital stocks for urban wealth creation and the mission statements of urban planning authorities (e.g. Urbanity, Smart City and the European City). Therefore, single sustainable real estate (development projects) can contribute to urban welfare in three internal and external (objective) dimensions (see Figure 12.1).

Due to the complexity of urban development in general and the change in the roles – mainly from the public perspective – not one single stakeholder, but only a cooperation of stakeholders is able to create sustainable real estate developments. The reason for the rising importance of partnerships is on the one hand that public funding for urban development retreats. Thus, projects more and more depend on inputs from the private sector. On the other hand, private developers also need public partners, since complex urban development projects are too capital intensive and too risky to be realized on their own, leading to market failure. Thus, public–private partnerships (PPP) are ventures where public authorities and private promoters share risks, responsibilities, but also benefits for the limited time of a development project. In general, private and public decision makers could use urban assets in all stages of their life cycle to achieve triple-bottom-line objectives (see Figure 12.2).

In this context, especially the land development phase is suitable for a PPP following the triple-bottom-line approach. Only these large-scale development projects can enable public authorities and private promoters to create jointly new urban districts, quarters or communities in a sustainable way. Here, public authorities as well as urban planners usually apply a

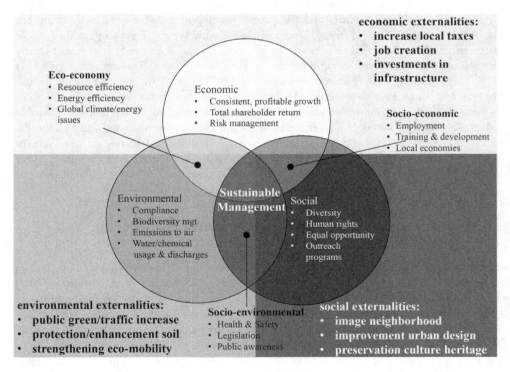

Figure 12.1 Real estate development process with triple-bottom-line objectives (internal versus external benefits)

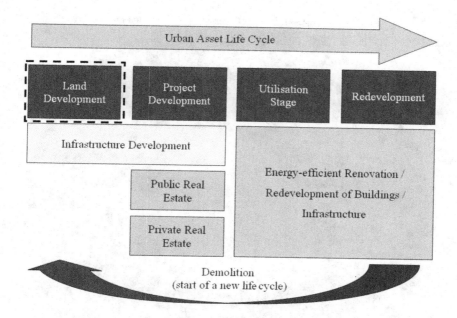

Figure 12.2 Classification of urban development projects according to the life cycle of urban assets

Source: Kreuz and Nadler, 2010: 16

(normative) guiding principle or "mission statement." In this context the widespread "Bristol Accord" identifies seven components of sustainable communities that name positive externalities achievable by large-scale development projects (see Figure 12.3).

Another crucial element of these projects is that private investors create sustainable communities through the process of land subdivision (Peiser, 2012: 72–145; Brueggeman and Fisher, 2015: 554–582; Owens, 1998). In subdivision development, promoters split a large land area into an appropriate number of lots (residential and commercial). Usually promoters can sell these lots more easily and more profitably to the property market because of the smaller size and the flexible timing of the sales process. However, the higher possible land development yield is only a compensation for the higher development risks. Promoters have to deal with substantial planning and approval risks resulting – e.g. from neighborhood complaints – with only a long-term amortization of capital and high start-up costs. These costs result from possible land contaminations (in case of brownfields) and extensive infrastructure required to prepare the lots for sale. Only through a strategic land subdivision process can private promoters establish new urban quarters as sustainable communities according to the wishes of public authorities, as in Egan's wheel. Therefore, Peiser (2012: 73) states accurately: "Subdivision of land is the principal mechanism by which communities are developed."

Therefore, we can sum up the status quo as follows: both public and private investors are highly interested in sustainable urban development projects that provoke positive externalities as laid out in Figure 12.1. In the context of rendering positive externalities, land development (see Figure 12.2) plays a special role due to its high leverage: only by realizing large-scale

Figure 12.3   Components of sustainable communities: Egan's wheel

Source: Office of the Deputy Prime Minister, 2004: 19

projects can promoters exploit all potential positive externalities leading to the development of sustainable communities (see Figure 12.3). On the negative side however, large-scale project development contains so many risks that a joint PPP approach is mandatory in order to avoid market failure. Since in the past PPP approaches have mostly proven to fail due to the absence of a common target system, poorly defined responsibilities, and wrong incentives in combination with organizational PPP forms, the aim of this chapter is to develop a financial innovation of a PPP solution in order to exploit all positive externalities connected to large-scale land development.

In order to pursue our research aim we will proceed as follows: in the next section we will present a large-scale land development project in order to identify positive externalities referring to the components of sustainable communities. In the following section, we will perform a dynamic investment calculation approach for the financial feasibility analysis of the given land development project to show that for a private investor alone the project is not viable, thus leading to market failure. Consequently, we develop an urban financial innovations approach on PPPs containing several instruments, before we apply these instruments to our specific project in order to realize all positive externalities connected to the development.

New urban quarters as classic example for a PPP in sustainable urban development

In the following section, we will analyze the internal and external benefits of a recent large-scale development project in Germany, the "Karolinger Yards," in order to illustrate our research approach. The mixed-use development project is currently under construction in Düsseldorf, the capital of North Rhine Westphalia (NRW). It fulfills all features of a large scale urban development project elaborated in the previous section. The site consists of 15,768 m² overall and has been in use for more than a 100 years, first by a paper mill and then by a (now insolvent) car dealer with (partially empty) buildings for sales, warehouse, and office spaces. Thus, with no actual use, the site had been a "typical" industrial urban brownfield for a few years with no private investor willing to invest (see Figure 12.4).

According to the zoning plan, the site is a commercial area with adjacent special housing. However, due to the inner-city location and the steadily rising demand for additional housing, the city of Düsseldorf recently converted it into a residential area to bring the site back into the urban asset life cycle (see Figure 12.2). The city developed a new land-use plan and zoning plan for the site and made use of the principle of land subdivision (see previous section). The actual plan redefines and subdivides the site into a general residential area and a mixed area. After the subdivision, a real estate company bought the site in order to redevelop it. According to the local development scheme, the promoter will create five building plots (land subdivisions) with five- to six-story buildings. The degree of building coverage for the site exceeds the upper limit of the zoning plan, but in this area it is justified in order to enable an efficient use of scarce land, which is also true for the surrounding buildings. The utilization concept contains housing, retail, services and gastronomy, parking, and green spaces. It also shows which interests − in terms of positive externalities − the public decision makers intend to accomplish with redeveloping the former brownfield. In Table 12.1, we have linked these positive externalities to the categories of the Egan's Wheel (see Figure 12.3) contributing to sustainable communities. The table shows that our reference project is able to contribute to five

Figure 12.4 Industrial urban brownfield in Düsseldorf, Germany

Source: bing.de

Table 12.1 Positive externalities of the large scale development project "Karolinger Yards"

Category of Egan's Wheel contributing to sustainable communities	Positive externalities due to land subdivision and development of "Karolinger Yards"
Places to live in an environmentally friendly way	Creation of three contiguous public green spaces of 600 + 600 + 600 + 900 m² (17 % of the total site) with playgrounds
Flourishing local economy	Due to the upgrading of the area and neighborhood the city (but also the county) expects significantly rising real estate taxes (property tax and real estate transfer tax) and rising income taxes (new residents) as well as trade taxes (new retail businesses)
Good transport services and connectivity with respect to accessibility	In addition to connecting to public transport the redevelopment provides additional parking spaces in connection with improved local supply (commercial areas)
High quality in housing and the built environment	High-quality apartments that achieve at least 12 €/m² on the open housing market
Vibrant and harmonious communities through social and cultural use	Mandatory to also offer *affordable housing* in new residential areas: every investor has to build – a fixed 20% affordable (social) housing: NRW county prescribes rents of (no higher than) 6.25 to 7.15 €/m² – and another 20% so-called "price-reduced housing": the city of Düsseldorf prescribes rents of 8.50 to 10.00 €/m² both depending on the household income category of the possible tenants. Investors have to maintain these below market rent levels for 30 years.

out of the seven components of sustainable communities that name positive externalities achievable by large-scale development projects.

Resulting demand for public support as a result of the feasibility analysis

As illustrated in the last section, large-scale projects in land development are able to induce high positive externalities presupposing that investors are willing to realize the project. In this section, we will perform a financial feasibility analysis of the given land development project to show that for a private investor alone the project is not viable.

In order to evaluate whether the given land development project is financially viable and thus feasible, we apply a dynamic investment calculation approach (Peiser, 2012: 72–145). In real estate economics, the discounted cash flow method (DCF) has been the dominating appraisal approach for many decades. It forecasts future project income (I_t) and expenses (E_t) and calculates their present value by using cost of capital (q) as a discount rate:

$$DCF = \sum_{t=0}^{n} \frac{(I_t - E_t)}{(1 + i)^t} = \sum_{t=0}^{n} \frac{(I_t - E_t)}{q^t} = \sum_{t=0}^{n} (I_t - E_t) \cdot q^{-t} = C_0 \qquad (12.1)$$

The resulting net present value (C_0) expresses the economic gain ($C_0 > 0$) or loss ($C_0 < 0$) of the investment and is able to integrate risk and sensitivity analyses next to delivering investment rates of return. The necessary input variables can be taken, e.g., from the guidelines of the German Society for Real Estate Research (Gif, 2006). In contrast to a "normal" land subdivision process (Peiser, 2012: 72–145) we integrate in our analysis not only the development phase but also the utilization and exit phase. Further, we apply an investor developer perspective by taking into account all estimated costs of the development phase next to all incomes and expenditures of the utilization phase as long as rents will be (partially) below market level. Therefore, the utilization phase amounts to 30 years in order to reach a uniform price level for all rental housing space. After this period, we calculate the capitalized value (Peiser, 2012: 173) for the exit phase by the net operating income (NOI) and a multiplier. The multiplier derives from the cap rates at sale (4.25% for apartments, 4.50% for mixed use) and the remaining life of the respective urban assets. In the exit phase, the residual value equals the selling price of the investor developer for the entire site.

In Table 12.2 (reduced to the development phase and the last year of utilization and exit phase), we first calculate the difference between project income and expenses leading to annual (free) cash flows. Second, we calculate the net present value (NPV) and internal rate of return (IRR) for the private developer. The IRR results when the present value of the income equals the present value of the expenses and displays the profitability of the capital invested in the project. By comparing the IRR with the return of an alternative investment or the capital costs of the investor, we can determine whether the project investment is favorable or not.

In the analysis, we identify three steps. First, at the level of the gross cash flow (before taking into account the financing structure), the IRR equals the return on assets of the project development. Therefore, the investor has to compare the IRR to the weighted average cost of capital (WACC). The latter is calculated by allocating total cost of investment according to the sources of funding. In our project, we assume that in 2011 the investor initially financed the purchase of the site by equity capital. The remaining construction costs are subject to loan financing. Here we assume no repayment until the utilization phase. Then the loan is paid back by constant annuities. The resulting equity ratio of approximately 23% is a common requirement of banks engaged in project financing.

Second, we derive the net or free cash flow for the investor developer by subtracting the annuities from the project cash flow. The free cash flow ("before taxes cash flow") is the base to determine the dynamic return on equity by calculating its IRR (IRR_{equity}). To decide whether the project development is favorable or not, the investor developer has to compare the equity IRR with his desired return on investment. When we apply this "hurdle rate" to discount the free cash flows, the resulting DCF_{equity} shows the financial feasibility of the project.

Third, we can also capture the income tax burden for the investor developer. Here we take the current NOI and deduct interest on loans and depreciation (simply assuming 2% of the property value). When we apply the current corporate tax rate for Germany (30%), the resulting free cash flow after taxes serves to derive a dynamic return on equity after taxes. Moreover, the approach is extendable to risk analysis, corresponding to state-of-the-art research on real estate investments (e.g. Peiser, 2012; Brueggeman and Fisher, 2015).

For our sustainable real estate development project we initially assume that the investor developer does not receive public funding. The following DCF calculation of the overall project reveals, however, that this leads to a non-viable real estate development project, since the return on equity is below 5.0% (4.90%). This is even lower than the average level of total return for the German housing sector (IPD/MSCI, 2016), plus it would imply a risk discount instead of a risk premium for the development risk. Thus, no private promoter would be willing to invest while the city of Düsseldorf is unable to realize the project itself due to limited financial resources. Thus, despite high potential positive externalities, the project is subject to

Table 12.2 DCF calculation for integrated urban real estate development project without public support (excerpt)

DCF	assumptions	2011	2012	2013	2014
(0) income housing with market rents !!!	60.00%				
(1) income affordable housing	20.00%				
(2) income social housing	0.00%				
(3) income price-reduced housing	20.00%				
income commercial rents	1,309.32 m²				
income parking (and additional fee) ###	366				
Sum potential income		0	0	0	0
vacancy-collection loss (year 1)					
vacancy-collection loss housing (year 2–30)	2%				
vacancy-collection loss commercial (year 2–30)	4%				
Sum missing income		0	0	0	0
Exit price after sales provision	3.57%				
Sum cash-inflow		0	0	0	0
KG 100 land acquisition —	980.0 €/m²	−16920668			
KG 200 site improvements	21.9–170.5 €/m²		−2688050		−871714
KG 300 hard costs construction housing	690.5 €/m²		−8493150	−8493150	
KG 300 hard construction commercial	774.4 €/m²		−5149760	−5149760	
KG 400 building technology housing	185.5 €/m²			−1949550	−1949550
KG 400 building technology commercial	180.4 €/m²			−1199660	−1199660
KG 300 + 400 hard costs parking spaces	20000 €/space			−3660000	−3660000
KG 500 hard costs outdoor facilities +++	69.2–173.1 €/m²				−1158206
KG 600 finish-out costs	19.1 €/m²				−254030
KG 700 soft costs	25%		−3410728	−5113030	−1702303
tax refunds/investment grants ***		0			379224
Sum development costs	−72643744	−16920668	−19741688	−25565150	−10416238
real estate taxes/building insurance §§§	1,54%/0,3%				
maintenance housing	8.62 €				
maintenance parking	82.60 €				
maintenance commercial (limited)	3.54 €				
management housing	279.35 €				
management commercial	2.00%				
management parking	36.43 €				
surcharge elevators	1.22 €				
modernisation risk	0.70%				
Sum operating expenses		0	0	0	0
Sum cash-outflow		−16920668	−19741688	−25565150	−10416238
Gross Cash Flow: IRR (total capital)	3.90%	−16920668	−19741688	−25565150	−10416238
Discounted Cash Flow with WACC	3.78%	1.1178	1.0771	1.0378	1.0000
	2067372	−18913516	−21262816	−26531793	−10416238
loan balance year end			19741688	45800380	57361628
interest %%%				−493542	−1145009
principal reduction				0	0
Debt Service/annuity		0	0	0	0
Free Cash Flow: IRR (equity = BTIRR)	5.48%	−16920668	0	0	0
Discounted Cash Flow with hurdle rate	8.00%	1.2597	1.1664	1.0800	1.0000
	−11552089	−21315168	0	0	0
depreciation &&&		0	0	0	0
taxable income		0	0	−493542	−1145009
tax	30%	0	0	148063	343503
Free Cash Flow: IRR Tax (equity = ATIRR)	4.90%	−16920668	0	148063	343503
Discounted Cash Flow with hurdle rate	8.00%	1.2597	1.1664	1.0800	1.0000
	−13066609	−21315168	0	159908	343503

Note: Symbols used in this table are explained in the text of the chapter

2015	2016	2017	...	2040	2041	2042	2043	2044	2045
2197502	2224971	2252783		2997817	3035289	3073231	3111646	3150542	3189923
381511	387233	393042		821907	821907	945193	945193	945193	1086972
0	0	0		0	0	0	0	0	0
518855	529232	539816		999272	1011763	1024410	1037215	1050181	1063308
197969	201928	205967		324789	331285	337911	344669	351562	358593
307440	313589	319861		504388	514476	524765	535260	545966	556885
3603276	**3656953**	**3711469**	...	**5648173**	**5714720**	**5905510**	**5973984**	**6043443**	**6255682**
	−62829	−63713		−96380	−97379	−100857	−101881	−102918	−106804
	−20621	−21033		−33167	−33830	−34507	−35197	−35901	−36619
−1080983	**−83449**	**−84746**		**−129547**	**−131210**	**−135364**	**−137078**	**−138819**	**−143423**
									91455304
2522294	**3573504**	**3626723**	...	**5518626**	**5583511**	**5770146**	**5836906**	**5904624**	**97567563**

0

0	0	0	...	0	0	0	0	0	0
−773949	−773949	−773949	...	−773949	−773949	−773949	−773949	−773949	−773949
−219242	−223407	−227652		−350976	−357645	−364440	−371364	−378420	−385610
−30232	−30806	−31391		−48397	−49316	−50253	−51208	−52181	−53172
−4635	−4723	−4813		−7420	−7561	−7705	−7851	−8000	−8152
−88833	−90521	−92241		−142210	−144912	−147665	−150471	−153330	−156243
−3959	4039	4119		6496	6626	6758	6893	7031	7172
−13333	−13587	−13845		−21345	−21750	−22164	−22585	−23014	−23451
−31030	−31619	−32220		−49674	−50618	−51580	−52560	−53558	−54576
−286330	−286330	−286330		−286330	−286330	−286330	−286330	−286330	−286330
−1451542	**−1450903**	**−1458321**		**−1673805**	**−1685455**	**−1697327**	**−1709424**	**−1721751**	**−1734312**
−1451542	**−1450903**	**−1458321**	...	**−1673805**	**−1685455**	**−1697327**	**−1709424**	**−1721751**	**−1734312**
1070751	2122601	2168402	...	3844821	3898055	4072819	4127481	4182873	95833251
0.9636	0.9285	0.8946		0.3810	0.3671	0.3537	0.3409	0.3284	0.3165
1031740	1970751	1939925		1464878	1431051	1440734	1406876	1373811	30328454
56510597	55638291	54744177		26714566	25097359	23439722	21740644	19999089	18213995
−1434041	−1412765	−1390957		−707308	−667864	−627434	−585993	−543516	−499977
−851030	−872306	−894114		−1577763	−1617207	−1657637	−1699078	−1741555	−1785094
−2285071	**−2285071**	**−2285071**	...	**−2285071**	**−2285071**	**−2285071**	**−2285071**	**−2285071**	**−2285071**
−1214320	**−162471**	**−116669**	...	**1559750**	**1612984**	**1787748**	**1842410**	**1897802**	**75334185**
0.9259	0.8573	0.7938		0.1352	0.1252	0.1159	0.1073	0.0994	0.0920
−1124370	−139292	−92616		210881	201924	207225	197741	188598	6931954
−1114462	−1114462	−1114462		−1114462	−1114462	−1114462	−1114462	−1114462	−1114462
−396768	−321177	−252271		2152599	2246939	2466287	2564105	2663715	23656753
119030	96353	75681		−645780	−674082	−739886	−769232	−799114	−7097026
−1095290	**−66118**	**−40988**	...	**913971**	**938902**	**1047862**	**1073179**	**1098687**	**68237159**
0.9259	0.8573	0.7938		0.1352	0.1252	0.1159	0.1073	0.0994	0.0920
−1014157	−56685	−32538		123570	117538	121462	115182	109185	6278914

market failure. Against this background, the crucial question to be answered in the next two sections is how it might still be possible to realize the brownfield redevelopment with an overall investment of more than €72.6 million in order to accomplish the desired positive externalities.

Urban financial innovations to promote feasibility of sustainable development projects

In general, public authorities can apply different instruments in order to stimulate projects rendering positive externalities that are subject to market failure. First, they could use law and regulations. However, this can lead to reduced total values for local economies if no private sector promoter is willing to develop such kinds of projects. A more market-based solution would be to use negotiation and incentives to allocate externalities. By applying different approaches in the context of PPP, we derive the following (combination of) financial instruments for all spatial areas. However, because of the high demand pressure and the resulting high land values these instruments are especially suitable for urban and metropolitan areas. Table 12.3 gives an overview of the presented urban financial innovations. Furthermore, we analyze the requirements but also the possible benefits/costs ("outcomes") for the participating stakeholders.

The first innovation, a special government promoter, is a government entity usually set up to fulfill a specific role and responsibility. Funding stems from a combination of fees for services and taxes. One important funding source for this public developer could be a development fee from the private sector developer (Bryant, 2015). This fee comes, for example, for buying the relevant land parcel, building, running, and maintaining the public green spaces. In our example, such a public development partner could realize all public spaces on his own land parcels. This would lead to a substantial reduction in the development costs for land acquisitions and construction of outdoor facilities (marked with +++ in Table 12.2). When the developer is able to separate clearly the buildings and relevant land parcels, this approach is extendable to the affordable (social) housing part of the project. In Germany, managing authorities have privatized such kinds of development companies (such as LEG NRW).

A second option for public support would be to sell public land (of the municipality) to the private sector developer at below market land prices. This kind of "in kind" subsidy requires the existence of extensive urban land areas, for example in the form of land banks. A land bank is a public–private fund that purchases and holds land with the purpose of realizing the most desirable use of the land. It earns money by receiving rents or profits for selling (now) redeveloped land, similar to (community) land trusts. This would lead to a substantial reduction in the development costs for the whole land acquisition (marked with — in Table 12.2). In NRW, the "land fund NRW" is such a public land bank.

Third, in case public authorities don't have any land available, they could award density or floor area ratio (FAR) bonuses. This development right might be tradable and separated from the ownership of the land itself. In Germany, local authorities can give higher density ratios to private promoters but it requires a decision on a case-by-case basis. In the example, the private promoter already had permission to exceed the site occupancy ratio (marked with !!! in Table 12.2).

A fourth innovation in case of public green spaces might be usage pricing. Here, users pay for their individual cost of using a collective service (like a park). In our development example, this would lead to additional income from the usage fee (marked with ### in Table 12.2). Introducing such a kind of financial instrument requires an often very controversial decision,

Table 12.3 Urban financial innovations for public support of integrated urban real estate
development projects: Qualitative evaluation

Urban financial instrument/innovation	Requirements	Classification of life cycle phase	Evaluation of outcomes of promotions
Special government entity	Development fee	Land development phase	Public cost risk
Subsidized land (land with subsidized ground rents)	Land banking	Land development phase	Land speculation risk
Density or floor area ratio (FAR) bonuses as tradable development rights (TDR)	Suitable planning regulations	Project conception phase	Political decision risk
Usage pricing for public/subsidized goods	(Demand for) improvement districts	Operating phase	(Neighborhood) solidarity
Project bond with tax incremental financing (TIF)/value capture financing (VCF)	Market value taxation; suitable capital markets	Project conception/ operating phase	Waiving of tax income in local budgets
Investment grants	Sufficient national and federal budgets	Project conception phase	Non profitable projects (C-projects)
Tax reliefs (increased amortization)	Efficient national tax system with high income/corporate tax rates	Operating phase	Risk of market distortions in non-demand areas/for low-income/for foreign investors
Soft loans	Promotional bank	Operating phase	Restricted to (subsidized) mortgage finance of objects
Equity for PPP companies (joint ventures)	SPV/CDC	Project conception phase	Project management capacities in public authorities
(Equity/debt) mezzanine capital for project finance	Urban development funds (UDF) with independent fund managers	Project realization phase	EU regulations and competition with OP-projects with grants

which policy makers in Germany refused so far. In addition, granting of rent allowances (also marked with !!! in Table 12.2) to bridge the gap to market rents (here: 6.25 to 12.00 € /m²) is no longer possible in Germany (due to the high public payment burden).

Fifth, managing authorities on municipal and federal level profit from the given land development project because of rising real estate taxes. Since the area is a former brownfield, all captured taxes (marked with §§§ in Table 12.2) in the investor calculation could be used, for example for the creation of public spaces. It would be possible to combine this kind of tax incremental financing (TIF) with a project bond. Investors in these project bonds would get the future real estate tax receipts of this new urban quarter as return. Like (housing/business) improvement districts, TIFs are a form of land capture financing, but with longer durations of up to 20 years. However, up to now they are nonexistent in Germany. The reason for this is simple: by looking at the development example, we can see that the ongoing real estate taxes are quite low (1.54% (= 0.0035*440%) of the real estate "value"). This would only lead to a rather small project bond volume (even in the current yield situation of very low interest rates). The low tax payment is because Germany still lacks a market valuation for tax purposes. The plan is to reform the German real estate tax system by 2022. A second problem might be whether managing authorities are willing to waive part of their future tax income in their local public budgets.

Sixth, the traditional "model" to promote an integrated development project with positive externalities is the investment grant. Public stakeholders provide this kind of promotion usually as "lost" equity participation based on the investment costs of the promoters (marked with ★★★ in Table 12.2). This means that the public partner does not expect a repayment of the equity participation. Most often, managing authorities also do not participate in the management of the whole development project.

Instead of giving a (cash-based) investment grant, public authorities could provide private investors also (indirectly) with tax refunds (marked with &&& in Table 12.2). The final equivalent value of this subsidy depends on the (corporate/personal) tax rate of the private investors. However, in Germany, public authorities abolished these kinds of tax subsidies because of distortions in non-demand areas and inequitable effects for low-income and foreign investors. Today, cash-based investment grants are quite common for most development areas in the form of the "urban development/regeneration promotion program," especially in NRW. However, funding is restricted to non-profitable development projects (so called C-projects in the CABERNET systematization for brownfield development projects). The relevant development project in Düsseldorf creates a limited profit. Therefore, the promoter must separate the creation of public green areas, for example, from the rest of the development project. This separated brownfield project (for public green space) would then fulfill the requirements of the urban development/regeneration promotion program because it is not profitable. However, the national and federal governments would require that the city of Düsseldorf take on up to 33% of the cost of the investment grants. Overall, this kind of promotion creates quite substantial payment burdens for public decision makers.

A further approach to support the development project would be to give soft loans after completion of the buildings, e.g. for the affordable (social) housing spaces (marked with %%% in Table 12.2). The NRW.BANK, a promotional bank, provides soft loans to completed housing units that provide affordable housing (with the already described maximum rents of 6.25 to 7.15 €/m²). Right now the interest rate conditions are quite low with an effective rate of 1.0% (0.5% nominal interest rate and 0.5% for overhead costs) for a duration of up to 20 years. However, this promotional loan program is restricted to ready to use real estate objects and not for development or project finance. The bank allocates loans on rent level, credit rating of the real estate owning company, and the possibility of mortgage financing. Given the fact that at present free market credit rates are only 2.5%, the promotional effect is limited. In the meantime, as a further incentive for affordable housing NRW.BANK loans have a 25% grant share lowering the repayment to 75% of the loan sum.

A final very important innovative instrument to stimulate a PPP approach is an urban development fund (UDF). In 2010, managing authorities, not just in NRW, realized that they needed an additional instrument to support development projects indirectly by revolving financial instruments (Nadler, 2010). They decided to use the Joint European Support for Sustainable Investment in City Areas (JESSICA). The most innovative element of JESSICA is the provision of financial instruments – loans, mezzanine, equity capital, and guarantees – on a revolving base. Instead of financing sustainable urban development projects with "lost" investment grants, revolving financial instruments for successful projects may generate a capital backflow enabling managing authorities to reinvest in new urban development projects. Furthermore, the funding should finance ("bridge") the development phases because this seems to be the "market failure" in capital supply (marked gray in Table 12.2). In order to channel funds effectively, the institutional framework of the JESSICA initiative intends to set up UDFs as financial intermediary (here as part of the NRW.BANK). The UDF will provide catalytic first-loss capital, meaning that it will bear a specific amount of loss, which is typically set and agreed upon up front. By improving the private promoters' recipient's risk-return profile, this capital catalyzes the participation of investors that otherwise would not have joined. The basic idea is that the reduction in risk as well as the provision of favorable financial conditions reduce the overall weighted average cost of capital for these stakeholders. If public authorities provide equity-based instruments, they can also participate in the financial benefits associated with the value enhancement of successful projects. However, only real estate development projects that generate some cash flow return (i.e. repayable investments) and create positive externalities according to an integrated plan for sustainable urban development are eligible and fundable. Since the presented quarter development project is not only part of an integrated land-use plan of the city of Düsseldorf but also creates substantial positive externalities, it would be a perfect example for a (not yet existent) NRW-UDF. Such a kind of an independent investment fund could provide equity capital (and create a joint venture company with private promoters) or mezzanine capital to development projects.

Applying urban financial innovations to large-scale quarter development projects to solve the feasibility problem

In this section, we will now apply all urban financial innovations presented in the last section to our reference project in order to quantify the added value of these urban financial innovations for private promoters and public authorities. For the private partners we use the already introduced IRR equity calculations and for the public authorities the (negative) NPV. The calculation here is simplified because public authorities in Germany have an interest rate of nearly 0% at the present time. Therefore, we can also calculate the benefit–cost–ratio (BCR) by combing the overall costs (absolute value of NPV) of the promotions with the benefit (overall investment volume in development project) to quantify the intensity of the public promotion.

The results of Table 12.4 reveal that not all urban financial instruments render a financially feasible project development for the private promoter (with at least 8.0% equity return after taxes). A special government entity (in this case for the public green space), the higher density, or floor area ratio (non-tradable), tax reliefs with increased amortization, and the soft loans in the operating period (even with an additional 25% immediate loan amortization) would not create sufficient incentives. This would still result in market failure of private development partners. A quite easy and comparable solution for the feasibility problem would be the investment grant, which public authorities could give as a cash payment (based on investment costs) and as a below market price land sale (if the public authority is the landowner). Here the subsidized land solution would be favorable because the multiplier effect of the subsidy is higher than the investment grants. A comparable subsidy intensity would be tax increment finance combined with 20-year project bonds. If public authorities use the bond capital for the

Table 12.4 Urban financial innovations for public support of integrated urban real estate development projects: Quantitative evaluation

Urban financial instrument/innovation	Financial feasibility of private promoter	NPV subsidy for public authorities	Intensity of public support (BC-Ratio)
Special government entity	5.51%	–3,087,789 €	22.5 but not feasible
Subsidized land (reduction land price)	8.00%	–10,372,695 €	6.0 and feasible
Density or floor area ratio (FAR) bonuses as tradable development rights (TDR)	4.90%	0	Not feasible (base scenario)
Usage pricing for public/subsidized goods (1 €/per quarter per sq.m floor space)	8.70%	0 (–3,209,204 €)	22.6 (by neighborhood not public authority)
Project bond with tax incremental financing (TIF)/value capture finance (20 years)	9.67%	–12,955,233 €	4.6 and feasible
Grants (for investment costs ex post)	8.00%	–31,821,177 €	1.3 and feasible
Tax reliefs (increased amortization 4% p.a.)	5.68%	–7,355,446 €	9.9 but not feasible
Soft loans (30 year operating phase)	5.11%	–3,933,119 €	18.5 but not feasible
Soft loans (in addition: 25% immediate amortization of credit volume)	6.20%	–8,249,671 €	5.0 but not feasible
Equity for a joint PPP company (until exit)	8.00%	–30,437,012 €	2.4 and feasible
(Equity/debt) mezzanine capital for project finance (7 years with soft conditions)	8.00% according DCF (11.61% IRR)	–11,001,586 €	7.0 and feasible

complete public green development, site improvements, and roadworks, the private promoter would already get a return on equity of 9.67%. As an alternative, managing authorities could reduce the bond volume (and therefore the equity return for the promoter) and thus raise their subsidy intensity and therefore the benefit–cost–ratio. If the public authority used equity (until the exit phase with a complete waiving of a market conforming equity return of 8.0%) this would lead to a high NPV for the subsidy and a low benefit–cost–ratio. Although this solution is still better than an investment grant (because of the revolving character), a much better alternative would be the use of (equity) mezzanine capital because the duration as well as the payment burden would be lower for the public authorities. A usage pricing would be even more efficient for them. However, it requires substantial neighborhood solidarity, which is not realistic in new urban quarters such as in our example.

Therefore, revolving financial instruments in the form of mezzanine capital are a reasonable strategy to create PPPs in the context of large-scale urban development projects. An institutional framework like the JESSICA initiative and an organization form like the UDF could provide added value for both private investors and public authorities. This is even more relevant bearing in mind the qualitative benefits. If the revolving financial instruments only finance part of the investment volume (as assumed in the example) private capital mobilization will be mandatory. When the duration of mezzanine capital is restricted up to seven years, more capital recycling takes place. At the same time, UDFs could be not temporary but permanent financing partners in complicated development projects, free from budgetary situations. On the other side, mezzanine capital could be as effective as equity but it does not require management capacities from public authorities. Furthermore, project discipline might generally be better. Establishing institutional project control, which is essential for a UDF, ensures that a feasibility study is carried out first for each potential project. It includes an analysis of location, markets, risks, profitability, and sustainability. This detailed project planning allows for a periodic target–performance comparison with the project's progress that can identify and eliminate causes of deviations. Through professional fund management, with expertise from institutional investors and commercial banks, the efficiency of projects will further increase. At the same time, the private sector could benefit from public knowledge in the field of urban policy, promotion programs, building regulations, and town planning. The bundling of expertise could create a win–win situation for both partners.

Conclusion

Both public and private investors are highly interested in sustainable urban development projects that provoke positive externalities. In this context, land development projects especially can render high positive externalities, but are often subject to market failure. This is due to the fact that land development often is too capital intensive and risky for one investor alone. Therefore, PPP solutions are extremely favorable to overcome market failure, but are also subject to criticism themselves. Therefore, our chapter addresses several urban financial innovations on a PPP approach. Here, we find that urban financial innovations, especially in the form of revolving financial engineering instruments, are meaningful for complicated, large-scale development projects where public and private stakeholders have to work together. This PPP approach is especially suited to enlarge the scope of investable development projects: in contrast to classical grant-funding, revolving financial engineering instruments should be used for those projects with below market returns ("B-projects" in the CABERNET systematization). For these projects, the introduced financial innovations could be a way to combine private return on investments with public favored externalities for urban citizens (Squires et al., 2016). With an efficient design, financial engineering could provide a permanent contribution to new public management and sustainable urban development (not only) in Europe.

References

Brueggeman, W. and Fisher, J. (2015) *Real Estate Finance and Investments*, 15th edn, New York.

Bryant, L. (2015) Who really pays for urban infrastructure? The impact of developer infrastructure charges on housing affordability in Brisbane, PhD Dissertation, Queensland University of Technology, Brisbane.

Elkington, J. (1997) *Cannibals with Forks: The Triple Bottom Line of 21st Century Business*, Oxford.

European Commission (2014) Integrated sustainable urban development cohesion policy 2014–2020, Brussels.

European Parliament (2014) Urban dimension of EU policies, European Parliament resolution of 9 September 2015 on the urban dimension of EU policies (2014/2213(INI)), Brussels.

European Union (2009) Promoting sustainable urban development in Europe: Achievements and opportunities, Brussels.

German Society for Real Estate Research (Gif) (2006) *Standardisierung des DCF-Verfahrens*, Wiesbaden.

IPD/MSCI (2016) *Germany Annual Property Index, Results for the Year to 31 December 2015*, Frankfurt.

Kreuz, C. and Nadler, M. (2010) *JESSICA – UDF Typologies and Governance Structures in the Context of JESSICA Implementation*, Brussels, Luxembourg.

Mankiw, N. G. (2014) *Principles of Economics*, 7th edn, Stamford.

Nadler, M. (2010) *JESSICA Initiative for North Rhine-Westphalia: Evaluation Study*, Brussels, Luxembourg.

Office of the Deputy Prime Minister (2004) Skills for sustainable communities, London.

Owens, R. W. (1998) Subdivision development: bridging theory and practice. *Appraisal Journal*, July, 274–279.

Peiser, R. (2012) *Professional Real Estate Development*, 3rd edn, Washington.

Squires, G., Hutchison, N., Adair, A., Berry, J., McGreal, S., and Organ, S. (2016) Innovative real estate development finance: evidence from Europe, *Journal of Financial Management of Property and Construction*, 21:1, 54–72

Thiel, S. and Nadler, M. (2015) External benefits of private property-led development projects. In: Michael Nadler (ed.), *Working Papers for Integrated Real Estate Development*, Issue No. 2, Dortmund.

13

The self-organizing city

An analysis of the institutionalization of organic urban development in the Netherlands

Edwin Buitelaar, Erwin Grommen, and Erwin van der Krabben

Abstract

Due to the impact of the global financial crisis on the Dutch real estate market, the poor financial position of municipalities that resulted from that, and their reluctance to continue with proactive public land development, and an ongoing shift from greenfield development to urban transformation, private sector-led organic urban development increasingly seems to replace the predominant comprehensive, integrated approach to planning and real estate development in the Netherlands. The question we address in this chapter is to what extent these "new" organic development approaches have been institutionalized in the Netherlands. We use insights from theories on institutional change to create a better understanding of the way new planning and development practices institutionalize, and to "measure" whether the new practices have exceeded the threshold of institutionalized behavior. The issue has been further explored in a study of the city of The Hague. We were able to study in-depth characteristics of all urban development projects initiated in The Hague between 2008 and 2015. Our research shows that urban developments in The Hague have indeed become more "organic," though there is still a prominent role for the municipality. Instead of "traditional" large private development companies, end-users and investors with a long-term interest increasingly appear to take initiatives for new developments. The results of our study may relate to debates taking place in many countries regarding shifting public and private sector roles in planning and development projects.

Introduction

Dutch cities have for a long time relied on a comprehensive, integrated approach to planning and real estate development, based on a public land development model (Needham, 2007; Buitelaar, 2010; van der Krabben and Jacobs, 2013). The defining feature of this approach is that the municipality purchases the land earmarked for development. The municipality then prepares the land for further development (including all necessary infrastructure works), before

169

it sells the serviced land to developers and/or housing associations. The system aims to provide both high-quality development locations and cheap land for subsidized social housing. For years, this policy has allowed municipalities to make profits with which other municipal facilities could be funded. Also, as landowners, municipalities were in a good position to negotiate with private developers about the desired development of the location. Private developers appreciated the model, because it reduced their financial risks, while still being able to make a profit (Faludi and Van der Valk, 1994), provided them with high-quality locations and helped to control competitive developments elsewhere. Recent years have shown the downside of this development model: the economic and financial crisis led to a downturn in the demand for new housing, private developers stopped buying building plots from municipalities, and municipalities lost hundreds of millions of euros, mainly because of unanticipated increased interest costs (van der Krabben and Heurkens, 2015). As a consequence, calls for a change in the practice of area development in the Netherlands have increased in recent years.

In response to these developments, some, such as Urhahn Urban Design (2010, pp. 1–3) call for an approach based on small-scale, process-oriented developments that are user-oriented and reflect common values. Buitelaar, Galle, and Sorel (2014), foreseeing a period in which growth is not guaranteed and urban area development has structurally changed, take over this plea and advocate a more organic approach to area development. A growing attention to self-organization, spontaneous order, and organic area development is not limited to the Netherlands, but can be found across many developed countries (Alfasi and Portugali, 2007; Andersson and Moroni, 2014; Moroni, 2015).

The question we address in this chapter is whether these "new" organic development approaches have been institutionalized in the Netherlands. The issue has been further explored in a study of the city of The Hague. We were able to study in-depth characteristics of all urban development projects initiated in The Hague between 2008 and 2015. This analysis provides insights into, among other things, the type and size of projects, the scale of the projects, the role of the municipality, and the role of private actors, the development approach and the management of the project. On the basis of institutional theory we first analyze the extent to which institutional change towards and institutionalization of organic urban development have taken place, and second, what the underlying determinants are. For the first part we make use of quantitative empirical data, for the second part information is used that is derived from interviews with the main stakeholders in the city of The Hague (both public and private) involved in urban development (the municipality, housing associations, developers, and investors). In the concluding section of the chapter we discuss what the impact of these changes may be on the outcome of land and real estate development in the Netherlands. In addition, we focus on possible implications elsewhere of our attempt to measure institutions, change and institutionalization in land development.

The self-organizing city

Many "mature" European cities increasingly focus on policies of urban transformation and renewal – often as part of broader urban regeneration policies – but face a common set of challenges with regard to the implementation of these projects. Typical urban transformation projects may concern the transformation of brownfield sites, the redevelopment of inner-city shopping areas, waterfront and dockland redevelopment projects, and the renovation of post-war social housing blocks. Some of the most significant obstacles to the implementation of such projects may relate to periods of economic downturn and reduced market demand, but various studies have also revealed more fundamental/structural obstacles, such as (institutional) barriers to the assembly of land and properties, shortcomings in existing value capturing mechanisms to cover public infrastructure costs, suboptimal public–private cooperation, and increased complexity of the projects themselves (Adams et al., 2002; Buitelaar et al., 2008. Typical for this kind of complex land and property development project is the wide range of stakeholders involved – such as landowners, municipalities, investors, property developers, construction companies, architects, infrastructure providers, and housing corporations – that must try to reach an agreement on planning, development, and financial issues, although they often have partly contradictory interests, which may prevent them from reaching such an agreement (Samsura et al., 2010).

Partly as a response to this changing context, many European cities are now embarking on a variety of innovative experiments with respect to governance arrangements and financial packages for urban transformation (van der Krabben and Needham, 2008; Nordahl, 2014; Adams, 2015). Particular interest goes to self-organizing urban governance, for one reason because it will reduce public sector responsibility and (financial) involvement. But also more principally, because some expect self-organization to create cities that are more dynamic and better meet people's demands, which are therefore more resilient and sustainable (Urhahn Urban Design, 2010; Buitelaar, Galle, and Sorel, 2014). Portugali (2000) refers to the concept of the self-organizing city as urban development that arises out of the spontaneous local interactions between stakeholders based on the initial conditions chosen/caused by the stakeholders themselves, without active and goal-oriented government intervention. The role of government is to create a basic legal framework (an urban code), without a particular urban order in mind, within which self-organization among local actors takes place (Moroni, 2015).

Promoting self-organization seems to go hand in hand – at least in the Netherlands – with a development strategy to which we refer here as *organic* area development, as opposed to *integrated* area development. In comparison with more plan-led integrated development, this organic approach aims for smaller scales and is more strategic, process-oriented, with a stronger role for end-users at the expense of large developer firms (see Figure 13.1). Especially in the Netherlands, with its strongly institutionalized planning culture (CEC, 1997; Alterman, 2001), this type of planning and development certainly deviates from development practices in past decades in Dutch cities. But also in many other advanced economies, albeit to a lesser extent, this type of development has received (renewed) attention in recent years (Tira et al., 2011; Squires and Lord, 2012; Alfasi and Portugali, 2007; Moroni, 2015).

Dutch urban development in an international context

Urban development in the Netherlands since the Second World War has been based on a kind of "blueprint" or "end-state planning" (Needham, 1997, 2007; Buitelaar, 2010; Mori, 1998).

> A collaboration between private and public actors was established with arrangement about the building program, the urban design as well as phasing of the various aspects of the area development, supported by a land account, upon which the plan would ultimately be turned into a legally binding land-use plan.
>
> *(Buitelaar and Bregman, 2016, p. 7)*

The tradition of large-scale integrated urban development – residential development in particular – seems to relate for some practical reasons to (social) housing policy, and to urban design principles. From a practical point of view, integrated land development seems to be efficient, particularly in the Randstad region, where development land usually is below sea level and requires huge investments to make it suitable for development (Buitelaar and Witte, 2011; Tennekes, Harbers, and Buitelaar, 2015). Also from a practical point of view, integrated land development enables the placing of infrastructure in an efficient way as an integral part of the phased development of an entire area. With regard to housing policy, the integrated development of land was effective in reserving sufficient land in residential development areas for social housing. Finally, with regard to urban design principles, one may argue that a tradition has developed in the Netherlands that appreciates an integrated design for residential areas based on a rather detailed blueprint for that area (Buitelaar, Galle, and Sorel, 2011).

The blueprint planning and development approach used to go hand in hand with a public land development model. One of the objectives of municipal land policies is to make sure that sufficient land becomes available for planned urban development, be it greenfield development or infills or brownfield development. Often this requires a form of land assembly, since the required subdivision for the new development does not match up with the existing ownership structure (van der Krabben and Jacobs, 2013). The land assembly policies adopted by municipalities depend on, among other things, sometimes long-standing traditions in both urban planning and land policies (Healey, 2004), the policy instruments that cities have at their disposal, municipal finance (e.g. how cities finance their infrastructure; legal restrictions for municipalities to invest in land development), and market conditions (i.e. a so-called public goods argument: public authorities "produce" building land, because the private sector does not produce it in time and/or sufficiently; van der Krabben and Buitelaar, 2011).

Van der Krabben and Jacobs (2013) distinguish between four main types of land development models: a public comprehensive model (public purchase and development of land), a public planning-led quasi market model (public purchase of land in order to enable a private sector-led (re)development program for a specific area), a private market model (private purchase of land, if necessary and when desirable supported by expropriation by a public authority), and an urban land readjustment model (landowners in a certain area engage in a joint venture; the readjustment of ownership brings all owners into the best position to (re)develop). Dutch municipalities have had a reputation internationally for applying a public comprehensive land development model on a wide scale (CEC, 1997; Alterman, 2001; Buitelaar and Bregman, 2016; Valtonen et al., 2017). While in many countries, local authorities do purchase land now and then, mainly in urban transformation areas, to support future private development in that area, only a few countries – as far as we are aware in Europe, besides the Netherlands only Finland and Sweden – apply such a proactive plan-led approach to land development. Dutch cities used to purchase and develop (almost) all (future) building land within a city "to guarantee building developments according to public policies, to realize full cost recovery of all public works via the sale of building plots and to capture at least part of the surplus value of the land (after a change in use), to use that for public use" (van der Krabben and Jacobs, 2013, p. 775). Though the public development of land may be risky for public authorities and does not guarantee future urban development, because this still depends on market conditions for real estate markets, Dutch cities nevertheless relied on it for many decades.

The tradition of large-scale integrated development supported by a public land development model may be long-standing in the Netherlands, but nevertheless came under discussion after the outbreak of the global financial crisis in 2008. The subsequent decline of property and housing markets induced a reduction in demand for new housing and building land, which brought both municipalities and private developers, both holding substantial land banks, into financial trouble (Deloitte, 2013; Buitelaar and Bregman, 2016). More or less at the same time, in many Dutch cities spatial planning strategies shifted from a dominant focus on greenfield residential developments towards a greater emphasis on urban transformation (Buitelaar, 2010).

Compared to greenfield development, a large-scale integrated end-state planning approach for urban transformation may often be more complicated and more risky, due to fragmented ownership structures and potential hold-out problems, for instance (Adams et al., 2002) and relatively high costs of buying out of properties that are still in use and/or related to contaminated land.

As a response, municipalities have become much more reluctant to apply the "blueprint" model to new development, while they seem to have left at the same time the idea of taking an "active" role in the actual implementation of plans (Buitelaar and Bregman, 2016). As an alternative, a new "organic" development approach (Figure 13.1) has been introduced. Whether this new approach has already been "institutionalized" remains to be seen.

Understanding and "measuring" institutional change and institutionalization

Actions by agents such as action within land and real estate development processes are not voluntaristic. They are constrained and enabled, thus influenced, by the context within which they take place. "Doing – the central thread of practice – is not just doing in and of itself … but is always doing in a historical and social context that gives structure and meaning to what we do" (Laws and Hajer, 2006: 411). That context is made of institutions, which can be defined as the man-made structures that guide and give meaning to human interaction (North, 1990).

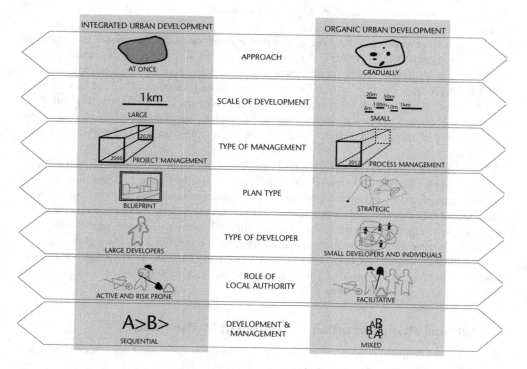

Figure 13.1   Integrated and organic urban development

Source: Buitelaar, Galle and Sorel, 2014

This shows that the relation between institutions and actors must be seen as reciprocal; institutions are actively created, changed, and maintained through action (Buitelaar, Galle, and Sorel, 2011). The reciprocity between actors and institutions has become known as "the duality of structure" (Giddens, 1984). Patsy Healey emphasized the relevance of this notion for the land development process long ago (Healey, 1992).

Institutions can be formal in the sense that they are written down in laws, zoning plans, ordinances, etc., and legally enforceable by the court, or they can be informal. Informal institutions are non-legal and unwritten rules such as taboos, conventions, and codes of conduct. Institutions emerge through social interaction and are the result of imitation and repetition of behavior (Zijderveld, 2000). This is especially the case with informal institutions but to a large extent also applies to formal institutions. Those are not designed overnight either. Law making, for instance, is an often cumbersome process. And when rules are finally adopted they often do not instantly become what we consider institutions. "Legislation, for instance, while formalized through a distinct decree, takes shape gradually, within evolving patterns of social expectations. Moreover, the formal act of commencement must be followed by practices of validation in social interaction" (Dembski and Salet, 2010: 618). Without that taking place, without really impacting on the behavior of those they target, formal rules are not institutions, but just a collection of words on paper.

In this contribution we are concerned about measuring the degree of institutional change (i.e. *from* integrated *to* organic development) or institutionalization (i.e. *of* organic development). We contend that the change and institutionalization can be measured by the extent to which a particular behavior occurs. Is it widespread or occasional behavior? Obviously, it is matter of degree and it is arbitrary to say when behavior has exceeded the threshold of institutionalized behavior (Buitelaar, Galle, and Sorel, 2011).

It is important at this point to emphasize that institutions and behavior are ontologically distinct (Hodgson, 2004). However, behavior provides an indication of the presence and influence of institutions. Repetition of behavior by one actor – in other words routine behavior – and imitation of it by others can be seen as the result of institutions. One-off behavior has then become institutionalized. In other words, when rules genuinely affect actors and their behavior in the sense that it shows repetition and imitation, whether in an intended direction or not, and becomes predictable to some extent, those rules can be said to have become institutions.

We consider organic and integrated development as types of collective behavior, as outcomes of the interaction between actors and institutions. In Figure 13.1 we distinguished between seven features of organic (and integrated) development. Based on what we outlined above, we can say that the more a development conforms to the seven features, the more it can be considered "organic." And when the scale at which this occurs – in terms of the number of practices – is increasing, we say that organic development is institutionalizing and that there is an institutional change away from integrated development. The case of The Hague is used to illustrate how this might work and to indicate, for one city, to what extent this process of institutionalizing actually has taken place.

Institutional change in urban development in The Hague

To analyze the degree to which "new" organic development approaches have institutionalized and the factors that have allowed for or hampered this, urban development practices in the city of The Hague have been studied in more detail. Both data on development projects and interviews with key stakeholders have been used to retrieve the necessary information.

The data

The empirical data have been derived from the Database Programmering Gemeente Den Haag (Gemeente Den Haag, 2015) (based on earlier work by Grommen, 2015). This database contains information on the status of developments taking place, in various stages, between January 2008 and January 2015 within the municipality of The Hague. For each of the intermediate years, the status of the different projects on 1 January has been registered in the database. The database contains over 3500 unique records (i.e. urban development projects). Each record covers one project for one year. A project thus occurs only once per year. In addition, the database contains information on the development of housing, commercial real estate, and public space, the stage the project is at, the developer involved, and the role of the municipality. This research does not take into account all seven dimensions from Figure 13.1, but focuses on the three dimensions for which empirical data are available: the scale of the project, the developer involved, and the role of the municipality.

The scale of projects

The scale of projects is measured by the number of houses being planned and the number of square meters of commercial real estate. Commercial real estate consists of spaces for leisure and catering, offices, businesses, and other services such as schools.

The developer type

The data allow for distinguishing between different types of developers that are responsible for real estate development or construction of public spaces. In this research, the following five categories are distinguished:

- Municipality: real estate development is carried out by the municipality or one of the municipal services.
- Housing association: a housing association or developer under direct control of a housing association is responsible for development.
- Real estate developer: develops a project aimed at making profits before selling its share to end-users or investors.
- Investor: involved in land development aimed at long-term property investment and returns. This includes investment firms with and without a development branch and other firms who realize property for their own investment portfolio.
- End-user: the developer is also the user of the property after completion.

The role of the municipality

To measure the role of the municipality, three types of municipal involvement have been identified without having data on the exact (financial) commitments made by parties.

a First, the municipality alone can take care of the land development or land and real estate development. In the latter case the municipality is fully responsible for the resulting costs and benefits of both the land and the real estate development stage.

b Second, in a public–private partnership (PPP) multiple parties take care of the development of an area, usually only the land development stage, with the municipality being one of them. The PPPs often take shape in a joint venture, a dedicated and separate legal entity, in which parties participate proportionately to the amount and value of the land they put in.

c   Third, the municipality takes no financial risks and plays no role beyond its public planning powers; it takes a facilitative or enabling role. Private parties take care of all costs and benefits associated with the development of land and real estate.

Interviews

For this study a number of interviews with professionals have been conducted so as to obtain a greater understanding of the causal factors behind the results of our data analysis. Municipal employees, people working for housing associations, real estate developers, and investors were approached. They were selected on the basis of the findings of the quantitative analysis; their role has often changed, as we will see in the descriptive statistics. We considered them to be the most appropriate actors to help us explain continuity and change in urban development practices in The Hague. This led to a total of seven interviews. The interviewees were asked to give their professional interpretation of the changes (if any) that have taken place. Interviews were conducted between 30 April 2015 and 16 June 2015.

Results

The data have been analyzed for the new projects that were initiated each year. We left out the projects that had already started before 2008 and might have changed in the face of the financial and economic crisis, since those are least likely to reveal potential fundamental shifts in behavioral patterns. Pragmatic decisions and reactions to unanticipated events are more likely to have steered those developments after 2008. A project only counts as new in the first year of appearance in the dataset.

The scale of projects

Figures 13.2 and 13.3 provide key statistics on the scale of new projects. Looking at the data, it is clear that the scale of new developments is decreasing for both the number of housing units and the square meters of commercial space. While the amount of commercial space is decreasing from 2010 on, the amount of housing units starts falling after 2011. This late response is caused by the nature of developments: it is often difficult to scale down planned developments immediately (interview: municipality). This might be related to the low number of projects. The average amount of public space increases substantially in 2010 before dropping sharply in 2012. After this, it recovers again. For all segments, the number of projects seems to be rising again from 2012 on. This trend towards smaller projects can easily be explained. Due to the economic crisis, demand has dropped while insecurity about the available means has grown. In order to minimize risks, developers decreased the size of projects. Smaller projects allow for easier adjustments and a reduced time lapse between investment and return (interviews: housing association; developers; investor).

The developer type

The next step is to analyze the type of developer. Figure 13.4 depicts the developer types for new projects. Combinations in which different developers participate have been

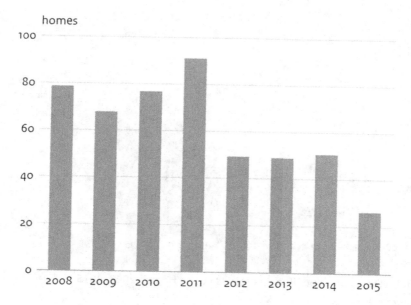

Figure 13.2 The average number of homes per new project in The Hague, 2008–2015
Source: Gemeente Den Haag, 2015

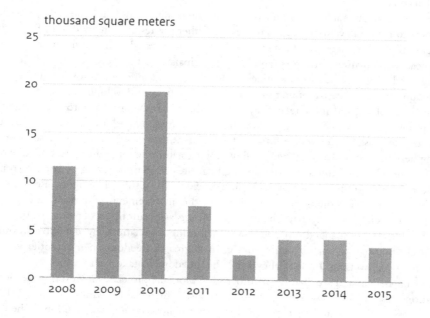

Figure 13.3 The average number of square meters of commercial real estate per new project in The Hague, 2008–2015
Source: Gemeente Den Haag, 2015

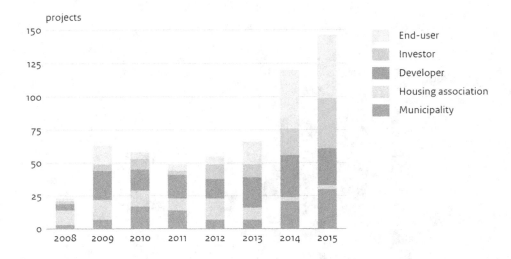

Figure 13.4 Developer types in new projects in The Hague, 2008–2015

Source: Gemeente Den Haag, 2015

deconstructed into individual developer types. This implies that some projects are counted more than once.

Due to the unfavorable economic situation, the ability and willingness of professional developers to realize developments has decreased. Other parties have stepped in to fill the void. The figure shows the increasing financial involvement (expressed in the number and the share of projects) of end-users, investors, and the municipality in the development process. The growing involvement of other types of developers means that different factors need to be taken into account. Investors are becoming more important to finance developments. These investors want to be involved, and in general, they want to be involved early so as to ensure they will have a product that fits their needs in the long run. The increasing involvement of end-users links to the decrease in the average scale of development that we discussed earlier and the shift of a suppliers' to a buyers' market, which allows for a greater role for consumers. End-users such as homeowners have no interest in developing large areas, but are focusing on their own needs (i.e. one-off housing) (interviews: municipality; developers; investor). The figure also shows that now the number of projects is growing again, the involvement of professional property developers is rising too. The participation of housing associations, as depicted in the graph, seems to reflect the changed attitude towards housing associations by central government. Historically, housing associations enjoyed a great degree of freedom, their activities were not limited to social housing. Initially, they used this freedom to take over in projects as private parties decreased their participation in urban development projects. However, some housing associations also suffered severe losses due to risky financial activities and in 2012 the Herzieningswet (Revision Act) changed the rules for housing associations. Since then, they have only been allowed to build, develop, and maintain social housing, below a state-defined rent level. In addition, they are allowed to develop projects that form a relevant, societal addition to their real estate, or to develop housing for private rental if the market parties do not do so (interview: housing association).

The role of the municipality

Over the years, the role the municipality plays during the land development stage has changed substantially. To assess this change, this research focuses on two indicators. First, the financial involvement in developments will be discussed. Second, the developments for which the municipality of The Hague is listed as the main land developer have been analyzed.

Figure 13.5 shows that the share of projects developed by the municipality alone is rising. However, more important is the number of privately financed development projects, many of which are developments of single homes by end-users. The number of PPPs gradually decreased, until it rose again. Nevertheless, in relative terms, the share of projects in which land development is carried out in cooperation between private parties and municipalities is decreasing.

Our research shows that urban developments in The Hague have indeed become more "organic," though there is still a prominent role for the municipality. Our data indicate a clear trend towards smaller-scale projects. Instead of "traditional" large, private development companies, end-users and investors with a long-term interest increasingly appear to take initiatives for new developments. Though our data do not allow us to analyze all the aspects of a development strategy (indicating either integrated or organic urban development; see Figure 13.1), we may nevertheless conclude for the city of The Hague that a shift in development practices occurs, in favor of smaller-scale projects and more active roles for end-users and investors with a long-term interest.

Conclusion and discussion

Whether organic urban development has been "institutionalized" or has even become mainstream in the Netherlands can of course not yet be derived from this single case study, made over a limited number of years. However, we believe the results of our study, combined

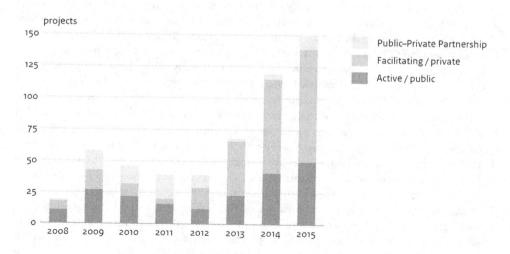

Figure 13.5 Role of the municipality in new projects in The Hague, 2008–2015
Source: Gemeente Den Haag, 2015

with ongoing policy debates in the Netherlands, do allow for some conclusions. First, we have no reason to believe that the changes in the approach to urban development that occurred in The Hague would be very different from what can be found in other Dutch cities. For instance, many cities have adopted a new land policy in which they state that public land development no longer forms the default development model, and instead decide between an active or a more facilitating role, depending on the situation. On the other hand, many municipalities and some of the big nationwide operating private developers still hold huge land banks that allow integrated developments in some locations and it is likely that they will continue with these developments as soon as market conditions allow them to do so.

Second, a new planning law is under preparation, including regulation supporting municipal land policies. This new planning law is expected to take away some current legal obstacles for municipalities – related to the recovery of costs for public works – to provide room for organic development approaches. We therefore assume that after the (expected) introduction of the new planning law in 2018, municipalities will have even better opportunities to apply an organic development approach.

Third, when referring to Portugali's definition of "self-organisation," it is obvious that the apparent shift towards organic development has certainly not entirely arisen out of the spontaneous local interactions between stakeholders without any government intervention. On the contrary, the local authorities in The Hague still play an important role in organic urban development, sometimes by participating financially in land acquisition or by initiating cooperation among private actors.

Fourth, we do not expect that in future municipalities will completely abandon the traditional comprehensive integrated development approach, but regional differences may appear. In highly urbanized areas expecting high demand for new housing, both municipalities and big nationwide operating private developers will probably still appreciate the traditional approach for a limited number of greenfield developments, because of the supposed efficiency of the model, next to an organic approach for urban transformation projects. In other regions, with less expected demand for new housing or even demographic decline, a return to integrated development is less likely.

We believe that the results of this study can be relevant for an international audience as well, for various reasons. Changes in (governance) approaches to land and real estate development take place – or at least are being discussed – in other countries as well (Lefcoe, 1977; Tan et al., 2009; Hartmann and Spit, 2015). As an example, we can refer to recent land reform debates in Scotland, quite in the opposite direction of what we have analyzed in the Netherlands. While Dutch cities now promote private stakeholders' initiatives for organic urban development as an alternative to public-led development, Adams (2015) has suggested in this land reform debate in Scotland a shift in the opposite direction, towards public-led development as applied in Sweden and the Netherlands, to overcome problems in Scotland (and elsewhere in the UK) with speculative land banking by big private developers. Discussions about paradigm shifts in urban development often concentrate on necessary changes in underlying planning laws – to enable the "new" development model. Such formal, legal changes are of course part of the process of institutionalization as well. However, the analysis of the extent to which organic urban development approaches have institutionalized in The Hague demonstrates the significance of informal processes of institutionalization. The introduction of organic urban development in Dutch cities does not require any changes in planning law in the short term; the main issue is whether both private and public stakeholders are prepared to work in that way.

Apart from the findings regarding development practices for the Dutch case and what this may imply for international land and real estate development practices, we believe that the research method itself, that is, to identify and measure institutions, institutional change, and institutionalization in land and real estate development, is appropriate to be used elsewhere as well. It helps to shed light on the development of institutions in a longitudinal way, which allows for monitoring institutions, and changes thereof, and therefore provides information upon which decisions to intervene or to refrain from doing that can be based. To be able to provide such information, a lot depends on municipalities' willingness to collect micro data and the accuracy with which they are registered.

Finally, we did not analyze in the The Hague case what the effect is of the introduction of an alternative development approach on market outcomes, in terms of the quantity, quality, and prices of the real estate "produced." For further study, both in the Netherlands and elsewhere, it would be interesting to analyze the impact of the institutionalization of a new urban development approach on this type of market outcome.

References

Adams, D. (2015), *Urban land reform briefing paper No 4: Explaining public interest led development*. Glasgow: Policy Scotland / University of Scotland

Adams, D., Disberry, A., Hutchinson, N., and Munjoma, T. (2002), "Land policy and urban renaissance: the impact of ownership constraints in four British cities," *Planning Theory and Practice*, 2: 195–217.

Alfasi, N. and Portugali, J. (2007), "Planning rules for a self-planned city," *Planning Theory*, 6(2): 164–182.

Alterman, R. (2001) *National-level planning in democratic countries: An international comparison of city and regional policy-making*. Liverpool: Liverpool University Press.

Andersson, D. E. and Moroni, S. (eds) (2014), *Cities and private planning, property right: Entrepreneurship and transaction costs*. Cheltenham: Edward Elgar.

Buitelaar, E. (2010), "Cracks in the myth: challenges to land policy in the Netherlands," *Journal of Economic and Social Geography*, 101(3): 349–356.

Buitelaar, E. and Bregman, A. G. (2016), "Dutch land development institutions in the face of crisis. Trembling pillars in the planners' paradise," *European Planning Studies*, 24(7): 1281–1294.

Buitelaar, E. and Witte, P. (2011), *Financiering van gebiedsontwikkeling: een empirische analyse van grondexploitaties*. The Hague: PBL Netherlands Environmental Assessment Agency.

Buitelaar, E., Segeren, A., and Kronberger, P. (2008), *Stedelijke transformatie en grondeigendom*. The Hague/Rotterdam: RPB/Nai Uitgevers.

Buitelaar, E., Galle, M., and Sorel, N. (2011), "Plan-led planning systems in development-led practices: An empirical analysis into the (lack of) institutionalisation of planning law," *Environment and Planning A*, 43: 928–941.

Buitelaar, E., Galle, M., and Sorel, M. (2014), "The public planning of private planning: an analysis of controlled spontaneity in the Netherlands," in D. Andersson and S. Moroni (eds.), *Cities and private urban planning: Property rights, entrepreneurship and transaction costs*. Cheltenham: Edward Elgar, pp. 248–265.

CEC (1997), *The EU compendium of spatial planning systems and policies*. Luxembourg: European Commission.

Deloitte (2013), *Financiële situatie bij gemeentelijke grondbedrijven 2013*. Utrecht: Deloitte.

Dembski, S. and Salet, W. (2010), "The transformative potential of institutions: how symbolic markers can institute new social meaning in changing cities," *Environment and Planning A*, 42: 611–625.

Faludi, A. and Van der Valk, A. (1994), *Rule and order: Dutch planning doctrine in the twentieth century*. Dordrecht: Kluwer Academic.

Gemeente Den Haag (2015), *Database Programmering*. The Hague: Municipality of The Hague.

Giddens, A. (1984), *The constitution of modernity*. Oxford: Oxford University Press.

Grommen, E. (2015), *From integral to organic development? A study of area development approaches in The Hague*. Walferdange: Faculty of Language and Literature, Humanities, Arts and Education – University of Luxembourg / The Hague: PBL Netherlands Environmental Assessment Agency.

Hartmann, T. and Spit, T. (2015), "Dilemmas of involvement in land management – comparing an active (Dutch) and a passive (German) approach," *Land Use Policy*, 42: 729–737.

Healey, P. (1992), "An institutional model of the development process," *Journal of Property Research*, 9: 3344.

Healey, P. (2004), "The treatment of space and place in the new strategic spatial planning in Europe," *International Journal of Urban and Regional Research*, 28: 45–67.

Hodgson, G. M. (2004), *The evolution of institutional economics: Agency, Structuralism and Darwinism in American institutionalism*. London: Routledge.

Laws, D. and Hajer, M. (2006), "Policy in practice," in M. Moran, M. Rein, and R. Goodin (eds), *The Oxford handbook of public policy*. Oxford: Oxford University Press, pp. 409–424.

Lefcoe, G. (1977), "When governments become land developers: notes on the public-sector experience in the Netherlands and California," *S. Cal. L. Rev.*, 51: 165–263.

Mori, H. (1998), "Land conversion at the urban fringe: A comparative study of Japan, Britain and the Netherlands," *Urban Studies*, 35: 1541–1558.

Moroni, S. (2015), "Complexity and the inherent limits of explanation and prediction: Urban codes for self-organising cities," *Planning Theory*, 14(3): 248–267.

Needham, B. (1997), "Land policy in the Netherlands," *Tijdschrift voor Economische en Sociale Geografie*, 88(3): 291–296.

Needham, B. (2007), *Dutch land use planning: Planning and managing land use in the Netherlands, the principles and the practice*. The Hague: SDU.

Nordahl, B. (2014), "Convergences and discrepancies between the policy of inclusionary housing and Norway's liberal housing and planning policy, an institutional perspective," *Journal of Housing and the Built Environment*, 29(3): 489–506.

North, D. C. (1990), *Institutions, institutional change and economic performance*. New York: Cambridge University Press.

Portugali, J. (2000), *Self-organization and the city*. Heidelberg: Springer.

Samsura, D. A., van der Krabben, E., and Van Deemen, A. (2010), "A game theory approach to the analysis of land and property development processes," *Land Use Policy*, 27(2): 564–578.

Squires, G. and Lord, A. (2012), "The transfer of Tax Increment Financing (TIF) as urban policy for spatially targeted economic development," *Land Use Policy*, 29(4): 817–826.

Tan, R., Beckmann, V., Van den Berg, L., and Qu, F. (2009), "Governing farmland conversion: comparing China with the Netherlands and Germany," *Land Use Policy*, 26(4): 961–974.

Tennekes, J., Harbers, A., and Buitelaar, E. (2015), "The institutional origins of morphological differences between the Netherlands, Flanders and North Rhine-Westphalia," *European Planning Studies*, 23(11): 2165–2183.

Tira, M., van der Krabben, E., and Zanon, B. (eds) (2011), *Land Management for Urban Dynamics: Innovative Methods and Practices in a Changing Europe*. Milan: Maggioli Editori.

Urhahn Urban Design (2010), *The spontaneous city*. Amsterdam: BIS.

Valtonen, E., Falkenbach, H., and van der Krabben, E. (2017), "Risk management in public land development projects: comparative case study in Finland and the Netherlands," *Land Use Policy*, 62: 246–257.

Van der Krabben, E. and Buitelaar, E. (2011), "Industrial land and property markets: Market processes, market institutions and market outcomes, The Dutch case. *European Planning Studies*, 19(2): 2127–2146.

Van der Krabben, E. and Heurkens, E. (2015), "Netherlands: A search for alternative public–private development strategies from neighbouring countries," in G. Squires and E. Heurkens (eds) *International approaches to real estate development*. London: Routledge, pp. 66–81.

Van der Krabben, E. and Jacobs, H. M. (2013), "Public development as a strategic tool for redevelopment: Reflections on the Dutch experience," *Land Use Policy*, 30: 774–783.

Van der Krabben, E. and Needham, B. (2008), "Land readjustment for value capturing: a new planning tool for urban development," *Town Planning Review*, 79(6): 485–506.

Zijderveld, A. C. (2000), *The institutional imperative: The interface of institutions and networks*. Amsterdam: Amsterdam University Press.

Part IV
Finance and investment for real estate development

14

A historical evolutionary and cyclical perspective on models of development finance

Colin Jones

Abstract

Speculative development (and hence its financing) in Western economies was very limited up to the end of the 1930s except in the residential sector. The focus of this chapter is therefore on the evolution of the different models of funding development from World War II. The traditional model takes the form of a property company receiving incremental debt funding from a bank to pay out outlays through the development period. The debt is then repaid on completion through selling or taking out a mortgage on the property. However, market conditions rarely enable this simple model to work, and property cycles dominate the availability and terms of debt finance from banks with easy credit in booms followed by droughts in downturns. One alternative is forward funding/equity sharing partnerships between developer and a financial institution. Other innovations include non-recourse or limited recourse loans, usually involving a separate property vehicle for the sole purpose of individual developments, and very occasionally mezzanine finance. Another source of finance for property companies listed on a stock exchange is the issue of shares, bonds, and commercial paper. This approach is seen to transmute into asset securitization of properties set within a single purpose vehicle. Global integration has meant that development finance is available on a competitive basis from banks around the world, a trend that has gone hand in hand with larger construction projects. Large-scale development projects are also often dependent on a funding partnership with a public agency.

Introduction

Availability of development finance is a crucial component of the real estate market as no finance means no building. This finance can take a number of different forms, partly linked to the future ownership of the property and any investment or funding partners in the development process. In the case of construction by owner occupiers, this finance is almost certainly intricately tied up with their own businesses, often internally generated, and is not considered here. The focus of this chapter is on development finance for development with a degree of risk, primarily speculative, defined in terms of being either for future occupation or ownership. Before looking at this issue in detail it is useful to set the context.

Looking back through time, property development is interwoven with the evolution of towns and cities. High street shops evolved from housing conversions to individual purpose-built entities by proprietors, with shopping centers a very new phenomenon. Offices emerged as a separate property form in the first half of the nineteenth century and although from the beginning there was some speculative development most were built for their occupiers (Scott, 1996); many are still known by the name of the original occupier. Factories were traditionally specialist bespoke buildings but the first industrial estates for tenants were built around the world in the 1890s. Even so they were on a small scale, with only 48 commercially built estates in the UK by 1939 (Scott, 2001). Overall the scale of speculative development up to the end of the 1930s was very limited, and almost exclusively in the residential sector. While the focus of this chapter is on the post-World War Two period, its starting point is a model of development finance that can be traced back well beyond this date.

A further focus of the chapter is on cycles as research on property development has identified booms and downturns back to before the Industrial Revolution (Lewis, 1965). Since the 1950s there have been many well documented cycles. This period was initially one of reconstruction for countries impacted by the Second World War, stimulating commercial property development booms. In the UK for example, Churchill announced in November 1954 the lifting of all building restrictions setting off a ten year boom (Rose, 1985). Japan experienced a parallel boom (Dehesh and Pugh, 1998). Elsewhere, post-war booms were later: in the late 1960s through to the early 1970s there were development booms for example in New York, Sydney, and Dublin, driven primarily by structural changes in the macroeconomy (Daly, 1982; MacLaran et al., 1987; Schwartz, 1979). Barras (2009) has identified three global office cycles since the 1980s beginning with the speculative boom of the late 1980s, followed by a more subdued upturn in the late 1990s and another speculative driven boom in the mid-noughties. In the housing market, Jones (2012) reports on parallel booms across many countries from the mid-1990s, but also individual cycles in the 1970s and 1980s. So, while much of the evidence presented in this chapter is drawn from the UK, the commonality of these dynamics within the property market means that it has a resonance across the developed world.

The period since the 1950s has also experienced the growth of a greater services-oriented economy that has led to a long-term rise in the demand for offices and a decline in the role of manufacturing in Western economies. Since the 1970s this trend has been augmented by decentralized forces fashioned by the motor age and information communication technology improvements that can be characterized as a new long-term urban development cycle. New property forms, such as retail parks, have been established, traditional city cores have been supplanted by a more polycentric urban form, and many existing buildings replaced as obsolescent because of new technology. These modern developments are larger in scale than their predecessors, for example individual high street shops are being replaced by shopping centers (Jones, 2010), and provide a context to innovations in development finance.

In addition, the world has seen economic globalization, and specifically the emergence of world capital markets with the growth of international financial services from the 1980s. Underneath these trends is the liberalization of capital movements that has also seen global real estate investment strategies (Lizieri, 2009). Development finance too is now an integral part of these international capital flows as banks and investors fund projects around the world. With foreign banks and investors competing with national banks to offer finance (see Bank of England, 2015), this has stimulated international knowledge transfer and commonalities of practice. This chapter therefore examines the evolution of development finance for commercial real estate against a backcloth of urban transformation and redevelopment, new property forms, globalization of the real estate investment, and a series of property booms and busts.

The chapter will briefly outline basic forms of finance and introduce the classic roll-up finance model (Fraser, 1993). It will then chart the evolution of the different models of funding development, explaining the significance of differential market conditions and investment sentiment driving innovations. These will include forward funding/equity sharing partnerships between developer and financial institutions, non-recourse or limited recourse loans usually involving a separate property vehicle for the sole purpose of individual developments, and very occasionally mezzanine finance. It will explain the potential benefits and disadvantages in terms of the balance of risk and return and partners in the development process. Property companies/REITs' issue of debentures and shares on stock exchanges, together with equity securitization will also be assessed.

A cyclical perspective will examine the balance of debt and equity capital in development funding through its phases including in particular the changing lending requirements of banks linked to perceived risks. It will also highlight the changing shape of development finance through cycles and present a chronology of innovations within a cyclical framework. Innovation is also considered in the changing context of the investment sentiment of financial institutions.

The fundamentals and logic of public–private partnerships will be similarly reviewed, encompassing the spectrum from local authority city center redevelopment partnerships, central government support via for example grants to developers as part of urban regeneration initiatives, through to tax incremental financing.

The basics of the development finance process

It is useful to start with some definitions, distinguishing between short-term finance for private development and long-term/investment finance once the building is complete but the two are inter-related. Short-term finance covers the financing of construction or more precisely the development period encompassing potentially the purchase of land through to the point at which the property is completed and let, or either sold or kept as a long-term investment. In the USA this is known as a "construction loan" (Fergus and Goodman, 1994). Long-term finance, sometimes referred to as funding, can be sought once the building is complete to pay off the costs of development including the short-term finance. Alternatively, the property can be sold to pay off the development costs and take any profit. Long-term finance, for example in the shape of a mortgage, will usually be applicable for the developer to retain an equity interest. In reality, as discussed later, this distinction between short- and long-term finance can be clouded as often short-term finance stretches into the medium term.

The traditional model of short-term finance is roll-up finance whereby the developer receives incremental funds, usually from a bank, to cover outlays at the different phases of the development process from land purchase, stage payment of the construction costs, through to the letting period. On completion, the developer repays all these outlays with the built-up interest. Short-term finance can be provided on a fixed rate basis but in the UK today for example it is normally on a variable rate basis linked to a set number of base points (bps) above the bank base rates set by the Bank of England. The precise rate paid is therefore a function of the level of interest rates. Credit facilities with variable interest charges may be subject to a "cap" or "floor" limiting upward and downward movement of the interest rate to be charged.

An unfinished building is of limited worth and hence a poor security for a bank. Finance terms offered by a bank are dependent on the financial standing of the developer, taking into account other assets and existing borrowing. Notwithstanding a developer's track record and financial standing, repayment is related to the ultimate "success" of the completed development.

As this is a function of the state of the macroeconomy and the property market these may have an over-riding influence on the availability and cost of finance. This is discussed in more detail below.

Historical perspective on development finance

These underpinnings of the development finance process provide a framework for an examination of the changes that have occurred since the 1950s. Useful insights can be gauged from the office development boom in London. First, the property market conditions were very benevolent. There had been few offices built since the 1930s, first because of the depressed macroeconomy and then because of the Second World War and subsequent building material shortages. At the same time, there were many potential sites because of bombing during the war and demand was expanding, which meant shortages of offices and rising rents and capital values (Marriott, 1967).

Furthermore, there were other factors that favored developers. The government was following a low interest rate for growth strategy. Contractors were normally paid on completion of the building, reducing the need for finance. Usually the final development value exceeded costs by 50 percent and developers on completion were able to receive mortgages based on two thirds of a property's value. As a result, the mortgage could pay off construction costs. Marriott summarizes the implications:

> Since all the money to buy the site was usually lent by the banks, all of the construction costs paid for by the contractor or the bank, and the total repaid from a long term mortgage borrowed from an insurance company, the developer seldom had to find any money at all, once his credit was established.
>
> *(Marriott, 1967, p. 5)*

Interest payments on mortgage finance were less than the rent received. In other words the whole process was "self-financing," a very lucrative business that stemmed from the specific market conditions of that time.

Self-financing of speculative development has never occurred again in the UK, not even in the property booms of the 1980s and 2000s. However, during the 1980s boom this form of development finance was subject to innovation. The length of the loan term was extended from the initial development to the first rent review point after five years. The logic was that rents would have risen sufficiently by then to permit long-term financing. Later in the decade, short-term debt finance became more sophisticated through the arrangement of general credit facilities/options that enabled a developer to tap into funds at prior agreed interest rates up to agreed limits. These arrangements were often syndicated across a panel of banks, expanding finance opportunities to the developer and reducing the specific risk of large developments for banks (Lizieri et al., 2001). These facilities could also last up to 35 years blurring the short and long term.

The late 1980s also saw a range of other innovations that still apply today. Traditionally, banks have had a ceiling of 75 percent of development costs. Specialist lenders emerged that topped up this finance with "Mezzanine Finance." This enabled developers to borrow normally up to 90 percent of their costs. This top slice is more risky and financiers require a higher rate of return and a share in any profit. Mezzanine finance in this instance is the junior debt and owners of the mezzanine finance will only receive the payments they are owed once the senior debt has received its repayment, including interest in full. Mezzanine finance incorporates interest charges and sometimes an equity share of the final value of the development. Mezzanine finance supporting development in this way applies primarily to a property boom, and then probably rarely. Many banks have subsequently required not just interest payments on the senior debt but similarly an equity return from the completed development. There are also "exit" charges in the UK – their form can vary but they are usually an additional charge based on either the total debt borrowed or the value of the completed development.

Another innovation introduced in the 1980s was "limited recourse" or "non-recourse" finance loan arrangements (Fraser, 1993). In these arrangements, the bank lends only to the development project which is set up as a subsidiary company of say a property company or two companies. Today the terms have changed and the equivalent are now special purpose vehicles (SPVs) and limited legal partnerships (LLPs). The former is a general term for these partnerships. LLPs set legal limits to the liabilities of the partners, although the law varies across countries.

Non-recourse or limited recourse loans usually involve the creation of a separate SPV or company/subsidiary for the sole purpose of the individual development. The collateral for the loan is secured (restricted) to the assets of this company with no recourse to the parent company or to its other assets if the project fails. The arrangement protects the developer from the project's major risks, beyond the initial capital required, but not the lender. The advantage to the lender is that the project is sheltered from any wider problems that may occur for the parent companies. In some instances the opposite may be true and the lender could require some form of guarantee from the parent company. A further advantage to the parent company is that in certain circumstances the subsidiary and its debt will not have to feature in its balance sheet, so the subsidiary is taken off the balance sheet thereby masking some of its liabilities.

Sale and leaseback schemes began to emerge in the late 1950s as an alternative to long-term mortgages for developers when the UK government introduced occasional freezes on borrowing – known as "credit squeezes" at the time – to stop the economy overheating (Marriott, 1967; Darlow, 1988). In fact, sale and leaseback have a long pedigree. Scott (1996) notes their existence in the 1890s and how they had been used by individual retailers to expand their business, particularly from the 1930s on. A retailer could enter into a sale and leaseback arrangement on one store, and use the finance to buy another, and by this process establish a chain of shops. A retailer could also enter into such an arrangement with an insurance company to finance the building of a store (Scott, 1996). The innovation of the late 1950s was their use for speculative development. Under these schemes a developer (pre)sells the property on completion (usually) to a financial institution and at the same time takes out a long lease of the order of 100 years.

The structures of these sale and leaseback arrangements vary but the developer would retain a proportion, say 20 percent, of the equity, selling sufficient to pay back the rolled-up development costs. The developer then sublets and manages the property, with the financial institution receiving an agreed (share of the initial market) rent, a guaranteed income so that it receives the "bottom less risky" element. This type of arrangement is often referred to as a top/bottom slice partnership. There were initially long rent review periods on these arrangements – up to 33 years – reflecting the lack of inflation in the 1950s. The essential drawback of sale and leasebacks in this form was that each interest was not easily sold.

By the end of the 1950s, financial institutions became more interested in taking equity interests in property. This was partly because of the high returns made in the boom and partly because the 1960s saw the beginning of inflationary pressures with consequences for their investment business to provide endowment payments and pensions to policy holders that kept pace with rising prices. Property as a real investment fitted this new paradigm. As a result, the

1960s saw a move from the use of elementary sale and leasebacks into complex equity sharing partnership models of development finance. A financial institution became involved from the outset of a development, providing the finance for the project including land purchase and construction costs. Initially the basic sale and leaseback partnerships did not involve arrangements about sharing future rental (equity) growth (Darlow, 1988). In the late 1960s these schemes began to include a proportional formula to share future rental income, often collectively referred to as "side by side schemes." A further development was a priority yield arrangement that combined both elements of the side by side and top/bottom slice. The development company prioritizes the financial institution to receive a set yield on its investment, with the next say 1 percent going to the developer, and the rest (if any) split equally. The breakdown of returns is shown by Figure 14.1 in which the priority yield is the bottom slice, the property company receives its share if there is sufficient profit, and the rest of the profit is shared in a side by side arrangement.

With the property market collapse of the mid-1970s the prioritized/guaranteed income proved an illusion. This led to the evolution of the "profit erosion" model in the 1980s, in which if the guaranteed return on completion is not forthcoming the developer is subject to penalties (Darlow, 1988). If the profit erodes to a certain point, the developer disappears and the fund is left with the building and to find tenants. A common feature of these partnerships is that they provide incentives to the property company in the development stage and its ultimate returns are based on its performance in completing the project on time and within budget. The funding institution shoulders most of the risk as it is its capital, but the developer risks losing the right to profit in the development process (Fraser, 1993). Generally, these arrangements include a buyout by the institution at an agreed yield on an agreed base rent, and a higher yield on surplus rents known as "overage." The ultimate step in the changing partnership forms between developers and financial institutions is the appointment of the property company as a project manager to the development. Payment then takes the form of a basic fee plus a performance incentive.

Listed property developers have always had another source of finance; in common with firms more generally they can issue shares, bonds, and commercial paper. Raising capital to

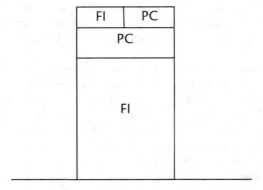

FI = Financial Institution PC = Property Company

Figure 14.1   Breakdown of returns in a priority yield arrangement

Source: Fraser, 1993

finance property company activities on a stock exchange is usually based on a prospectus explaining what it plans to use the additional funds for to justify the issue and persuade investors to purchase. An evolution from debentures is asset securitization of properties set within a "single purpose vehicle." Essentially they are bonds secured against rental income, but rather than being issued on the stock exchange they are placed by a bank with investors supported by a risk grading from rating agencies. Both can be cheaper than borrowing debt capital from a bank, especially when share values are high, and can be used to repay development loans. They can also raise up to 90 percent or even 100 percent of real estate value compared to a bank loan of up to 75 percent (Lizieri et al., 2001). However, any conditions placed on securitization by investors may bring constraints on restructuring a company's real estate portfolio or its management such as subsequent lease terms.

An extreme example of the use of securitization is Olympia and York, a private Canadian property company that just before it filed for bankruptcy in 1992 owned trophy buildings in Toronto and London, together with 75 percent of Manhattan's office space. It was then revealed that it had non-recourse loans owed to 91 creditors across the globe secured against the company's real estate and public stockholdings. Over $1 billion alone was securitized against Canary Wharf in London, its main underperforming asset and the primary instigator of the company's downfall (Ghosh et al., 1994). Despite the problems of Olympia and York, the 1990s saw the slow rise of asset backed securitizations, many of them trophy buildings or offices to be occupied by a blue chip company with an excellent credit rating.

Real estate cycles and the role of development finance

Upturns in the property cycle are normally associated with readily available bank finance, while the reverse happens in the downturn. This can be explained in terms of relaxed bank lending criteria involving the under-pricing of risk in a boom followed by a more conservative approach engulfed by fear (over-pricing of risk). Banks who typically lend up to 75 percent of development costs with low mark-ups over base rates in a boom turn volte-face during the bust. This debt finance then may not be available at all or only under strict conditions, for example only for developments that are pre-let with tenants waiting, and even then banks may offer less generous sums based on a lower loan to cost ratio, the order of say around 60 percent. Such a short-term development finance "famine" for example occurred in the years following the global financial crisis after 2008, and also can be seen in previous major downturns.

The inter-relationship between development finance and the performance of the economy/property cycle goes much further than this relationship with bankers' lending criteria and funds available. The historical perspective above also demonstrates that **forms** of development finance available also vary with the state of the economy. Table 14.1 summarizes the cyclical nature of finance forms and contrasts the funding droughts during recessions and the availability of loans in upturns. Corporate finance in general is most attractive during periods of economic growth when property and share prices are rising. For example, 1987 was a record year for share issues in the UK when both were soaring. These rights issues can also be issued to support a company in downturns when external funding is not available. In this case, the company's shareholders are called upon to safeguard their existing investment by providing more capital.

However, superimposed on these relationships are the portfolio investment strategies of non-bank financial institutions. Property as an investment in comparison with other investment modes – equities and government bonds – has had periods when it was seen as relatively attractive, such as the 1960s and 2000s, but also when sentiment moved in the opposite

Table 14.1 Cyclical perspective on forms of development finance

Period	State of economy/ property cycle	Forms and innovations in finance	Real estate institutional investment sentiment
1950s and early 1960s	Low inflation/low interest rates/economic growth/property boom	Fixed interest loans for speculative development	Very limited institutional investment
Late 1960s and early 1970s	Rising inflation, slow economic growth	Corporate finance, sale and leasebacks, variable interest loans for speculative development	First substantial wave of institutional investment
Mid 1970s	Recession	Funding drought	No investment
Late 1970s and early 1980s	High inflation followed by recession	Partnership with FIs	Build up of institutional property portfolios
Late 1980s	Rising economic growth, modest inflation/rise in stock market values until 1987/property boom	Speculative development loan finance with variable interest rates, mezzanine finance, corporate finance (until late 1987) sale and leasebacks	Positive institutional investment sentiment
Early 1990s	Recession	Funding drought	No investment
Mid/late 1990s	Modest economic growth/ low inflation, rise/fall in stock market values	Corporate finance, speculative development loan finance difficult	Property out of favor with financial institutions
2000s	Continuing and strong economic growth/low inflation/property boom	Speculative development loan finance, equity securitization	Large flows of institutional investment funds into property
2008 on	Post Global Financial Crisis/low interest rates	Funding drought London	Weak investment outside

direction such as the 1990s. These opinions influenced the nature of the development finance available and certainly when positive stimulated investment partnership models.

Property booms also brought innovations in development finance. For example, the late 1980s saw the arrival of non-recourse lending and mezzanine finance. The busts often exposed the flaws in these innovations, such as the lack of transparency in equity securitization and unexpected high levels of gearing. In the distress of the bust, energies focused on addressing the financial consequences, not stimulating further evolution. One such novelty in the post credit crunch period is the use of mezzanine funding for refinancing completed buildings (Giostra, 2011).

Public–private partnerships and development finance

Modern commercial (re)development is no longer a small incremental process, and individual schemes have increased in size. From the late 1950s, UK local authorities were also conscious of the importance of large-scale city center development, especially shopping centers, to the local economy and engaged in entrepreneurial partnerships with the private sector to attract investment (Shapely, 2011). Local authorities had a critical role in these developments contributing through planning support, already owning the land central to the project, or acquiring it compulsorily, so unlocking land's development potential. As a result, an authority also has an equity interest in the development. Sometimes a public authority or agency may also seek a joint venture with a private sector partner to provide finance for a project that is not commercially viable. In these cases, the financial return models above (see Figure 14.1) were reformulated at the time with up to three partners to reflect the different roles and risks (Fraser, 1993).

Local economic development is arguably even more important today for communities. Public authorities continue to initiate and take a role in major real estate projects, often reconfiguring city centers or reviving key sites. The evolving urban economy as noted earlier, with its closure of traditional businesses, brought fundamental policy challenges from the 1980s in terms of a legacy of large tracts of vacant/derelict buildings and land. The schemes are now generally mixed use, a potential combination of commercial, residential, and leisure. The precise mechanics vary between countries depending on planning systems and the local fiscal regime/freedoms (Munoz-Gielen, 2014). They may also be underpinned by significant transport links supported initially by public funding.

The scale and location of the projects often mean a lack of private sector confidence to invest and there are market doubts about the viability of regeneration. There are a range of solutions that encompass state subsidies/tax incentives and a variety of types of area strategies in which the state focuses and coordinates resources to set a direction and vision for the future. Enterprise zones are one model that has been applied around the world offering tax incentives for development (Jones, 2006, 2013). Another approach is the use of "development corporations" for defined areas that buy up extensive land tracts, provide modern infrastructure such as new access roads, and provide subsidies for development and market user-friendly land parcels (Jones, 2013). These are pump-priming strategies in which initial public sector expenditure/tax incentives, ultimately for a limited period, are provided to stimulate private real estate investment.

These schemes provide public support as one element of a broader development funding package. Government involvement offers positive and reassuring signals to private financiers for what are generally more risky projects. The initial public support is also crucial to the viability of such development, not only in terms of crude profitability but also by reducing the required equity required to meet the available finance in the development equation. There are a range of ways to seek to repay the public funding via any rise in land or property values. This can be simply through land sales, forms of development land gains tax, or tax incremental financing in which local rises in rental/capital values are translated into higher occupancy property taxes (Squires et al., 2015).

Conclusions

Development finance schemes are products of the contingent market conditions applying at a particular time. Notwithstanding a gamut of sophisticated innovations over the last 50 years, the long-standing financing model of a property company receiving incremental debt funding to pay out outlays through the development period is still the mainstream approach. However, property market conditions rarely enable this simple model to work with the repayment of development funding having to stretch into the medium term until rents/capital values rise.

The distinction between short- or long-term finance is therefore arbitrary, reflected in the use of general credit facilities from banks and equity securitization. Global integration has meant that much of this finance is available on a competitive basis from banks around the world and standardization of approaches in different countries. This trend has gone hand in hand with a move toward larger development projects as noted earlier.

Property cycles dominate the terms and forms of development finance, with booms stimulating innovations and the busts dealing with the ramifications of their shortcomings. The changing attitudes of financial institutions toward real estate in any particular decade have also been a factor in the prevalence of the form of development finance of that period. More generally the availability and terms of debt finance from banks are very pro-cyclical with droughts in downturns.

Large-scale development projects are often undertaken in partnership with a public agency seeking to promote local economic development. Public support is crucial to the viability of many of these projects not just simply in terms of profitability but also to the provision of private development finance. More generally, finance is the key to successful individual developments, but as we have seen, its form has gone through many guises and its availability is very much dependent on the economy, general real estate sentiment by investors and ultimately the risk of the project itself.

References

Bank of England (2015) The Failure of HBOS: A report by the Financial Conduct Authority (FCA) and the Prudential Regulation Authority (PRA), Bank of England, London.

Barras, R. (2009) *Building Cycles: Growth and Instability*, Wiley-Blackwell, Oxford.

Daly, M. T. (1982) *Sydney Boom Sydney Bust*, George Allen & Unwin, Sydney.

Darlow, C. (ed.) (1988) *Valuation and Development Appraisal*, Second Edition, Estates Gazette, London.

Dehesh, A. and Pugh, C. (1998) Property cycles in a global economy, paper presented to RICS Cutting Edge Conference, London.

Fergus, J. T. and Goodman, J. L. (1994) The 1989–92 Credit Crunch for real estate: A retrospective, *Journal of American Real Estate and Urban Economics Association*, 22, 1, 5–32.

Fraser, W. D. (1993) *Principles of Property Investment and Pricing*, Second Edition, Macmillan, London.

Ghosh, C., Guttery, S., and Sirmans, C. F. (1994) The Olympia and York crisis, *Journal of Property Finance*, 5, 2, 5–46.

Giostra, N. (2011) *The Funding Gap: Is Mezzanine Lending the Solution*, CBRE, London

Jones, C. (2006) Verdict on the British Enterprise Zone experiment, *International Planning Studies*, 11, 2, 109–123.

Jones, C. (2010) The rise and fall of the high street shop as an investment class, *Journal of Property Investment and Finance*, 28, 4, 275–284.

Jones, C. (2012) Introduction: The housing economy and the Credit Crunch, in C. Jones, M. White, and N. Dunse, *Challenges of the Housing Economy: An International Perspective*, Wiley-Blackwell, Chichester, 1–24.

Jones, C. (2013) *Office Markets and Public Policy*, Wiley-Blackwell, Chichester.

Lewis, J. P. (1965) *Building Cycles and Britain's Growth*, London: Macmillan.

Lizieri, C. (2009) *Towers of Capital: Office Markets and International Financial Services*, Wiley-Blackwell, Oxford.

Lizieri, C., Ward, C., and Lee, S. (2001) *Financial Innovation in Property Markets: Implications for the City of London*, Corporation of London, London.

MacLaran, A., MacLaran, M., and Malone, P. (1987) Property cycles in Dublin: The anatomy of boom and slump in industrial and office sectors, *The Economic and Social Review*, 18, 4, 237–256.

Marriott, O. (1967) *The Property Boom*, Hamish Hamilton, London.

Munoz-Gielen, D. (2014) Urban governance, property rights, land readjustment and public value capturing, *European Urban and Regional Studies*, 21, 1, 60–78.

Rose, J. (1985) *The Dynamics of Urban Property Development*, E & FN Spon, London.

Schwartz, G. (1979) The office pattern in New York city, 1960–75, in P. W. Daniels (ed.) *Spatial Pattern of Office Growth and Location*, Wiley & Son, Chichester, 215–237.

Scott, P. (1996) *The Property Masters*, E & FN Spon, London.

Scott, P. (2001) Industrial estates and British industrial development, 1897–1939, *Business History*, 43, 2, 73–98.

Shapely, P. (2011) The entrepreneurial city: The role of local government and city-centre redevelopment in post-war industrial English cities, *Twentieth Century British History*, 22, 4, 498–520.

Squires, G., Berry, J., Hutchison, N., McGreal, S., Adair, A., and Organ, S. (2015) *Innovative Finance in Real Estate Development in Pan-European Regeneration*, RICS, London.

15

The future of finance and investment for real estate development and investment

Changing approach for a new structural era

Christopher B. Leinberger

Abstract

Real estate is structurally shifting toward walkable urban development. This structural shift affects every aspect of the industry from site acquisition, construction, design, property, and asset management to finance and place management, a new field. This chapter focuses on the financing and investment in real estate in this structurally new era, contrasting it to the previous era (drivable suburban), in which most financial professionals learned their skills and gained their experience. It also touches on the needed skill set changes for the real estate development and investment industry to adapt to this new era. In essence, the skill sets and experience learned in the drivable suburban era have to be completely modified to the new walkable urban era. Applying the previous era's skill sets to the current will lead to substantial under performance and even failure.

Introduction

Real estate development and investment is one of the oldest asset classes in economics and business history. Real estate is a basic industry essential to human existence, where we live, work, and socialize. It has consumed a large portion of the investment capital of economies since cities emerged over 6,000 years ago. Today in developed economies, approximately 35 percent of all assets are invested in real estate (residential, retail, office, industrial, institutional, government, university, etc.) and the infrastructure needed to support this real estate. Infrastructure includes transportation in all of its forms, water, sewerage, electrical, cable, wireless, parks, etc. Hence, real estate and infrastructure – the built environment – is probably the largest asset class in the economy.

Real estate has always been a long-term asset. Historically, buildings last 25–50 years until they either need to be torn down or need fundamental redevelopment. It is the reason the United States Internal Revenue Service mandates that real estate should be depreciated over 39 years for non-residential property and 27.5 years for residential rental property (land is not included since it does not depreciate) (Department of Treasury Internal Revenue Service 2014). Yet, real estate investment is generally financially underwritten like every other investment asset; that is, it is required to generate cash flow in the short term under conventional discounted cash flow methodologies. The result is that this long-term asset class is evaluated for initial development or redevelopment using short-term evaluation techniques, generally with a 3–7 year time frame.

This chapter reveals and challenges the mismatch between this long-term asset class being evaluated and underwritten using short-term techniques. These techniques include the various forms of discounted cash flow, particularly the prevalent use of internal rate of return (IRR) in real estate underwriting. Far more research is needed to prove this mismatch in long-term investment horizons and short-term underwriting is harming investors and the quality and form of the built environment, but the hypothetical case is proposed here. However, many private individuals and families, and increasingly publicly traded companies and institutions with long-term cash flow needs, are recognizing, once again, like their forbears a century ago, that a long-term asset such as real estate is best served with long-term investment underwriting.

Recent structural changes in real estate investment and market demand

Prior to the 1990s, real estate investment was generally locally financed. Local banks or regional offices of national banks provided construction financing to primarily local developers, regional offices of national developers and investors. Long-term financing was provided by local and national institutions, such as insurance companies and pension funds predicated on the understanding of local market fundamentals. Private investors, corporations, government, and non-profit institutions also undertook substantial real estate investment, many times using internal financing.

Furthermore, real estate has been an illiquid asset class for millennia. It has been assumed for many centuries that real estate generally takes a long time to sell, depends upon "windows of opportunity," and that it has high transaction costs, slowing the sales process. There were times during the real estate cycle when selling at anywhere near the cost of acquisition or replacement was not possible and an investor just had to wait for better economic conditions. Real estate continually went from boom to bust, such as boom times in the late 1970s, mid-1980s, mid-1990s and mid-2000s, always followed by the busts of the early 1980s, early 1990s, and 2007–10. Most of the busts led to general economic recessions given the large asset base of the real estate industry in the economy as a whole.

In the early 1990s, during the worst real estate bust up until then since the 1930s, the beginning of a revolution in real estate finance started. For the first time in history, a significant portion of the real estate industry was made liquid and instantly tradable. Taking the form of secondary residential and commercial mortgage packages and publicly traded companies, particularly real estate investment trusts (REITs), it began to fundamentally change how real estate financially functioned. Buying and selling tradable stock in a REIT, which owns and manages billions of dollars of real estate assets, was as easy and quick as any other common stock in a household or institutional portfolio. Secondary mortgage-backed securities hit a major downdraft and in fact helped cause the Great Recession in 2007–8 due to lax oversight of underwriting and lack of understanding of the assets held in those packages. Yet, residential and commercial mortgages being sold into a worldwide market has returned as of 2016 and one hopes the recently enacted reforms, such as better underwriting and minimum long-term ownership stakes by the fund sponsors, will bring needed discipline for continued expansion.

However, there was a price required to easily, cheaply and quickly trade in financial markets.

That price was commodification. Financial markets required commodification of any product being traded, whether a specific class of General Electric common stock, the grade of pork bellies, or currencies. Without commodification, financial markets cannot trade at high volume and low trading costs. Financial markets are not auction houses, which trade unique assets that require high transaction costs and specialized knowledge.

As real estate companies approached Wall Street to float the "class of 1993" for REITs, there became a demand for commodification of real estate products. Wall Street investment analysts knew relatively little about how real estate would perform financially in the early 1990s, even though the REIT structure had been around since 1960. The question was whether REITs were a utility (low risk with predictable dividends), a growth hybrid stock (high risk with high dividends), a cyclical investment (high returns in good times and implosion in recessions), or generating cash flow in good times and bad. There were few publicly traded REITs in the USA until the recession of the early 1990s drove formerly privately owned real estate companies to seek this "new" funding source, most listing on the New York Stock Exchange in the USA.

Wall Street analysts and institutional stock owners had a lack of experience with the REIT management. Many REITs were led by high profile company founders, such as Sam Zell (Equity Office and Equity Residential), Bill Sanders (multiple REITs in different product categories), and various spin-offs of Trammel Crow Company, the largest developer in the USA in the 1990s. Wall Street analysts have experience with professional managers in conventional corporations, not high profile "promotors" in real estate.

Many other countries have recently enacted legislation allowing the REIT form of publicly owned real estate, including the United Kingdom (2006), Germany (2007), Japan (2001), Canada (1993) and Hong Kong (2005). The investment analyst understanding of how these companies work and behave has to go though the same learning curve as the US markets have since the early 1990s.

Nineteen standard real estate product types

In the USA, the commodification of real estate led to what is referred to as the "19 standard product types" comprised of silos of single product types such as rental residential, entry level housing for sale, office, warehouse, etc., shown in Table 15.1. Without a public track record, Wall Street investment bankers wanted REITs to focus on one product type. Mixed-use projects and mixed-use product public companies were generally not allowed to go public. In the case of two REITs in the 1990s, Federal Realty (neighborhood retail centers) and Post Properties (suburban apartments), the founders and CEOs began to experiment with mixed-use projects. Both firms started walkable, urban, high-density developments with destination retail on the ground floor with apartments in the upper floors. However, investment analysts soon turned on these new strategies, leading to the firing of both CEOs. The message was clear: "stick to your knitting."

In addition to mandating single product types, Wall Street and major investors had a strong preference for low density, "drivable suburban" products, which were the "conventional," well-understood form of development that the industry had perfected in the late-twentieth century. This product form was segregated from other development, generally had surface parking and was connected to the rest of the region only by car and truck. Once again, the naturally conservative financial markets were not interested in the newly emerging forms of high-density walkable urbanism, such as downtown redevelopment, "New Urbanism," and "transit-oriented development." Lack of REIT track record, little experience with real estate management teams, and the conventional wisdom being that Americans only wanted drivable

Table 15.1 Nineteen standard real estate product types, 1990–2006

These real estate products are the easiest and most acceptable to the conventional investment community. They are generally single product type, stand-alone developments with self-contained parking, though some mixed-use developments are now possible.

Income products

Office	Build-to-suit office
	Mixed-use urban office/retail/restaurant
	Medical office
	Multi-tenant office
Industrial	Multi-tenant bulk warehouse
	Build-to-suit industrial
Retail	Grocery anchored neighborhood center
	Big box anchored power center
	Lifestyle center
Rental apartments	Garden apartments
Miscellaneous	Self-storage
	Mobile home park
Hotel	Budget hotel
	Business hotel

For-sale products

Housing	Entry level
	Move-up housing
	Luxury housing
	Retirement (includes a variety of segments, e.g., assisted living, independent, etc.)
	Resort/second home

Source: Author

suburban forms of development, led institutional investors to keep REITs on a "short leash" for the last 20 years.

Wall Street wanted as much predictability as the real estate industry could deliver. That meant well defined, well understood, drivable-only locations and low density, single purpose products. For example, a "neighborhood retail center," pictured in Figure 15.1, has all the same characteristics, as shown in Table 15.2. Generally speaking, a neighborhood center anywhere in the country follows this formula and trades as a commodity. It also explains why the country tends to look the same, aside from the superficial architectural theming.

The financing of the 19 standard product types became the defining feature of US real estate investment, whether publicly traded companies using Wall Street capital or private companies building with commercial bank financing or institutional investors. Hundreds of billions of investment capital poured into publicly traded real estate companies and secondary mortgage backed securities, providing a fundamental financial infusion of capital into the industry, allowing for much more liquidity, low cost trading, and more predictability.

Figure 15.1 Standardized neighborhood retail center with national credit grocery store anchor, Sacramento, California (USA)

Table 15.2 Neighborhood retail center as an example of the 19 standard real estate product types

- 12–15 acres (5–6 hectares) in land area
- 80,000 to 350,000 square feet in total size (7,500 to 32,500 square meters)
- 80% of the land will be covered with surface parking quite visible from the street traffic
- 20% of the land will be covered by the buildings set back from the street with significant store signage
- Loading and trash in the back lane behind the stores
- Located on the "going home" side of a major arterial
- Minimum of 25,000 cars per day passing the site
- Minimum household demographics within a 1–3 mile radius (2–5 kilometers)
- Anchored by national credit grocery and drug stores at opposite ends of the center
- In-line stores also national credit including card (example, Hallmark), fast food (Subway), fitness (LA Fitness), office supplies (Staples) and occasionally local tenants, such as nail salon, dry cleaner, etc.
- Outparcel pad sites immediately at the entrance from the arterial for banks and casual chain restaurants (TGIFs, Chipotle's, etc.)
- Architecture as an after thought, depending on the locally accepted popular theme, such as Mediterranean in Southern California, Williamsburg in Virginia and Cape Cod in New England.

The market changes

Just as the finance industry had commoditized the industry in the 1990s, the consumer market began to change, slowly at first and only in selective metropolitan areas in the USA. There arose a demand for the opposite of drivable suburban development, which is walkable urban, where high-density concentrations of various product types are clustered in a relatively small amount of land. It allowed for multiple ways of getting to the place (car, truck, rail and bus transit, biking, walking, etc.) but once there, the place was walkable for all destinations, as the example in Figure 15.2 shows. Unlike the USA which abandoned its city centers, most other developed nations maintained the traditional walkable urban development of their city centers even as they were adding the new American-style drivable suburban development on the expanding metropolitan fringe.

European walkable urban real estate values in the city centers were maintained even as drivable suburban development occurred along the arterials and highways. However, from the start of the twenty-first century, walkable urban real estate in European and Asian city centers became *the* most expensive in most metro areas, as housing and commercial space in the center of Paris, Munich, Singapore, London, along with selective US center cities, appreciated well beyond any historic experience, to become the most expensive real estate in the world.

Starting in the mid-1990s, the market for walkable urban, mixed-use development started to revive across the USA, gaining speed in the 2000–7 real estate cycle and increasing even faster during the current cycle, depending on the metropolitan area. This has resulted in high rent/sales prices and valuation premiums, as demonstrated in the GWU research *Foot Traffic Ahead* ("FTA") (Center for Real Estate and Urban Analysis 2016), released in June 2016. FTA ranked the 30 largest metropolitan areas in the USA (54 percent of US GDP) on three criteria: current walkable urbanism percentage for rental office, retail, and multi-family occupied space; a forward-looking "Development Momentum" ranking; and "Social Equity" (the percentage of low-income household spending on housing and transportation and accessibility to employment). The key findings are summarized below (Foot Traffic Ahead, 2016):

- All 30 metro areas achieving walkable urban rental rate premiums over drivable suburban space in the three product categories, averaging 74 percent. This is the first time all 30 have achieved walkable urban rent premiums in probably 60 years.

Figure 15.2 Mixed-use, walkable urban place: downtown Detroit, Michigan, USA

- In 26 of 30 metro areas, rental premiums increased between 2010–2015, indicating the premiums have not leveled out yet.
- All 30 metros are seeing walkable urban market share gains over the base period (1st quarter 2010) for the six years ending 4th quarter 2015, and 28 of the 30 metros have seen market share gains over 2010 of between 77 percent and five times. This is the first time all 30 have achieved walkable urban market share in probably 60 years.
- In the six most walkable urban metros (New York City, Washington, DC, Boston, Chicago, San Francisco and Seattle), 81 percent of total office and multi-family absorption over this six-year period went to "regionally significant" walkable urban places that used only 0.4 to 1.2 percent of the metropolitan land. This is an indication that drivable sub-urban sprawl may be over in these product types in these metro areas.
- The most walkable urban metros have the most educated workforce (40 percent of the population over 25 years of age have a college degree) and the highest GDP per capita ($72, 110) compared to the least walkable urban metros. The most walkable urban metros had 33% more college graduates in the workforce than the least walkable urban metros. In addition, the most walkable urban metros have a 49 percent higher GDP per capita than the least walkable urban metros. This 49 percent GDP per capita premium is the difference between high performing Germany and moderate performing Russia, Croatia and Latvia.
- In spite of these rental premiums, the most walkable urban metros surprisingly have the highest social equity ranking. While low-income households in high walkable urban metros spend slightly more of their income on housing, they spent a much lower percentage on transportation, more than offsetting the housing premium. Also, all metro households also have two to three times the accessibility to employment in a walkable urban metro for the same commuting times.

Required changes for the walkable urban real estate era

Institutional investors and Wall Street are under increasing pressure to finance walkable urban, mixed-use development and investments. One would think this would be an easy concept to understand for the investment community. After all, many of them live and work on Manhattan Island, which is the highest density, most walkable urban mixed-use place in the country, generating the absolute highest rents, sale prices, and valuations. Yet, there is a disconnect between how they view real estate underwriting on their small island and the rest of the country.

How can the real estate investment community understand this changed market dynamic playing out in most metropolitan areas in the country? What needs to be done to allow for the financing of this market-driven type of development? There are a number of things to address this structural shift in financing, as outlined below.

Education of the investment community

Commercial and investment bankers and institutional money managers are obviously bright professionals. If a case could be shown to them for investing in mixed-use walkable urban development, they will certainly shift their underwriting. In fact, many conferences, such as the Urban Land Institute, have focused attention on the growing market demand and rental/sale and cap rate premiums for walkable urban real estate.

However, walkable urban development is much more risky and complex than developing and managing the 19 standard real estate products of drivable suburbanism. Many investment managers and bankers only have experience with one product type, say neighborhood retail or entry level housing. In addition, finance is by its very nature conservative. Increasing risks or investing in a "new" real estate product type is not what the majority of investment managers want to do. Staying with the tried and true, even if the market is slipping away, seems like the right thing to do from a fiduciary perspective. Plus, most funds have proscribed what type of real estate product they invest in, so mixed-use, high-density walkable urban would probably not fit the official proscription.

Finally, as will be described below, the cash flows of walkable urban development perform differently than drivable suburban. Walkable urban development often takes longer to stabilize its cash flow and it performs better in the mid- to long-term. The 19 standard drivable suburban product types have been crafted as formulas that fit any place in the country and generate cash flow in the short term; the mid- to long-term cash flow potential is generally less. Most real estate funds are set up with a 3–7 year life, hence, just when the walkable urban development is maturing, it is time to sell and liquidate.

Upgrading of development/asset management skill sets

Developers/asset managers have to develop different skills sets and track records to engage in walkable urbanism. Investment managers correctly note that developers/asset managers need fundamentally different skill sets for walkable urbanism. The market for this form of development in the USA has only resurfaced since the beginning of the twenty-first century – depending on the metro area – so there is only a minority of developers/asset managers with much experience. Walkable urban development and management are fundamentally different to drivable suburban development. Developers/asset managers have to retool from the ground up. Walkable urban requires fundamentally different skill sets in finance, site acquisition, architecture, construction, phasing, marketing, project management, and place management (many have never heard of this new field). Transitioning from drivable suburban to walkable urban has many pitfalls, such as falling back on old rules of thumb that do not apply in the walkable urban world.

For example, a major publicly traded REIT was developing a high-rise rental apartment for the first time, across the Hudson River from Wall Street. Financial analysts could actually watch its progress from their windows. It was located in a former industrial area that was redeveloping as a mixed-use, walkable urban place. This area was in the early phase of the redevelopment process, which increases the risk. The REIT "stuck to their knitting" by not including any other products aside from multi-family rental but it had no experience with high-rise construction or marketing a walkable urban product. There was a significant construction overage; however, there was an even more significant rental rate premium than projected, leading to a better than forecasted return on investment. However, the Wall Street analysts were disappointed in the company on two points: their inability to manage construction costs *and* forecast rental rates, even though it turned out financially better in the end. If another product type, such as retail, was added, the financial analysts would be even more nervous. The learning curve for walkable urbanism is steep and the investment community is still skeptical of developers/asset managers who do not have the skill sets or track record.

Better national place-based inventory, trends, and performance metrics

The emergence of real estate as a separate asset class, joining cash, bonds, and stocks, only occurred in the 1990s with the surge of REIT initial public offerings and the rise of residential and commercial mortgage-backed securities. Only then did private data companies with national datasets emerge, such as Zillow, CoStar, Redfin, WalkScore, etc. Yet these databases are in product silos, reinforcing the 19 standard product types. Therefore, the typical Wall Street investor would say that if you want to invest in a mix of uses, invest in an office REIT, a retail REIT, and a multi-family REIT. However, the economic premiums referred to above in the *Foot Traffic Ahead* research (74 percent rent premium per square foot for walkable urban) only happens when there is a high density, walkable mix of uses or a mixed-use project.

Therefore, there is a need for far better datasets and performance metrics of walkable urban versus drivable suburban development. In other words, there needs to be a "Bloomberg" for real estate that is national in scope and dissects the data into walkable urban versus drivable suburban. Only then will the investment community have a clear picture of economic performance.

Increased equity in capital stack, including "patient" equity

There is much to be learned from developers from the early twentieth century, such as J. C. Nichols (Country Club Plaza in Kansas City), Henry Flagler (Miami), and the Rockefeller family (Rockefeller Center). They have become role models to be emulated in the walkable urban revival of downtowns, suburban town centers, new urbanist projects, transit-oriented developments, and lifestyle centers throughout the USA. While attention has focused primarily on the urban design lessons they employed, there are financing lessons they can teach us as well. The most important financing lesson is to recognize that real estate has always been a long-term asset class. One reason downtowns lately are reviving so quickly is the positive market response to the rehabilitation of historic structures – buildings whose quality of construction could never be matched today. However, over the past half-century real estate has changed. Today, most real estate projects have a seven- to ten-year life as a Class A property – the result of a reduction in the construction quality of projects and the building of a commoditized, single-purpose product. To build projects like Country Club Plaza or Rockefeller Center – mixed-use projects erected in walkable environments – something in short supply today was undoubtedly employed: patient equity. Patient equity is the capital committed to a development or redevelopment budget that does not have a defined payback. To many today it is an oxymoron – since equity is the most expensive and, therefore, most impatient of all capital. However, patient equity in the past was generally required to move a project forward, but also used because of the pride taken in building something that J. C. Nichols called of having "enduring value."

Increasing the equity in the capital stack, especially patient equity, is more difficult to do but not as outrageous as most investors and developers think. The sources of patient equity include:

- Land: All projects start with land control and often, the existing landowner could be a contributor of that land in a patient equity manner. Especially when there is a common philosophy of the landowner and developer (walkable urbanism, sustainability, revitalization, etc.) and the landowner has owned it for many years, if not decades, possibly at zero basis and possibly it is not generating cash flow at present, it could be painlessly contributed to the venture. It is important to note that the land would have to be subordinated to any debt and possibly much of the additional equity invested in the project.
- Developer fees: The developer should be willing to invest any fees for development, marketing, brokerage, asset management, etc. into the project for an increased share of ownership in a patient manner.
- Professionals: Development professionals (architects, planners, lawyers, etc.) have been known to contribute their fees in the development in exchange for ownership.
- Mezzanine equity: This is a high cost equity but it also increases equity in the capital stack.

- Mission-driven investors: Many walkable urban developments are of interest to foundations, university endowments, public sector players, etc. They may have financial resources that could be invested in a patient manner.

The benefits of increasing equity in the capital stack, especially patient equity, is that it allows the project the required time for it to mature, it can weather an economic downturn that can sink a conventionally financed development, and makes construction lenders much more interested in the project since there is more money in front of the debt. In fact, having 30–50 percent of the capital stack as equity might mean the construction lender may negotiate a non-recourse loan, a major benefit to developers.

Build/buy and hold for the long term

One of the key differences in the financial performance of walkable urbanism versus drivable suburban is the length of economic performance. Drivable suburban financial performance tends to peak over a 5–7 year period. During the late twentieth century, sprawl had many times pushed demand further out to the fringe and inner suburban developments may have suffered. In addition, the construction quality was not of the highest so as to minimize front-end costs. As a result, drivable suburban projects would be re-evaluated at the end of the 5–7 year period to see if additional investment for re-habitation was justified or would the asset be left to unwind economically over time. Finally, the competitive market area (CMA) was generally very large, dictated by driving distance. As more competition was built, especially further out toward the fringe, the trade area might become overbuilt with newer projects. In other words, drivable suburban can suffer from "more is less": as more competition is built the threat of overbuilding and demand moving further towards the fringe would degrade financial performance.

Walking distance – between 1500 and 3000 feet defines a walkable urban place – puts a governor on competition. Walkable urban places tend to cover a relatively small amount of acreage (Pi × R squared). In metro Washington, the average walkable urban place is 406 acres; about three times the size of a regional mall, including the surface parking lot. Not only does this limit overbuilding to a degree (obviously not entirely) but also another crucial factor kicks into financial performance: "more is better." As the next rental apartment is built next to your project that may have retail on the ground level, the sales, overages, and rents all go up. Even if the adjacent development includes retail, that creates more critical mass of retail that increases traffic for all of the retail tenants, increasing sales, overages, and rents. The more urban vitality makes the place perform better for all participants.

"More is better" changes the holding period of real estate investment. Rather than the 3–7 year holding that has become prevalent for investment portfolios, it makes more sense to hold long term. That was the way the Nichols, Flaglers, and Rockefellers financed a century ago. It seems to be affecting portfolio managers today as it appears there has been the beginning of a sea change toward long-term holding of walkable urban developments. Certainly the entire REIT industry is predicated on long-term holdings of assets, especially if the companies are focused on walkable urban development, which include highly regarded companies such as Equity Residential, AvalonBay, Federal Realty, and Forest City.

Investment in place management

Walkable urban places must be managed 24/7 to achieve their maximum economic performance.

Place management is a missing level of governance in society and the extent of services provided has barely been tapped as we learn how walkable urban places grow and prosper. Even more intense than ordinary property management, place management includes elements outlined below:

- Clean and safe: increasing the safety and cleanliness provided by the local municipality to deliver a higher level which the market for walkable urbanism demands.
- Economic development: crafting the product mix for every real estate product type within the walkable urban place, whether it be retail tenant mix, restaurant concentrations, clusters of companies and employment, tourism mix, cultural assets, etc.
- Festivals: developing and managing festivals, in essence temporary product types, that occur periodically such as ethnic events, sporting events, farmers' markets, cultural festivals, charity events, seasonal events (Christmas shows and markets, New Year's celebrations, etc.), parades and public celebrations (winning the professional or university championships).
- Parking and transportation: managing the parking of the place, lobbying for or even providing transportation to or circulating around the place.
- Social equity: ensuring there is a mix of housing for all employees working in the place.

Summary and conclusion

The rediscovery of walkable urban real estate changes everything, including how it is financed. This structural shift gets the industry back to how it was financed a century ago; actually since cities were first built thousands of years ago. It should transform real estate finance from a transactional business, continuously buying and selling over a short time period, into building economically sustainable, cash flow generating assets worthy of holding long term. Given the price and cap rate premiums we are already discovering, this is a profitable way of building value in real estate.

The need to retool finance and investment skill sets is essential for success in this new walkable urban era. It is also essential that the development and asset acquisition community that the finance and investment community is funding needs to completely retool their skill sets. Underwriting standards need to be fundamentally adjusted to this new reality. Those developers and investors being financed also need to be informed of these new underwriting standards and that they require a fundamentally different business strategy and tactics in the future.

References

Center for Real Estate and Urban Analysis 2016, *Foot Traffic Ahead*. Available from: http://business.gwu.edu/wp-content/uploads/2016/06/CREUA_Foot-Traffic-Ahead_2016.06.14.pdf [accessed August 18, 2016].

Department of Treasury Internal Revenue Service 2014, *How to Depreciate Property*. Available from: www.irs.gov/pub/irs-pdf/p946.pdf [accessed August 18, 2016].

Foot Traffic Ahead 2016, http://business.gwu.edu/wp-content/uploads/2016/06/CREUA_Foot-Traffic-Ahead_2016.06.14.pdf [accessed August 18, 2016].

16

Mechanisms for financing affordable housing development

Graham Squires

Abstract

It is important to improve understanding of how cities can be better financed to meet affordable housing challenges over the short and long-term; be they economic, social and/or environmental. The lack of adequate affordable housing is a major challenge in the development of human settlements. With rapid urbanization, governments are increasingly having difficulties to meet the growing demand for affordable housing. The lack of revenues is one of the biggest problems facing most cities administrations all over the world, which makes them one of the vulnerable layers of government, with increasing responsibilities and small share in the allocation of affordable housing.

This research project critically analyses the financing of cities by looking at San Francisco City and Bay Area examples of city finance that engage with real estate development of affordable housing. This research is needed to improve the way finance in urban spaces can maintain quality affordable housing in economically constrained circumstances. In turn, this research into the financing of affordable housing in cities facilitates a greater quality of urban places and provides a more sustainable and resilient economic platform with which urban areas can thrive.

Introduction

This chapter critically analyses the financing of affordable housing development using the case study of San Francisco and the Bay Area in the United States (Squires, 2014; Wallace, 1995). Further, it aims to help understand the way finance in urban spaces can maintain quality resources in economically constrained circumstances (Squires, 2013; Harvey, 1974; Kim, 1997; Murphy, 2015). The study is set within an era of austerity for city administrations, which provides challenges for all stakeholders whether locally, nationally, or internationally (Gibb, 2011). Lessons to be learnt from the case study are of relevance and importance given the difficult affordable housing issues being faced. The key and important focus of the chapter is to put forward an improved understanding of how affordable housing development finance challenges can be met.

The insight also provides several opportunities: the opportunity for improved management of real estate and financial resources in housing; to provide more value for money for all stakeholders; and subsequently the possibility of creating a better social and environmental condition. Furthermore, it helps to support those who are seeking to find sustainable solutions to the needs and wants of rapidly urbanizing cities, and in some cases de-urbanizing, given increasingly scarce resources – particularly in scarce affordable housing supply.

Findings and analysis consider the more technocratic discovery of which specific mechanisms used to finance housing development in cities are most prominent. Different "types" of mechanisms are put forward in the shaping of affordable housing development markets (Warnock and Warnock, 2008) – with "types" being broadly direct, fiscal, and monetary (Calza et al., 2013). These "type" mechanisms cover a multitude of acts, bonds, investment trusts, fees and levies, agreements, zone and districts taxes and credits, regulation, recapitalization, and trust funds. Final discussion points draw conclusions that a "Weighted-Blended" Finance Mechanism is the most appropriate framing to consider the financing of affordable housing in the development process. Plus a "tiered" level finance generating structure is important in order to begin to start claiming the strength of weight with which mechanisms have the greatest power.

San Francisco and the Bay Area case study is used to draw lessons for individual city nuances with respect to their use of financial mechanisms in affordable housing development. The City of San Francisco and the Bay Area was chosen as a case study as the area has been trying to overcome the challenges of high volume of affordable housing development in difficult market circumstances. Furthermore, the City of San Francisco and the Bay Area region has managed to lever in innovative finance approaches to tackle affordable housing where there is a property boom for some cities; whilst some other city administrations continue to deal with a stagnant property market (Gibb et al., 2012; Squires et al., 2016).

Case study context of mechanisms in financing affordable housing development

Mechanisms of finance in affordable housing development need to be placed in the context of the case study, although many of the mechanisms' fundamental principles can provide insights for other places. The aspects of discussion (such as types, weights, and blends) can be considered more conceptually for this case and further cases elsewhere. First of all, the occupation–rent ratio is important when considering the finance for affordable housing development in the case. What is interesting about San Francisco is that it is comprised of 70 percent renters. It is different from the rest of the Bay Area cities, with a composition where the city of San Francisco is polarized largely with extremes of wealthy owners and poor renters (or homeless). Moreover, those in the middle are predominantly renters rather than owner-occupiers.

There is a significant number of non-profits for affordable units in cities of the Bay Area region (Bratt et al., 1998). The Bay Area has very strong non-profits that have been in business 35–40 years, with cash of USD 20–30 million on their balance sheet. They have become so strong because public policy has driven projects towards affordable housing. In the 1960s and 1970s, the federal government under the direct provision of public housing development mainly provided affordable housing. More recently, affordable housing development has transitioned more towards private enterprise, where non-profits provide the local jurisdictions with an alignment of public policy along with a mission. Despite affordable housing development "enterprise," the sector still interlinks with more direct provision of funds for affordable provision, albeit in a more sophisticated way to encourage mixed and cohesive communities, such as via the San Francisco (SF) Hope VI and National Hope VI projects. Housing authorities are still heavily involved, and they still are government entities with federal organizations such as Housing and Urban Development (HUD).

"Type" mechanisms for financing affordable housing development

Given these broader contextual considerations in the case study and beyond, the specific mechanisms that channel finance are now put forward. They are conceptualized as "type" layers that work simultaneously, with incentives and selection finance being taken from different types. Here we consider type 1 as direct regulation such as shaping by acts and legislation; type 2 as the fiscal incentives of taxation and spending; and type 3 as monetary effect such as the influence of bonds and land value capture. These three types are seen as fundamental additions to conventional private financing in order to make the development of affordable housing supply "affordable." Private financing could be, for instance, in the form of institutional funds, loans, trusts, and donations. These three types (see Table 16.1) are now unpacked to get a great handle on what could be used as a mechanism in the financing of affordable housing development.

Table 16.1 Affordable housing finance "type" mechanisms

"Type" mechanisms	Key policy and concepts
Type 1: Direct regulation – acts, legislation	Community Reinvestment Act (CRA)
	Transit-Oriented Development (TOD)
	Metropolitan Transportation Committee (MTC)
	Priority Development Areas (PDAs)
	TOAH (Transit-Oriented Affordable Housing)
	Housing Community Development (HCD) Agency
	Up-zoning
Type 2: Fiscal – tax and spend	Impact fees and levies – Rational Nexus of External Cost
	Communities Benefit Agreement (CBA)
	Developer Contributions and Planning Gain
	Inclusionary Zones (IZ)
	Inclusionary Housing Ordinances (IHO)
	Inclusionary Housing Fees (IHF)
	Proposition 13 – Capped Property Tax
	Low Income Housing Tax Credits (LIHTC)
	Housing Trust Funds (HTF) – "pooling" general funds
	Community Development Block Grant (CDBG)
	Housing and Urban Development (HUD) Grants
	Re-Development Agencies (RDAs)
Type 3: Monetary – bonds and land value capture	Tax Increment Financing (TIF)
	Value Capture Bonds (VCB)
	Infrastructure Financing Districts (IFD)

Source: Author

Type 1: Direct regulation – acts, legislation

With regard to the more technical mechanisms at play in affordable housing development in the Bay Area, the Community Reinvestment Act (CRA) is a fundamental underlying legal instrument that directs the sector. The bulk of multi-family or apartment construction, whether it is new or renovation in the affordable housing sector, is done primarily to satisfy a federal regulatory requirement called the Community Reinvestment Act (CRA). CRA is a banking regulation passed in the 1970s at the federal level to prevent red-lining, where low-income neighborhoods are disadvantaged by being labeled so, and thus not invested in. The CRA means that banks are incentivized to reinvest back into the communities that they make their deposits to.

Cross-sectoral appeal for affordable housing development finance is particularly found within incentives provided by transport and regional planning (Austin et al., 2014; Beer et al., 2007). Transit-Oriented Development (TOD) dollars are available through the Metropolitan Transportation Committee (MTC) as a regional planning body. TOD will often involve a sustainable community strategy that has the goal of reducing vehicle miles travelled and of getting the cars off the road in order to reduce greenhouse gas emissions. TOD is planned to concentrate more development in areas that either have transit stations, or are along transit corridors or transportation corridors. To do so, cities were invited to identify in their general Bay Area plans Priority Development Areas (PDAs), which are pre-identified zones where city plans and permits would encourage development.

To explore PDAs (Priority Development Areas) further, there is resistance among voters to agree development in their own locality. This is largely due to NIMBYism (Not In My Back Yard) behavior, but also due to a lack of surrounding infrastructure and fear of further congestion. The business community also has reservations, as the priority areas will not immediately be where economic growth is actually going to happen. Even more so, there is an attitude that it is unlikely to achieve the desired greenhouse gas reductions as an aggregate. Further issues are that the majority of construction is for 1–2 bedroom apartments that are not the preferred choice for most families. This means that the plans may or may not actually have much influence on where people choose to live. Despite this, as congestion gets worse, more people may have the incentive to trade off commuting for family space, and be forced to take up demand for TOD accommodation. The theory is that people will pay a higher value (a premium) to rent a unit or buy a unit within walking distance of transit.

An affordable housing requirement of TOD often involves value capture concepts. Value capture in TOD is based on assessing properties to capture some of its value and pay for the initial transit outlay. In doing so, the large funds generated are used to pay for only Transit-Oriented Development in affordable housing. This type of affordable housing requires a minimum density, depending on where the development is. It also tries to increase densities and obtain a minimum percentage of affordable housing. Most projects are mixing current market rate and affordable housing. The policy here is the affordable housing development financing model known as the TOAH (Transit Oriented Affordable Housing) fund, which is a regional model directed by the State of California's Housing Community Development (HCD) Agency.

Changes in regulation can generate economic circumstances to strengthen the financial viability of affordable housing development. As well as horizontal zoning, regulatory circumstances can be provided to change viability vertically and in design. Under regulation, local governments have responsibilities to shape the design of buildings, the uses of buildings within certain areas, and the design in terms of height and bulk, and the impact on the street for example (Glaeser and Gyourko, 2003). Vertical regulatory changes are referred to as "up-zoning" and the development process can "up-zone" a building or a lot to create a greater developable parcel. City administrations can up-zone or down-zone, and that affects whether a developer may come to a community to build on a particular site. This regulatory process is also connected to economic factors such as a strong demand for real estate, both for office and residential. Obviously, larger buildings are more profitable than smaller buildings if there are more units per lot area, even though they may incur higher unit costs. Interestingly, some city administrations have facilitated the up-zoning of many districts in exchange for exacted inclusionary housing units.

Type 2: Fiscal – tax and spend

City administrations can adopt exactions that operate as a subsidy paid by the developer to an administration (Gibb and Whitehead, 2007). These could be, for example, impact fees or inclusionary fees. Inclusionary fees often apply to all new development in a particular land-use category at a uniform rate (Monk et al., 2005). These inclusionary fees are charged once for operating costs, but largely for capital costs. In turn, this fee is meant to offset the costs a city administration will incur to build, for instance, a road by using a project's funds. To incur a fee there has to be a direct connection between the project and the external cost, often referred to as having a "rational nexus." Impact fees are also similar to exaction although impact fees are like sewer fees or transit fees that go to cover "the impact" of additional density or use in a particular community. Affordable housing could also be part of the impact on a building-by-building basis, so there can be an attempt to exact some of the capital for this sector.

The Communities Benefit Agreement (CBA) is used in California and the Bay Area for exaction of fees and levies, and it often exists in specific zones. Within these agreements there is often tension, particularly as developers are struggling with how to deliver the housing, mainly as there are not enough subsidies to make the housing work if the lower affordable housing rents, as part of the agreement, are included in the viability model. Exactions as fees are often referred to as "in lieu fees" connecting to developer contributions as planning gain (Crook and Whitehead, 2002). Using San Jose in the Bay Area as an example, it demonstrates the issue of external demand and costs on public goods and services that can be extracted from real estate development, particularly if there is a direct rational nexus. San Jose mushroomed in its building of commercial property in the heart of the booming Silicon Valley. As a consequence of this boom, traffic has increased and generated congestion that in turn will need to be dealt with to address concerns over the environment and economic efficiency.

To implement exaction fees and levies in the housing sector of real estate development, Inclusionary Zones (IZ) have been set up to deal with the affordable housing problem. In these zones more specifically the fees are referred to as Inclusionary Housing Ordinances (IHO). Many jurisdictions around the Bay Area have IHOs that require market rate developers to put the new below market rate units on site within their development. Alternatively, the developer has to pay into a fund that is then leveraged to build affordable housing. Furthermore, developers can build affordable units off site but within approximately a mile of the principal permitted project. In essence, the inclusionary zoning process indicates that the developer pays a fee, to either provide onsite units or provide offsite units. Alternatively, in certain locations there may be the option to provide affordable housing development in exchange for permission to build at a slightly higher density than otherwise. So the IZ – in contrast to collections of housing units – are regulatory in nature with an exaction (e.g. a fee) for public benefit in exchange for providing developers with a private benefit.

Inclusionary Housing Fees (IHF) relate more to housing units, rather than the zone, and the fee or levy that is attached to the housing units developed. For example, the city administration can use these fees to generate a "pool" of finance. A developer could be building 100 units at the market rate, but needs to provide a certain percentage (a formula is used to calculate the percentage) within the development that is affordable. Alternatively, the developer can pay a fee – known as an inclusionary housing fee. This fee rarely gets waived because it is a very targeted approach to solving affordable housing. Note that the IHF is on top of impacts to the infrastructure, which generate a separate levy. To note the significance of the IHFs within San Francisco, developers cannot build market rate apartments without having inclusionary housing requirements. Developers either have to set aside units within their development, or developers have to pay a fee or work with a non-profit developer and build affordable housing off site.

Another key fiscal financing mechanism of affordable housing development for the Bay Area and California state-wide is through Proposition 13 (or Prop 13). Prop 13 is a law from 1978 that limits and caps the amount of property tax collected on a homeowner to 1 percent of the value of the house at the time of sale. The property tax bill only increases 2 percent a year until the property is sold again. This cap keeps the tax bill low relative to the appreciation value of the house, plus the city does not see the revenue until the house sells. Furthermore, at the outset, taxes raised by local governments for a designated or special purpose via Prop 13 need to be approved by two-thirds of the voters. As a result of Prop 13, all housing is seen as a loser to most cities, because the administration cannot collect a lot of property taxes to pay for public goods and services – as collection is at the point of sale. Furthermore, the inequities in housing values mean that proportionally the 1 percent capped tax rate is not progressive and cannot extract higher value amounts in tax on higher rate properties and cannot capitalize on rising property prices. As such, local governments are extremely restricted in terms of their ability to generate tax revenue because of Prop 13.

With affordable housing development, Low Income Housing Tax Credits (LIHTC) are the primary fiscal financing vehicle that allows affordable housing to be built, and allow some modicum of return for the bank. There is a purchasing of tax credits by banks for distribution to a project, and there is also a lending against credits as a form of equity. The LIHTC industry started in 1986 when the tax credits were formally codified within the IRS (Internal Revenue Service) regulations. All the major banks participate in this industry, although it is a specialized field because the process involves dealing with government regulations and IRS and tax implications. In the 1990s when it first started there were a lot of economic investors, because it enabled a dollar for dollar write-off of a company's tax liability. Today, the market is heavily dominated, not by industry economic investors, but by financial institutions that want an economic return and also want CRA benefits. The credits have a lifespan, similar to a bond, due to the agreement that the affordability has to last for 20 years. Each state is also allowed to have a state credit to add onto the federal tax credits, and federal credits are offered for up to 60 percent of average median income.

A critical point with respect to LIHTCs was that they have a limited lifespan (say 20 years) and do not run in perpetuity. Affordable properties are granted a lifespan as designated affordable, but this means that there is a problem into perpetuity. The affordable housing industry is now doing approved development work that is largely recapitalizing existing affordable developments as they come to the end of their LIHTC life. Developments could be in tax credit developments that were undertaken 15 or 20 years ago, and that are coming to the end of their initial compliance period, or are even older developments. The problem is if all of those resources are going towards redevelopment of older tax credit developments, the industry is not supporting new affordable housing. Furthermore, the tax credit program has a difficulty in that it does not work well for buying existing market value apartments. Essentially from the start of the LIHTC project there is approximately a year until all of units have to be occupied, meaning there is no strict mechanism to phase in the credit.

Of the city funds available, an emerging fiscal mechanism of funding for affordable housing development is via Housing Trust Funds (HTF) that are administered by a city government. The HTF is a general fund set aside for the purposes of supporting affordable housing development, and is valued at USD 1.5 billion over a 30-year period. Local HTFs collect dollars in different ways. Sometimes cities are putting their own federal HUD money into the fund, as well as their own Community Development Block Grant money from HUD. Sometimes cities also charge a transfer tax on each market rate sale (e.g. Prop 13) that can go into a housing fund pot. Berkeley is one city that does this independent of redevelopment agencies, as was carried out in the previous system prior to the Re-Development Agencies (RDAs) being dismantled. San Francisco city administration has got a housing trust fund set up and approved by voters (via Proposition C). This means that city authorities can capitalize finance themselves and have a set-aside property tax fund. The San Francisco city and county HTF is used for people on moderate incomes, such as police and fire fighters. The focus will be to replace some 3,000 public housing units that need rehabilitating, particularly those that house people living in extreme poverty.

Type 3: Monetary – bonds and land value capture

Government-backed lending of money can be considered a monetary type of finance mechanism that can stimulate affordable housing development. The raising of finance via lending to project districts on the basis of future clawback in taxes is the basis of tax increment financing (TIF) (Squires and Hutchison, 2014). TIF money in the Bay Area was instrumental in gaining match-funding grants from the federal renewal program – with TIF enabling affordable housing development via the redevelopment agency. The termination of RDAs in California on January 2012 effectively ended state-wide TIF being used as a major funding mechanism for dealing with local "blight" in the real estate (re)development process. Commentators noted the underlying reason for RDA abandonment being due to the large budget shortfall and in order to protect funding for core public services at the local level. The RDAs (funded largely by TIF) played a significant role in San Francisco and their loss will be felt in the remediation of "blight," although investment at the high end of the housing development market will no doubt continue.

Given the demise of TIF bonds to deal with blight, other mechanisms have been introduced or are being made more use of to enable leverage of finance against future uplift of taxes – namely the use of Value Capture Bonds (VCB) in Infrastructure Financing Districts (IFD) without a necessity to focus on blight and hence redevelopment. These have already been available for a long time in California, and it is a diversion of existing property tax (see Prop 13) to a special financing district. The IFDs have a limited set of purposes that are allowed, and there is a requirement for a certain percentage of the property owners affected to approve development. This generates some difficulties for city administrations, as there are always multiple property owners. As a special case to highlight these districts, the Port Authority in San Francisco have their own IFD, as they control all of their own property. As a result, they create ground leases and control a large amount of San Francisco's waterfront, with which special legislation helped create their own IFD. This means that the Port Authority does not need to get property owner votes, because they are the property owner. As such, the use of IFDs to finance affordable housing development is weak, as the priority will be to pay off initial bonds related to infrastructure rather than incentivizing less than market rate housing.

Tiered weighting and blending types of mechanisms for financing affordable housing development

The mechanisms highlighted in the previous section only serve to demonstrate financial forces that shape the market. Often it is the process of development by affordable housing providers that select and implement the appropriate mix or "blend" of finance mechanisms, and in various different degrees or "weights." As such, it is worth highlighting here the process of affordable housing development, to demonstrate that the blend and weight of finance will be one of the developer's choices; the other being ensuring that the development is realized beyond the financial choices. A concern of the overall affordable housing development process is making sure that all the typical planning issues are dealt with first (Type 1 – direct mechanism). Following this priority will be submitting an application for tax credits (Type 2 – fiscal mechanism), in addition to engaging with an architect and the city administrators to prepare a building permit (Type 1). A selection of financing partners would be next, with the first one as an equity investor – the institution that is actually going to buy the tax credits – the debt part of the finance is then sought. Often, local cities and counties will also put up and control targeted dollars in the project, attached to a number of program funds that are made available at the federal level and the state level (Type 2). Developers will then start to bring together other sources of financing – government loan financing, such as city bonds (Type 3 – monetary mechanism). Once this is in place and the equity is confirmed, then the affordable housing developer (or partnership) can obtain an interim construction loan and start to lever in the permanent debt from mainstream financial sources such as high street banks (Type 3).

During the development process, different mechanisms will be blended in the finance mix. Further, the weighting of importance for each mechanism is also of interest to the affordable housing developer. To put some weighting into this blend that occurs during the process over time, affordable housing development finance operates over several "tiered" finance generating levels. Here it is put forward to consider the financing of affordable housing development as a three-tier finance generating system (see Table 16.2).

At the first tier, federal–state–regional finance is generated, and devolved to the city and sub-city scale via government programs. A significant weight provided in this tier is by the LIHTC system, developed by the CRA, which encourages commercial lending in the affordable housing sector. Other first-tier finance generating levels are those such as the IFDs that operate as value capture instruments, albeit less focused on blight, as was the now disbanded (in California) TIF districts. Non-value capture bonds also contribute to this first tier, with mechanisms such as the CFDs that are bonded to finance to pay for public needs (not specifically housing) then paid back via taxation. At tier 1 federal–state–region finance generating level, state-led regional PDA grants also have been selected to encourage TOD affordable housing located at key transport hubs.

Second-tier finance generating levels are those highly powered within city and county jurisdictions for affordable housing development. City financing from tax extraction is via instruments such as special property taxes (e.g. Prop 13), and affordable housing incentivized finance via inclusionary housing fees and levies, plus inclusionary zones that encourage cross-subsidization of finance for affordable units. Exaction in this second city tier includes the use of impact fees on developers that contribute to public costs occurring due to development, or from a community agreement that will pay on the benefit principle, a CBA. At this city extraction and exaction second tier, funds can be pooled by the city administration, such as in the form of HTF in the example of affordable housing development.

Table 16.2 Affordable housing finance "weighted tier" mechanism power and associated key concepts and policy

"Weighted Tier" mechanism power	Key concepts and policy
1. Federal–State–Regional	Low Income Housing Tax Credit (LIHTC)
	Community Reinvestment Act (CRA)
	Infrastructure Financing Districts (IFD)
	Tax Increment Finance (TIF) Districts
	Community Facilities Districts (CFD)
	Transit-Orientated Affordable Housing (TOAH)
2. County–City–Neighborhood–Point	Proposition 13 – property taxes
	Inclusionary housing fees and levies
	Impact fees – external cost rational nexus
	Community Benefit Agreements (CBA) – benefit principle
	Housing Trust Fund (HTF) – pooled funds
3. Donations	Philanthropic
	Voluntary
	Charity
	Private syndicates and payments
	Mission funders
	Crowd funding
	Not-For-Profit (NFP) Organizations – for surplus

Source: Author

The third tier of finance generating for affordable housing development involves philanthropic, voluntary, private, and charity donations. Donations are not necessarily from government shaping type mechanisms (direct, fiscal, or monetary types) but will in turn provide less than market rate, and thus affordable housing developments. These could be philanthropic and as private syndicates or private payments. Donations in this way may in some instances provide a way that no individual affordable (or more specifically not-for profit) housing developer could finance. More philanthropic contributions come from several charitable trusts that often provide gap financing and seed-corn money for mission-based projects to be realized. The emphasis in this chapter was to explore more government shaping affordable housing development finance mechanisms, but with the continued rollback of state funding, available alternative donation-based "tiers" could be ever more weighted to drive affordable house development. Especially given the rise of crowd sourcing and the potential of crowd funding mission-based projects such as affordable housing developments. That is not to say that the most significant weight of affordable housing development finance generation, by tier one federal–state–regional power, is not to be dismissed, particularly by those that wish to see a fair housing allocation system, based on the greater public good and not just unfettered market forces and charity alone.

Conclusion

In conclusion, this discussion has argued that mechanisms for financing affordable housing development are indeed important and have varying degrees of power and influence. The case study can draw out many mechanisms set within the context of high ratios of renters in booming cities, a leading and high volume of mission-based, not-for-profit affordable housing developers, and an increasingly enterprising approach that was formally a large remit of direct provision. It was found that the most significant mechanisms could be considered as direct, fiscal, or monetary "types" in shaping affordable housing at less than market rate. For instance, the Credit Reinvestment Act as a direct mechanism, the Low Income Housing Tax Credit program as a fiscal mechanism and Value Capture Bonds as a monetary mechanism, are ways in which sense can start to be made of the plethora of mechanisms on offer. Thinking of such types as acting within the development "process" by affordable housing developers further crystallizes understanding.

The weight of the blend in financing affordable housing development can also be considered in relation to the "tiered" finance generating level. The strongest weighted tier for mechanism power is the federal–state–regional tier that devolves many significant programs and acts that enable the affordable housing development sector to exist. Second, the county–city–neighborhood second tier is argued as providing power to generate finance, with some autonomous powers for city and county to exact, extract, and bond, although largely controlled by the federal–state–regional tier. The third tier of finance generation is that from philanthropic, voluntary, private, and charity donations; these donations arguably generate a weaker weight of mechanism power, although paradoxically without such contributions at site level many developments may not be kick-started.

Just as with the many sources of finance, partnering to provide affordable housing development (such as private equity or debt finance) may have the most influence. Influences by the sources of finance, rather than the mechanism of finances, are for further study – especially as the willingness for different sources of finance to be forthcoming could depend on many heterogeneous reasons, such as institutional power or the specific site location of the affordable housing development.

References

Austin, P. M., Gurran, N., and Whitehead, C. M. (2014). Planning and affordable housing in Australia, New Zealand and England: common culture; different mechanisms. *Journal of Housing and the Built Environment*, 29(3), 455–472.

Beer, A., Kearins, B., and Pieters, H. (2007). Housing affordability and planning in Australia: the challenge of policy under neo-liberalism. *Housing Studies*, 22(1), 11–24.

Bratt, R. G., Vidal, A. C., Schwartz, A., Keyes, L. C., and Stockard, J. (1998). The status of nonprofit-owned affordable housing: short-term successes and long-term challenges. *Journal of the American Planning Association*, 64(1), 39–51.

Calza, A., Monacelli, T., and Stracca, L. (2013). Housing finance and monetary policy. *Journal of the European Economic Association*, 11(s1), 101–122.

Crook, A. T. D. and Whitehead, C. M. (2002). Social housing and planning gain: is this an appropriate way of providing affordable housing? *Environment and Planning A*, 34(7), 1259–1279.

Gibb, K. (2011). Delivering new affordable housing in the age of austerity: housing policy in Scotland. *International Journal of Housing Markets and Analysis*, 4(4), 357–68.

Gibb, K. and Whitehead, C. (2007). Towards the more effective use of housing finance and subsidy. *Housing Studies*, 22(2), 183–200.

Gibb, K., Maclennan, D., and Stephens, M. (2012). *Innovative financing of affordable housing international and UK perspectives*. York: Joseph Rowntree Foundation

Glaeser, E. and Gyourko, J. (2003). The impact of building restrictions on housing affordability. *Economic Policy Review*, 9 (2), 21–39.

Harvey, D. (1974). Class-monopoly rent, finance capital and the urban revolution. *Regional Studies*, 8(3–4), 239–255.

Kim, K. H. (1997). Housing finance and urban infrastructure finance. *Urban Studies*, 34(10), 1597–1620.

Monk, S., Crook, T., Lister, D., Rowley, S., Short, C., and Whitehead, C. M. E. (2005). *Land and finance for affordable housing*. York: Joseph Rowntree Foundation

Murphy, L. (2015). The politics of land supply and affordable housing: Auckland's Housing Accord and Special Housing Areas. *Urban Studies*, 53(12), 2530–2547.

Squires, G. (2013). *Urban and environmental economics*. Abingdon: Routledge.

Squires, G. (2014). *Future financing of cities for real estate development*. London: RICS (Royal Institution of Chartered Surveyors).

Squires, G. and Hutchison, N. (2014). The life and death of tax increment financing (TIF) for redevelopment: lessons in affordable housing and implementation. *Journal of Property Management*, 32(5), 368–377

Squires, G., Hutchison, N., Berry, J., Adair, A., McGreal, S., and Organ, S. (2016). Innovative real estate development finance: evidence from Europe. *Journal of Financial Management of Property and Construction*, 21(1), 54–72.

Wallace, J. E. (1995). Financing affordable housing in the United States. *Housing Policy Debate*, 6(4), 785–814.

Warnock, V. C. and Warnock, F. E. (2008). Markets and housing finance. *Journal of Housing Economics*, 17(3), 239–251.

17

The value investment approach to real estate development

A case study from Berlin, Germany

Stephan Bone-Winkel and Karim Rochdi

Abstract

The standard development approach all over the world is to "build new" on vacant sites. However, the refurbishment or redevelopment of existing buildings is becoming increasingly important, and not only for ecological and sustainability reasons. Buildings have to be adapted at a faster pace to changing uses and tenant needs for flexibility. With the "internet of things" and the growing digitalization of so many processes, our work environment has to adapt rapidly and without large capital expenditures. By presenting a case study from Germany, this chapter highlights an investment story from redeveloping former industrial buildings into modern uses for the growing tech and internet industry in the urban center of Berlin. In doing so, the chapter deals with the fundamentals of value investment in real estate. Based on the model that Dodd and Graham (1934) developed for analyzing securities in the 1930s, the authors apply the same principles to analyzing and investing in real estate. They compare the value investment approach to the prime/core investment approach of institutional investors and conceive it as a contrarian strategy for developers, independent of location and cycle.

Introduction

It is obvious that buildings are very different from cars. Yet both are durable goods, so we can draw some meaningful parallels. Just like a new premium-class car straight off the production line, from a technical perspective, buildings can also have a very long useful life if properly maintained and cared for. In terms of holding their price, however, the situation is completely different. As we all know, the value of a car declines continually, even if it is not being driven. Indeed, cars currently lose their value at an even faster pace, as vehicle manufacturers accelerate the rate at which they launch new models with enhanced features, and buyers perceive them as an improvement on previous models. Furthermore, only very few car owners hold on to their vehicles long enough for them to acquire a value among enthusiasts as vintage models, at which point, their value will slowly increase. A similar pattern can be observed for core real estate, that is, newly built or completely redeveloped and well-rented office and retail premises in prime city locations. Although car owners are presumably all aware that new vehicles in particular lose much of their value in the first few years, real estate investors appear to assume, when buying core properties, that they are economically sustainable, in the sense that they will retain their premium investment status and continue to generate premium rents or returns on sale. In fact, however, just like cars, these properties also age and lose their utility value. Signs of wear and tear become apparent, and in many cases, both the fixtures and technical features of a building very soon fail to conform to the latest standards and the requirements of tenants with constantly changing preferences.

Project developers bring new properties onto the market in regular cycles, in the form of buildings that more effectively meet the needs of tenants than their older counterparts. These trends in tenant expectations manifest themselves in terms of comfort (e.g. air conditioning, lighting), IT equipment (e.g. centralized/decentralized and integrated/external server rooms, wiring systems), workplace design (from cell offices to business club), and sustainability.

As an alternative to building new real estate, developers also take over empty, ex-core properties from investors with no development expertise. Such properties are then renovated with the aim of letting them and offering them again as premium or core properties, with a low-risk return strategy and predictable cash flows. The rather florid term occasionally used within the industry to describe this scenario is "broken core." This strategy follows a process of the original owner first suffering a heavy loss on what was a premium-ranking property, and then selling it on cheaply to a developer, who may subsequently make a substantial capital gain.

One of the main reasons for this widespread belief in the ever-appreciating value of prime real estate is a market-based approach that works to some degree as a method of interpreting the capital market. However, it is then often applied by inexperienced investors without proper consideration of the specifics of the real estate market and its various segments. They expect the value of their property to move in line with the market as a whole, but buildings are not a homogeneous commodity. No single building can "track the market" permanently. Even if an investor has a large portfolio in a particular location, that investor's property holdings will not correspond exactly to the market. He may be able to influence the market to some degree with his pricing policy, but he cannot monopolize or replicate the market.

To put it another way, if a new building or a completely refurbished one can be described as "core," it can only be classed reliably as such at the time it is first occupied. Over time, it is no longer representative of the market, but merely of its own specific rental situation. Using the example of a ten-year rental contract with an index clause, the cash flow generated will usually grow along with inflation, whether rents in the broader market are rising or falling. It is not until the landlord needs to find another tenant that he has to deal with the market again. An example of how this can work in practice is presented below.

Core investing in real estate

Movements in the value of core real estate can be demonstrated by reference to the well-known case of "Taunusanlage 11," a prime office tower in Frankfurt, Germany. In the early 1990s, an institutional investor purchased an office building in Frankfurt am Main, regarded as a "landmark property" in a prime location in the city's banking district. At the time, the tenant of the building (built in 1972) had already moved out, and the property was completely refurbished. Subsequently, it was the first property in Germany to achieve a monthly rent (for two floors of the building) of more than €50 per square meter. The property could probably have been valued thereafter at around 20 times the annual rent – €9,600 per square meter, assuming a sustainable consistent rent of say €40 per square meter. These figures are based on the assumption that the rent can be increased repeatedly, at least keeping pace with inflation.

On this basis, after 20 years, the rent would have risen from €40 to €56 per square meter (based on average inflation of 1.75 percent). Today's value would be around €13,580 per square meter, if the building were valued at the same initial return of five percent.

However, the situation evolved very differently. The new principal tenants moved out as early as 2007, and the property then remained permanently empty. Over the last 20 years, the return is likely to have been well below five percent, especially if we factor in non-apportionable ancillary costs such as maintenance. Thus, less than 20 years after its construction and again after its later refurbishment, the property was no longer competitive. It was then sold a number of times and completely refurbished once again by a developer, between 2012 and 2013 (hp developments, 2014; Wiederhold, 2011). According to Jones Lang Lasalle (2016) the premium rent for offices of this kind in Frankfurt is now roughly €36.5 per square meter, with the cap rate at around 4.4 percent. From today's perspective, the property's theoretical selling price is likely to be approximately €8,800 per square meter. However, this would only apply, if it were to be let at a premium rent in a completely refurbished state and could be sold at the core cap rate. Our example demonstrates that this property had core status when it was built, and again when it was refurbished in 1993 and again in 2013, but (after the first two of these occasions) it subsequently declined to the official reference land value, plus shell construction costs, within less than 20 years (see Thomas Daily, 2016).

Many other real-life examples could also be cited to illustrate the point that in economic terms, the residual useful life of tenanted office premises in the premium segment, and even of those occupied and used by the owner, is subject to constant decline. Consequently, the refurbishment costs that owners need to factor in every five to ten years, increase considerably. Yet, it seems likely that very few investors actually do factor in this expense – which can, from evidence, be as much as several years' rental income. Otherwise, they would not be satisfied with an initial return of 4.4 percent. Instead, investors assume that it will always be possible to let such office premises, simply because the location itself is outstanding. That is, it cannot be duplicated, and the premises are tailored to the tenants' current requirements. When calculating the purchase price, most investors do not appear to give sufficient thought to the fact that the premises will age, that tenant needs and preferences will change, that technology will progress and that the site itself will evolve for better or for worse.

Incidentally, this case study from Frankfurt is not an isolated example. The same pattern recurs not only for this building, but also for many others, whether in the office or retail segments, city-center shopping malls or luxury hotels. Figure 17.1 shows uniformly low cap rates for core property in the retail, office, and residential segments in Germany's top seven cities. Since 1996, there has been no observable increase in values as a result of higher purchase prices. As long as nearly all institutional investors such as pension funds or insurance companies focus exclusively on core real estate investments as outlined by Metzler Real Estate (2016), this situation is unlikely to change. For office premises, even the rise in premium rents has not produced any growth in capital values – in fact, the opposite is true. In real terms, we have seen a significant decline in the value of new buildings, although, as we shall now see, this decline is much less pronounced in the case of existing real estate than it is for new properties.

Intense competition driven by new buildings in core segments

From a management perspective, all premises undergo their own product life cycle, and in some cases, this may diverge considerably from the broader market cycle. Even if, for example, the premium market rent remains constant or increases, the level of rent is not always achievable for a specific building that was regarded as a core property perhaps ten years before. A series of

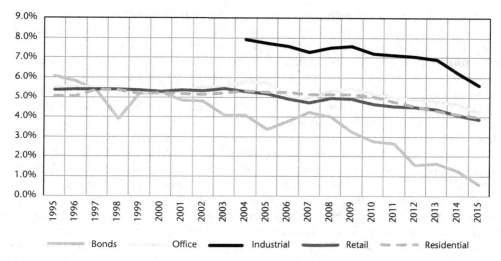

Figure 17.1 Low cap rates in Germany's top seven cities – commercial premises much cheaper

Source: BulwienGesa (2015) and Statista (2015)

filter processes tend to lead to a situation in which the original premium rent gradually declines to an average level over several rental cycles. The set of tenants also changes over time. The market itself filters demand for premises, with the most sought-after tenants moving to newer properties and being replaced by less demanding tenants. Good asset management and high levels of investment can slow down this process, but cannot terminate it.

Alongside the ongoing competition their properties face from new buildings, the situation for owners and investors is made more difficult by the fact that, while the core segment tenants they wish to attract are few in number, they are nevertheless highly mobile. What motivates such tenants is relatively clear; any new premises must be either cheaper or better than their current one. Whether a lawyer works in one location or in another 50 meters away is irrelevant, as far as his or her work is concerned. The marked readiness of such tenants to switch location gives them a strong negotiating position. In most stages of the market cycle, tenants can dictate the terms of any follow-on lease contract. Usually, the landlord takes the view that the risk and damage incurred if the tenant moves out would be higher and that he would therefore rather accept a lower rent. At this point, the difficulties associated with classic core investments become obvious (see also Figure 17.2).

Value investing in real estate

There are ways around the problem mentioned in the previous section. Most companies in Germany work on the basis that the higher the rent, the less attractive the premises. That is particularly the case for the German "Mittelstand," the small and medium-sized enterprises (SMEs) that make up 99.6 percent of all German businesses and account for 59.4 percent of all employees (Institut für Mittelstandsforschung, 2016). This group of tenants rarely looks for the kind of properties that most institutional investors currently regard as attractive. Apart from low rents, what matters most for such tenants are the productivity and flexibility of the leased property. For SMEs, business premises have a functional character, and status is less important.

To date, institutional real estate investors have tended to focus almost exclusively on prime sites with a high level of market transparency in terms of open and accessible data. Moreover, they have avoided low-rent properties in second-tier locations (not second-tier cities). Nevertheless, such "class B" sites can present an attractive long-term investment, if they can be acquired at low cost, following a thorough analysis of the competition and with good asset management. The key lies in adopting a value-based approach that is referred to in the industry as value investing. It is important not to confuse "value investments" with those in the "value added" risk category, which involve buying slightly higher-risk properties (some vacant units, less desirable locations, tenant credit ratings, refurbishment potential).

The value approach to real estate investing is based on long-term fundamentals, such as the available supply of premises, and growth in employment. The primary focus is on the user's perception of the supply situation, such as whether there is an over or an undersupply. It is the user's search profile that defines the relevant market in which landlords compete against each other. Value investors concentrate on a property's immediate competitive environment and intrinsic value. For example, they will examine, from the tenant's perspective, whether other landlords are in a position to offer better or cheaper premises. The value approach to real estate investing therefore presupposes a detailed assessment of the micro-location, based on a primary analysis of comparable factors such as rents, purchase prices, and refurbishment costs.

Value investing takes us logically into second-tier locations, often existing buildings that have already experienced a number of rental cycles, as can be seen from Figure 17.2. This approach

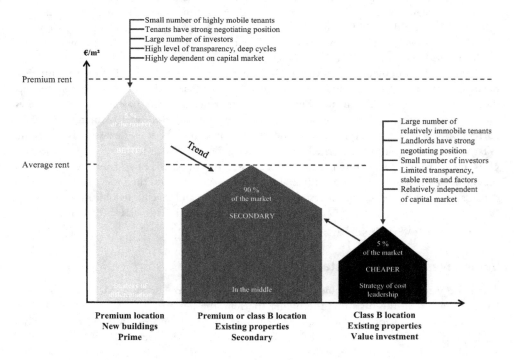

Figure 17.2   Value-based competition strategies generate benefits for their users and allow values to rise

Source: Authors

can be applied to any segment of the real estate sector, be it offices, retail, hotels, or logistics/industry. The fundamental aspects of this approach are outlined below by reference to mixed-use commercial properties, which are also referred to as corporate or industrial real estate. These are mainly buildings used as offices or for services, storage, production, and distribution. For the most part, the tenants in question are SMEs, so that the use of and demand for such industrial real estate is correspondingly diverse. Investing in mixed-use commercial buildings in second-tier locations enables investors to adopt a strategy of cost leadership, as this segment does not necessarily depend on volatile premium or average rents. since the rents are at a low, but constant level. The reason is that this segment is, in comparison to the two other segments that cover the majority of the properties, dominated by large numbers of less mobile tenants. Their commitment to a particular area places the landlord in a comparatively strong negotiating position. Furthermore, the segment attracts only a small number of institutional investors. A lack of market transparency, along with the need for active and specialized asset management, make it possible to acquire such properties more cheaply – assuming the investor has a good knowledge of the market. This price discount, when the purchase price ranges below the property's intrinsic value, is considered as a margin of safety, as popularized by Dodd and Graham (1934). This margin substantially reduces the overall downside risk of the investment.

Another point in favor of this market segment, in addition to stable rents and purchase-price factors, is a reduced dependency on the capital market. This is particularly clear in the risk–return ratio for different asset classes, as shown in Figure 17.3. The "German Property Index" published by BulwienGesa shows far less pronounced fluctuations in total returns for

Figure 17.3 The changing pattern of total returns for individual asset classes suggests a convergence between different cycles in Germany

Source: BulwienGesa, 2015

industrial real estate than for retail or office. Even new residential properties follow a more cyclical pattern. The only asset class to record consistently positive total returns over the last 20 years is industrial real estate, which underpins its suitability for value investing.

If the risk of declining property values is lower for industrial real estate than for other asset classes, investors should increase their weighting of this sector on account of the potential higher returns. As Figure 17.4 shows, since 1991, investors in the industrial segment not only have experienced much less pronounced volatility in capital real estate values, but also have achieved gross rental yields roughly 200 base points higher.

The apparent contradiction that higher returns can be achieved despite lower volatility, is explained by the fact that the market for industrial real estate is relatively intransparent in terms of available data and service provided covering the sector. Furthermore, industrial real estate is dominated by owner-occupiers who focus on cost and specific needs.

Industrial properties

In order to provide more detailed insight into our understanding of the abovementioned multi-use and multi-let commercial properties, the four distinct types of industrial properties according the "Initiative Unternehmensimmobilie" (BulwienGesa, 2016) are explained below, which have proven to be an excellent asset class for a value-based approach, namely: converted properties, logistics properties, business parks, and light manufacturing properties.

Converted properties are often former manufacturing sites with a building structure that evolved historically in line with changing business needs. Some of them have a campus

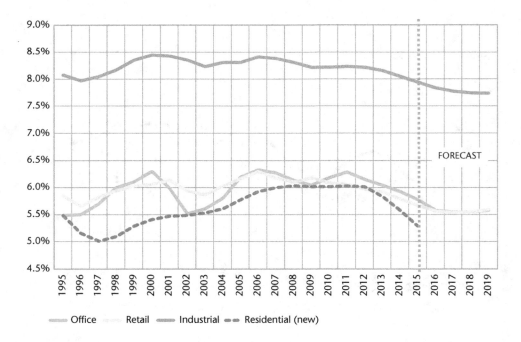

Figure 17.4   Individual asset classes vary considerably in terms of average gross annual rental yields in Germany

Source: BulwienGesa, 2015

character and are found in comparatively central locations in urban areas. During the conversion process, the existing rental income facilitates alteration, supplementary, and redevelopment measures aimed at turning a single-occupier homogeneous-use property into a multi-tenant property characterized by a variety of uses. The historic nature of period buildings adds a unique charm to the premises. Young service providers especially identify with the historical "Made in Germany" image of industrial brownfield sites.

Converted properties are highly versatile in their use options, because of their very heterogeneous and comparatively large and adaptable structures. Accordingly, they qualify for almost all classic types of use. Depending on the type of redevelopment undertaken, converted properties may continue to serve business purposes for another 50 years on average, assuming sound use concepts. Their often central urban locations justify elevated rent levels that match or even exceed the local reference rent, especially when properties of this type are redeveloped into office or retail accommodation. Accordingly, leases tend to be signed for extended average terms, with ten-year leases not uncommon for the major tenants.

Against the background of supply chain management, modern logistics properties serve purposes beyond the business of storing, trans-shipping, and order-picking goods and commodities. Existing infrastructure and technology (for instance, annexed offices and social areas, warehouses, conveyor technology, IT, service areas, shops) facilitate rapid adjustment to new requirements, while also lending themselves to mixed use by several occupiers and having a modular structure. Logistics properties constructed after 2000 generally display these characteristics. Most of them are operated by logistics firms on the basis of service contracts (contract logistics), and are frequently located near transport hubs. Their standardization and high alternative-use potential make properties of this type extremely flexible.

The best-case scenario for vintage stock properties is to be located in a region subject to great logistics demand, such as near a major center or port. A serious shortage of floor area, combined with a high level of demand, boosts their attractiveness as investment property. Due to their architectonic structure, dated legacy properties are rarely suitable for multi-tenant use, because they were often custom-developed for their former primary user. Accordingly, these vintage properties do not qualify as typical value investments in the sense explained above.

Factors such as globalization and the segmentation of manufacturing processes have greatly favored the logistics industry in Germany, not least because of the country's advantageous location in the heart of Europe. The momentum is further accelerated by the rise of e-commerce, and has recently generated strong demand among occupiers and investors. Among the industrial real estate types, it is the asset class with the fastest-growing footprint in Germany. The combination of high demand with a supply of limited floor space also makes it reasonable to expect considerable upward rent potential.

By their very nature, business parks are intended for multiple occupancy. A combination of office, service, storage, and clearance spaces is let in a building cluster, together with professional management. Distinguishing features of modern schemes include central locations and smaller dimensions, whereas older business parks were often located in suburban sites characterized by convenient transportation access.

The history of business parks can be divided into several generations. The first generations were defined by a very low share of office accommodation. Over time, office spaces became increasingly dominant, in some cases accounting for a share of up to 80 percent. More recently, the trend has shifted again in favor of other types of floor space.

While business parks are inherently designed as multi-tenant properties, the concept of business parks has repeatedly been adjusted to the needs of the economy, by changing the pro-rata floor space contingents. This has led to the creation of a wide variety of different business park types. Depending on the location, rents in business parks are comparatively high. This rent level can be explained primarily by the higher office share. Another factor that comes into play is their fairly central locations, because they justify higher square meter prices. The multipurpose nature of interconnected complexes means that these properties are safe and profitable vehicles. Older schemes with a high office share and located in peripheral sites, however, need to factor discounts into their office rents. The spread of achievable rents in business parks is therefore comparatively high for office space.

The fourth type of industrial properties are light manufacturing properties that represent buildings used for light, non-nuisance (low-pollution) manufacturing purposes. Present-day light manufacturing properties tend to be found in location clusters with excellent access to their constituent cities. They frequently combine several floor area types, such as manufacturing, laboratories, logistics, and warehouse space, along with a modest share of office units.

Light manufacturing real estate is located in commercial or industrial areas, either within or close to cities. Due to the relatively high degree of specialization and customization for specific manufacturing processes, lease terms signed by established businesses often exceed the ten-year mark. This is explained, inter alia, by the substantial proprietary investments that occupiers tend to make, and that presuppose long periods of occupation to allow for amortization. Another factor encouraging strong ties to a given location is the local commitment of some companies.

From an investor perspective, the above-average lease terms for light manufacturing properties are attractive. Then again, rent revenues rarely exceed the mid-range. The reason for this is in many cases, the relatively simple building standard that could alternatively be used for warehouse, service, and logistics purposes. Modern light manufacturing schemes are often designed for a flexible adaptation of these property types from the start, because of the close synchronization of production and logistics processes within the framework of supply chain management. This can be changed as soon as light manufacturing property is customized to serve the purposes of a certain occupier. Rental rates are therefore much higher for high-tech installations, for instance, whereas their alternative use potential is compromised.

In summary, it can be stated that the industrial real estate segment certainly provides a good example of the potential of value investing. However, the same approach could reasonably be adopted in other segments and niches within the real estate market. This would allow investors to implement a cost-leadership strategy, even in specialized markets such as budget hotels, retirement homes and student accommodation, provided they are able to acquire inexpensive property in the right location, manage it effectively and develop a close relationship with customers.

Asset management requirements

Asset management is generally a key factor for any type of asset. In the case of industrial real estate, asset manager qualifications are even more significant than would be the case with other single-use property types, because of their multi-use character and their frequent letting activities involving a larger number of occupiers. Industrial real estate is often associated with a high churn rate, but this association is rarely based on hard evidence. The prerequisite for tenant loyalty is first-hand asset manager knowledge of tenant requirements, and the capacity to meet changing floor-space needs. After all, the demand for space can fluctuate considerably over time. Depending on the order book balance, company premises may have to be expanded or reduced at short notice. The tenants of a landlord who demonstrates flexibility in this regard are more likely to remain long term. Yet, the flexibility comes with increased administrative costs, which are needed to ensure full occupancy of the floor areas.

Asset managers should communicate openly with their tenants, and place them squarely at the center of their activities. Of key importance here is direct communication on the ground, including regular visits of the asset manager to the occupiers or to the decision-makers of the various companies. This means one-to-one meetings, preferably at periodic intervals, such as quarterly. Between these on-site dates, the asset manager should tap other sources to remain up to date and be able to respond quickly. Other asset management tasks in this context include the periodic reappraisal of a tenant's creditworthiness and the monthly monitoring of rent payments. If a tenant falls into arrears, the asset manager is well advised not to respond immediately with a written reminder. Rather, the sensible thing to do is to engage the tenant in an informal dialogue and to hold back on tougher measures, such as legal threats, until an amicable approach has clearly failed.

It is essential that an asset manager understand the nature of the tenant's core business. This is in fact the only way to adapt the lease as much as possible to the respective occupier's needs. Familiarity with the tenants also enables an asset manager to respond to changing floor-space requirements. Equally important is that the tenants know who to contact if strategic issues arise concerning the premises.

Case study of value investing: Zeughof, Berlin

The following case study shows by way of example, how the proposed value investing approach can be successfully adapted individually in real estate development. The project was initiated and developed by BEOS AG, an independent developer and asset manager headquartered in Berlin. Founded in 1997, BEOS employs interdisciplinary teams in Munich, Hamburg, Frankfurt, Cologne, and Berlin to manage and develop properties throughout Germany. The main investment focus is on multi-use commercial properties that offer office, production, service and logistics space to cater for the needs of their heterogeneous tenants. Through personal conviction and as a contribution to sustainability, BEOS invests primarily in existing buildings and attempts to create new usage concepts by pursuing creative and economically prudent solutions.

Project summary

The Zeughof is a former single-tenant property in the vibrant Berlin-Kreuzberg area that has been restored, adapted, and transformed into a multi-tenant and multi-use business campus. The project area comprises 11 buildings and features high quality production, storage and office space, attracting emergent and aspiring companies within the technology, media and telecommunications sectors. The project was initiated and developed by BEOS, with the objective of revitalizing a large part of the existing space, developing a campus brand, and participating in the dynamic trend of Berlin's growing economy in general and the start-up scene in particular. With a total leasable area of about 51,000 square meters, a high occupancy rate, as well as an attractive tenant mix, the project can be considered highly successful, both operationally and financially.

Development background and the site

In the nineteenth century, the area surrounding the Zeughofstrasse was incorporated into the city boundaries of Berlin, following the industrialization and economic upturn in the German capital. Shortly after the turn of the century, the property was purchased by a German producer of telephone equipment and served as its headquarters. The existing buildings date back to the years between 1958 and 1982, and originally comprised office and production space for its single user (Figure 17.5 shows the main building in the year 1972). The project is located on a 30,269 square meter, half-block site along the Zeughofstrasse, Köpenicker Strasse, and Wrangelstrasse in Berlin-Kreuzberg, on the south side of Berlin, close to the Spree river (see Figure 17.6). It includes 31,254 square meters of office space, 12,605 of service and warehouse space, 4,112 of retail space and 2,546 of production space, amounting to a total leasable area of about 51,000 square meters. In 2013, when the property was purchased, the economic situation in Berlin had already improved extensively over the preceding few years, reflecting a decline in unemployment and a GDP increase. Even though Kreuzberg can arguably be considered as a relatively economically weak district, it is characterized by a very young population living predominantly in single households. However, the economy of Kreuzberg has steadily been changing from the historical "industrial-image" with factory workers, to a focus on creativity and culture with clubs, restaurants, and start-up fashion labels. In recent years, the Spree banks surrounding Friedrichshain-Kreuzberg have also undergone a shift towards media services with the establishment of MTV Europe and Universal Music. This has also been underpinned by a couple of ambitious development projects in the surrounding area, with the aim of attracting telecommunication and media companies. The "Zeughof" compellingly illustrates this very shift in appearance of the district, which is surely one of the key factors leading to the success

Figure 17.5   Zeughof: Main building, 1972

Source: BEOS AG

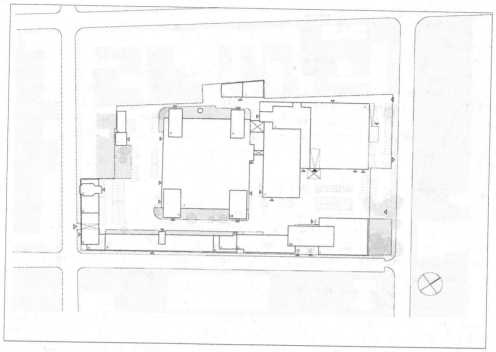

Figure 17.6   Zeughof: Overview of buildings
Source: BEOS AG

of the project. In summarizing the key points in the historical background and site character-istics, it appears that many proposed value investing requirements are met. First, in terms of city location, the Zeughof can historically clearly be classified as class B, even though the attrac-tiveness has risen constantly since 2010. Second, the property offers a broad range of usage types allowing a multi-use structure. Third, due to its historic multi-use, single-occupier, but flexible structure, the property qualifies for a multi-use, multi-tenant structure. In terms of industrial property classification as presented above, the Zeughof can clearly be classified as a converted property.

The idea and the concept

When the property was taken over by BEOS, the producer of telephone equipment vacated a large part of their former space, and various tenants from distinct sectors occupied the buildings with no uniform or efficient concept. In addition, there was a substantial vacancy rate amounting to 8,523 square meters. Most buildings faced deferred maintenance and the overall condition of the property was poor. Due to the partly resulting outdated and obsolete space, important anchor tenants intended to vacate the buildings after their lease contracts ended shortly. These circumstances created an opportunity to acquire the project at a modest price, well below replacement cost. Moreover, the project featured some of the abovementioned important aspects regarding value investing. Specifically, the property comprised various usage types and a flexible and functional layout. While the property is located in a typical class B location within a class A city, the area is characterized by economic growth and prevailing moderate rents.

Simply expressed, the basic idea was to gradually invest in the buildings in order to create a modern and open work environment for existing and future tenants. Further effort was put into establishing a property brand, identifying and attracting discerning companies from the TMT sectors and German Mittelstand, some of which had already located in the surrounding area. This entire process was intended to increase the very low rent level within the property that had prevailed up to that point, while still participating in the ongoing cash flow generated from the existing leases.

Development and management

As a first step, right after the purchase, all existing tenants were approached and involved in discussions about the future development of the "Zeughof," in order to demonstrate the genuine interest of the landlord in the needs and requirements of his tenants. One important point emerging from the discussions, and also from the expectations of typical investors, was the visual appearance of the Zeughof that had suffered over the years. The buildings were not considered as a harmonious entity and the age of the buildings was clearly visible on the facades and the entire courtyard. This circumstance shows one of the typical challenges that have to be faced when converting a property. Thus, a large part of the facades was cleaned (sandblasted) and painted, in order to harmonize the overall image. The courtyard was completely redesigned, by providing an open space with benches, as well as bike racks. By planting trees and shrubs, a green and recreational space was created that now emanates a welcoming, relaxed, and open atmosphere. In the course of redesigning the courtyard, the accessibility of the buildings was further improved.

In order to implement the architectural concepts in the buildings, several tenant units underwent major structural improvements aimed at providing a more open architecture. This included the construction of two additional floors, the reconstruction of the original entrance on the Wrangelstrasse, a new staircase, entrance, and lift serving existing and new floors, and the transformation of basement storage space into office and parking space. In addition to architectural improvements, considerable effort was put into important reconstruction measures, such as modernizing the central heating and the structural fire protection of the buildings.

Since the image of the buildings featured an industrial and aged character, marketing played a crucial role in repositioning the property. Apart from the new architectural concept, an overall brand concept was designed, focusing on the name "Zeughof," which was inspired by the street name and the history of the area. This marketing strategy entailed a new exterior and interior guidance system with markings for main axes, parking spots, and loading area. In conjunction with the overall redevelopment of the buildings and after completing the improvements, the "Zeughof" finally constituted a homogeneous and uniform property. As a consequence, the property has evolved from a single tenant one into a functioning and visually united multi-tenant property. The efforts made in order to take advantage of the property's value investing potential show the superior role of active and good asset management with regard to both the building and the tenants. Figure 17.7 gives an impression of the aforementioned key measures, for example, the new architectural concept and the brand concept.

Another important issue that often has to be handled when developing or managing industrial sites is that of the environment. Even the inner-city located Zeughof is, or has been,

Figure 17.7   Zeughof: Visualization
Source: BEOS AG

subject to some environmental problems primarily relating to contamination. As a consequence, the property is listed in the register of contaminated sites, which generally discourages institutional investors in particular. Therefore, sound historical research and ongoing drilling had to be conducted in order to resolve this issue and bring about a delisting of the site from the register. This whole process requires long-term commitment and ongoing investigation, but is likely to be rewarded by increased investor interest in the property, subsequently leading to higher pricing.

Conclusion

Convinced by the overall concept and willingness of the landlord to create a functioning and flexible working space, most anchor tenants extended their long-term leases. However, many new tenants also occupied the vacant or vacated space, which significantly increased the rent level. Through the entire transition, the property has not only become an outstanding location for tenants to produce, develop, distribute or offer their services, but also an attractive investment for those with an appetite for efficiently functioning and well diversified assets that are ready to face future challenges. In addition, even though the development yields a stabilized cash flow and already constitutes a viable investment, further opportunities are being provided, for example, by continuing to transform existing storage or production space into office space. The project shows clearly and convincingly, how the principles of value investing can be implemented successfully in the real estate market and offer a profitable alternative for real estate developers and investors who aim to be independent of location and cycle.

References

BulwienGesa (2015) "German Property Index (GPI)," BulwienGesa AG. Available at: www.bulwiengesa.de/en/node/250

BulwienGesa (2016) "Initiative Unternehmensimmobilien, Market Report 4, Second Half-Year of 2015," BulwienGesa AG. Available at: www.bulwiengesa.de/sites/default/files/marktbericht_ui_2._hj._2015_englisch.pdf

Dodd, D. L. and B. Graham (1934) *Security Analysis: Principles and Technique*, McGraw-Hill.

hp developments (2014) "T 11, Revitalizing the high-rise office building at Taunusanlage 11," hp developments GmbH. Available at: www.hp-developments.de/en/t-11-revitalizing-the-high-rise-office-building-at-taunusanlage-11/

Institut für Mittelstandsforschung (2016). "Overview – SMEs," Institut für Mittelstandsforschung. Available at: http://en.ifm-bonn.org/statistics/#accordion=0&tab=0

Jones Lang Lasalle (2016) "Office market profile, Frankfurt, 1st Quarter 2016," Jones Lang LaSalle Inc.

Metzler Real Estate (2016) "Individual vs Institutional," Metzler Real Estate Inc. Available at: http://metzlerna.com/metzlerre/servlet/linkableblob/start_mre_us/36864/data/Institutional-data.pdf

Statista (2015) "Yield on ten-year German government bonds from 1995 to 2014," Statista.com. Available at: www.statista.com/statistics/275265/yield-of-ten-year-state-bonds-in-germany/

Thomas Daily (2016) "Taunusanlage T11," Thomas-Daily.com. Available at: www.thomas-daily.de/en/project/detail/id/ee41d59f-577e-4c08-a82b-89a59b677803/lt/Frankfurt-am-Main-Deutschland/t/Taunusanlage-T11

Wiederhold, L. (2011) "Freo erwirbt Frankfurter Hochhaus T11," *Immobilienzeitung*, January 13. Available at: www.immobilien-zeitung.de/1000001812/freo-erwirbt-frankfurter-hochhaus-t11

Part V

Environment and sustainability in real estate development

18

Sustainable transformation in real estate developments through conversions

Hilde Remøy and Sara Wilkinson

Abstract

Of the buildings we will have in 2050, 87 percent are already built. If predicted climate changes are correct we need to adapt existing stock sustainably. Reuse is an inherently sustainable option, which reduces the amount of waste going to landfill and limits the use of raw materials. Inevitably, settlements and areas undergo change, whereby land uses become obsolete and buildings vacant. At this stage, the options are either to demolish or to convert to another use. Although office to residential conversions are still few in number in various CBDs, cities such as Sydney show an emerging trend in conversion. Some 100,000 m² of office space is earmarked for residential conversion as demand for central residential property grows and low interest rates create good conditions. With the Sydney market about to be flooded with the Barangaroo office supply in 2017, the conditions for residential conversion are better than ever.

However, what is the level of sustainability in these projects? This chapter investigates the nature and extent of conversion in Sydney, as well as the political, economic, social, environmental, and technological drivers and barriers to successful conversion. Through international comparisons between cases in the Netherlands, Belgium, and Australia, this chapter identifies some key lessons that are applicable to other market and urban contexts worldwide. There is substantial potential to change the nature of the CBD with residential conversion of office space and this chapter explores this potential.

Introduction

To date, conversions of office stock have not been widely undertaken in Sydney, whereas in cities outside Australia, such as in the Netherlands and other European countries, there is an established history of office to residential conversion. As cities transition through change, buildings become redundant and obsolete. The Sydney central business district (CBD) may see an increase in office to residential conversion towards 2020. Office supply is increasing along with an increasing demand for residential property in Sydney, which has led to withdrawals of non-CBD office stock for residential conversion (CBRE, 2015a). Of total Sydney office stock, 3.5 percent will be removed for residential conversion before 2020 as demand for central residential property grows (CBRE, 2015b), and low interest rates create good conditions. Furthermore, by 2031 the CBD population is projected to increase by 4 percent, requiring 45,000 new homes, and this increase coincides with a stock of ageing offices; some with conversion potential (City of Sydney, 2010). In addition, the Sydney office market is predicted to be saturated in 2016 and 2017 from the Barangaroo redevelopment, thereby making conditions for residential conversion better than ever.

A number of terms describe conversions, and these include adaptive reuse (Langston, 2014) and change of use, which essentially have the same meaning; that is to change the existing land use from one type to another, whilst updating the building. Interestingly "adaptive reuse" is the term favored in Australia whereas "conversion" is favored in Europe.

The UNEP report *Buildings and Climate Change* (2009: 4) concluded that the built environment has the "potential for delivering significant and cost effective greenhouse gas (GHG) reductions," and that nations will not meet emissions reduction targets without supporting energy efficiency gains in the sector. Furthermore, failure to deliver low carbon new build and retrofits (including conversions) will lock countries into the disadvantages of poor performing stock for decades. In total, the built environment contributes around a third to global GHG emissions (UNEP, 2009). Often, new sustainable building is perceived as the best way of meeting these aims, however 87 percent of the buildings we will need in 2050 are already built, and therefore adaptation and adaptive reuse is a necessity (Kelly, 2008). Adaptive reuse (or conversion) is an intrinsically sustainable option, as it lowers the amounts of landfill waste, and with the focus on redevelopment in the existing built environment, reduces land take for new buildings and infrastructure. On a pragmatic level, with population growth and increasing rates of urbanization, reusing existing buildings allows for a faster build time compared to demolition and new build (Bullen 2007).

Economic and demographic changes drive transformation in urban settlements. As a result, some land uses become obsolete and some buildings become vacant. In some regions demographic and economic decline causes obsolescence and vacancy, whereas in other regions a spatial shift occurs, with high demands in specific markets and changes of land use. At this point, options for existing real estate are demolition or conversion. This chapter examines sustainability in the context of office to residential conversion in Sydney, and is based on literature and interviews with Sydney stakeholders including real estate experts and developers. Illustrative case study examples are used to show real world practices.

Background – aim of this chapter

The potential to convert offices to residential land use has been established (Geraedts and Van der Voordt, 2007; Remøy, 2010; Wilkinson et al., 2014) and is explained by understanding the attributes of the residential and office markets, as well as the location and the building itself. This chapter focuses on evolving practices in the Sydney CBD in Australia, and examines the prevailing residential and office markets, before exploring key location and building attributes.

The residential property market

The residential property market in Australia, particularly Sydney, was again very heated in 2015 although there are some predictions that the "bubble will burst." Overall, the conditions are favorable for growth with low interest rates and reasonable employment conditions. In major Australian cities such as Melbourne, Sydney, and Brisbane there has been high demand by foreign investors, particularly Chinese buyers for CBD apartments (EC Harris, 2014; CBRE,

2015a). These investors are seeking to invest their money in markets they perceive as stable and safe. Ironically, residential property shortage is not addressed as many investment apartments purchased by Chinese buyers remain un-let as it is considered unlucky for Chinese people to occupy buildings that have been already occupied. Having these properties unoccupied exacerbates residential property affordability problems for many Australians. The pension or superannuation system in Australia also encourages citizens, excluding first time buyers, to buy investment properties for an income in retirement, which drives up prices further.

In recent history, residential supply in the Sydney CBD, along with other Australian capital cities, has been very low, and the stock has been restricted predominantly to office use or mono-functional use (JLL, 2014). There has been a shift since the 1990s onwards from urban planners who seek to reintroduce vitality and mixed use into CBDs, by permitting more residential land use. The provision of new apartments with amenities such as gyms and pools, and easy access to work, retail facilities, and entertainment, are attractive to some buyers and investors. Together, the combination of low interest rates, foreign investor demand, wealthy immigrants, and urbanization are driving up residential property prices at high rates and this creates the economic viability to convert older, vacant or partially vacant, office stock into residential land use. This is coupled with the migration of office tenants into the newer Barangaroo stock and the increasing vacancy rates in lower grade stock.

The office market

Australia's biggest office market is in Sydney, and the CBD has the largest portion, with a total of 4,961,728 m^2 (m3property, 2015). Australian office space is categorized using the Property Council of Australia's (PCA) office matrix categories, where premium is the top quality followed by A, B, C, and D grade stock. Over the last few years, demand for office space in Sydney has been high, especially in the CBD. Even after the great financial crisis of 2008, the office market showed growth with overall vacancy decreasing slightly to 9 percent in 2015, from 10 percent in 2011 (CBRE, 2015a). Furthermore, the demand for premium sustainable office buildings is high, and shows a higher value and lower vacancy rates for rated green buildings (Newell et al., 2011). Currently the Sydney office market is performing well and has decreasing vacancy and yields, and increasing absorption rates and capital values, and as a consequence, the office market supply is increasing also.

In 2014, the CBD supply pipeline was 460,000 m^2, 9.27 percent of the current stock. A large proportion of this, approximately 250,000m^2, is the Barangaroo development. Barangaroo is a large area adjoining the CBD previously used for maritime land use, which had become redundant when the maritime activity was relocated. It is highly unusual to get such a large amount of office stock coming to market in such a short space of time. This Barangaroo supply will come onto the CBD office market in 2016 and 2017, and is predicted to lead to movement of existing tenants in the CBD in prime stock (premium and A grade) into this new stock (CBRE, 2015a). In turn, the existing tenants in mid-grade CBD office stock are expected to relocate into the vacated premium stock and in this way, a replacement market develops. A form of relative obsolescence will develop with the result being increased vacancies in the mid- and lower-grade stock (Investa, 2014; Savills, 2015a, b).

Drivers for conversion

Conversion is defined as a change of use adaptation and usually requires major changes of the building. Conversion, as such, contributes to the continued use of historical cities and buildings

237

that are treasured by society; an example is the canal houses in Amsterdam, that were originally constructed in the seventeenth century. Over 400 years, the uses of the buildings have changed numerous times, from warehouse to residential to offices and back to residential and retail, inflicting many changes to the buildings (Leupen, 2006; Remøy, 2010). Several authors (Barlow and Gann, 1993; Beauregard, 2005; Bullen and Love, 2010; Coupland and Marsh, 1998; Heath, 2001; Langston et al., 2008; Remøy and Van der Voordt, 2007a, b, 2014; Tiesdell et al., 1996; Wilkinson et al., 2009) describe similar conversions of vacant office buildings in obsolete urban areas or downtown locations.

Heath (2001) describes office to residential conversions as a successful strategy for inner-city redevelopment in London and Toronto. During the 1990s, the Toronto city core was a mono-functional office district, which was depopulated after six o'clock in the evening. Office construction booms in the late 1980s and an economic recession in the early 1990s resulted in high vacancy rates, rent reductions, and tenants moving to newer accommodation with comparable rents (Barlow and Gann, 1995). Whereas the London planning authority was supportive though not proactive and conversions were mainly market led, the Toronto munici-pality introduced a planning policy to stimulate redevelopments. In Toronto, conversion and redevelopment contributed 9000 additional dwellings to the downtown in the 1990s. By 2000 the office vacancy had fallen back to acceptable rates and the buildings most suitable for residential use had been converted. Drivers for conversions in Toronto and London included demographics and household compositions with changing attitudes and housing demand, and the increased popularity of city-center living. In addition, new use was needed to activate obsolete offices (Heath, 2001). A third and most important driver was the rent gap between offices and housing: in some situations the return on housing was estimated to be 90 percent higher than for commercial property (Barlow and Gann, 1993).

Between 1992 and 1995, the New York downtown vacancy rate was 20 percent, caused by an economic downturn (Barlow and Gann, 1995). Reacting to this development, the New York City government initiated the Lower Manhattan Revitalization Plan to enable and subsidize residential conversion (Beauregard, 2005). Subsidies were given for conversion of office buildings constructed before 1975. The government encouraged conversions into studios and small apartments, targeting first-time renters. The low rents made the apartments popular for other groups as well, although the area lacked basic services and facilities. The most important drivers for conversions were the tight housing market, a high supply of obsolete office buildings, and governmental policy. From 1995 to 2005 more than 60 office buildings were converted, and the number of residential inhabitants in the area grew.

In Tokyo, the office market collapsed in 2002–2003, and oversupply and economic decline were the drivers for conversion. Older offices in secondary locations became obsolete and were converted (Ogawa et al., 2007). As tenancy perspectives for new, large office buildings were still good, redevelopment was generally a more interesting option than conversion. The local government had little control over the urban developments, though recent focus on urban conservation might enhance conversion potential in the future (Minami, 2007).

The drivers for conversion in the Netherlands are similar to the Japanese drivers, as office vacancy has risen since 2002 (Remøy and Van der Voordt, 2007b, 2014). With a fundamental demand for housing, especially in the larger cities, residential conversion has proved successful. Sustainability is mentioned as a driver, although developers focus on the intrinsic sustainability of conversion: few specific sustainability measures are taken (Remøy, 2010). Prolonging the use and lifespan of heritage buildings is another important driver (Remøy, 2014). Keeping and developing the characteristics of a building or an urban area are found to be important for sustainable urban development.

In Australia, although sustainability is a key driver for building adaptation, economic considerations are also very important. Upgrading the existing building stock to improve sustainability and reduce CO_2 emissions before 2020 is a target for the City of Melbourne (Wilkinson and Remøy, 2011) and before 2030 for Sydney. The aim is shared by Perth in Western Australia, where high office vacancy and increased residential construction activity has been another driver for building conversion since 2000 (Bullen, 2007). The governing authorities in many Australian cities seek to encourage sustainability in adaptations to deliver emission reduction targets.

In the described cases, sustainability aims, urban policy, office obsolescence, and a tight housing market were the most important conversion drivers. These relate to political, economic, social, technological, legal, and environmental drivers. Political, economic, and social drivers consider residential conversion as a strategy to introduce housing in CBDs that have historically been mono-functional office locations. Moreover, residential conversion in central urban areas is seen as a possibility for realizing affordable housing in city centers. In large cities, housing affordability in central areas has become problematic for lower income groups and for the middle classes. Technological and economic drivers are most important in cities where the value of residential property is higher than the value of offices. Due to technological and economic changes, and quantitative and qualitative mismatches in demand and supply, several cities have struggled with high office vacancy and obsolete office locations. In these cases, market forces drive residential conversions; conversion is less expensive and faster than demolition and new build, and existing obsolete office buildings occupy central locations. Changes in building acts or legislation can lead to legal obsolescence and are another driver for converting offices into new use. Changes in floor heights and fire escape demand, and increased Energy Performance Certificate (EPC) norms are examples that lead to legal obsolescence. Within use adaptation is a possibility, but conversion for new use is often chosen, especially in locations with a high market demand for housing. Finally, environmental drivers are increasingly important. Office users demand sustainable offices, and older property is left vacant and obsolete. Major adaptation or conversion is needed to accommodate new use.

Barriers for conversion

The barriers for conversion are categorized as political, economic, environmental, social, technological/physical, and legal. One of the obstacles for conversion is the specialized nature of the work and the competence of the actors in the real estate market. Developers and investors work within their own areas of expertise, and may have little understanding of related disciplines (Remøy and Van der Voordt, 2007b). Moreover, the market is sectorial; office investors do not invest in housing and vice versa, and moving from the office to the housing sector is therefore difficult. Socially, the infrastructure to support residential land use may not exist in a former, or predominantly, commercial area (Heath, 2001).

Legislation in the form of zoning plans and building laws is an important conversion barrier. In most countries, the building laws for housing are stricter than those for offices, especially in respect of fire escape, daylight admittance, and energy efficiency (Remøy and Van der Voordt, 2007b). Regulations can require structural alterations be undertaken that lead to higher costs or make conversion physically unfeasible (Bullen, 2007). In existing buildings, deleterious materials, such as asbestos, are a barrier where removal involves compliance with strict health and safety rules as well as incurring high costs (Remøy and Van der Voordt, 2007a).

Another issue arises when the original construction drawings of older office buildings are not correct; although this is not a technical barrier as such, it does make thorough inspection of the structure vital (Remøy and Van der Voordt, 2007a). The main structure or fabric of older buildings may be aged and experiencing decay, for instance the concrete may be deteriorating. Repairs can be costly, and secondary construction may be required. Physically, apartments require more vertical shafts for electricity, water, and plumbing services than offices (Remøy and Van der Voordt, 2007a). In newer European construction, pre-stressed concrete is commonly used, which loses strength when the steel is cut and thus forming voids for services shafts is problematic (Remøy and Van der Voordt, 2007a). Overall, several technical barriers are revealed that threaten the economic performance of the building and the financial feasibility of the project.

Location and building attributes favorable for conversion

Location

Of the attributes that influence the conversion potential of buildings, the location of the property is significant in terms of accessibility and public transport, access to amenities such as food retailing, other retailing, leisure, and entertainment facilities (Wilkinson et al., 2014). Furthermore, access to services such as education, healthcare, and childcare is important for residential stock (Wilkinson et al., 2014).

Building

In addition, the opportunities and risks of conversion are very closely related to the physical characteristics of the existing building (Remøy and Van der Voordt, 2014; Wilkinson and Remøy 2015). Remøy and de Jonge (2007) defined building type characteristics that influence conversion potential, such as structural form and floor structure, façade, type, floor layout, and the length and depth of the building, as well as the number and location of stairs and lifts.

Typically, office buildings have high conversion potential when characterized by a wide span or bay width, with few columns, high floor to ceiling heights, and a high load bearing capacity. Conversely, poor acoustic insulation, high beams, and, in older properties, a dense structural grid reduce conversion potential.

In addition, interventions in the façade lead to substantial costs and reduce the conversion potential and economic feasibility. Features such as cantilevered floors and curtain walling reduce the possibility to add balconies and to subdivide the façade, to accommodate interior partitions and walls. On the other hand, well-maintained façades in good technical condition, and with a dense grid, increase the conversion potential.

It has been found that large floor plates and building depth increase the conversion potential of office buildings (Remøy and de Jonge, 2007). The location of lifts and staircases has a high impact on which layout is possible, because relocating stairs and lifts adds significantly to building costs. A high number of lifts in offices adds to a high conversion potential and elevator shafts can be reused as shafts for HVAC, water supplies, and sewerage. Table 18.1 summarizes the characteristics affecting office to residential conversions.

International comparison

Conversion, driven by various forces, occurs all over the world. In markets worldwide, such as New York, London, Brussels, and the main Dutch cities, the focus on conversion has been high since the 1990s, with increased importance after the global financial crisis, driven by high office

Table 18.1 Summarizing typological characteristics that affect conversion capacity

	Positive	Negative
Structure and floors	• Large floor spans • Columns; free plans • Constructed for heavy carriage	• Dense grids • Low ceilings under existing beams • Thin floors: acoustic insulation insufficient
Façade	• Small grid • Good technical state	• Inadequate technical state, no attachment points for interior walls • Cantilevered floors: complicates adding balconies
Floor layout, length and depth	• Deep buildings	• Location of lifts and staircases
Stairs and elevators	• Excess number of lifts	• Insufficient number of escape routes • Excessive space occupied by cores

vacancies and high housing demand. Although office to residential conversions are still few in number in various CBDs, cities, including Sydney, show an emerging trend in conversion. Through international comparisons between cases in the Netherlands, Belgium, and Australia, some key lessons are identified that are applicable to other market and urban contexts worldwide.

The Netherlands – Rotterdam

"The Admiral" is the name of an office to housing conversion in a central location in Rotterdam, the Netherlands. The office building of more than 30,000 m² was built in 1989, and was vacant for several years. The building was converted into 600 rental studios and small apartments of 20–55 m², with commercial space on the ground floor. Moreover, 400 places for car parking were provided and 700 bike parking places. The project started in 2013, and was completed in 2015. Housing was not in the zoning plan for the area, and it had to be altered before construction could start. Neighbors filed a complaint about the zoning plan change, however, the plan fitted very well with the municipality's aims to attract more young professionals to live in the city center. Hence, the municipality agreed to alter the zoning plan, but set some quality requirements for the housing, and required a reduction in the number of studios provided and an increase in the average size of the apartments. In addition, the appearance of the original façade, was "out-dated" and too "office like" and that was altered also. Furthermore, the fire safety requirements for housing led to substantial unforeseen, additional building costs. The municipality was closely involved in the whole conversion process; they drew up the original agreement, put the zoning change procedure in motion, supported the developer in the environmental permit application, and assisted in obtaining the construction site permit.

Belgium – Brussels

In 2009, the Thon Hotel Group decided to convert three obsolete office buildings in the Quartier Leopold (the EU area) into a hotel and apartments, with some retail space on the ground floor (Laserre et al., 2011). Thon owned all three buildings, which together form a block, and could have decided upon demolition and new build, i.e. to increase the density of the site. However, after analyzing the possibilities, conversion was chosen as it was estimated to take less time, and required lower investment. Demolition and new build would take at least 5 years, whereas the conversion took half the time, just two and a half years; the conversion was completed in 2012. The original façades were removed, and replaced by a new homogeneous façade. To connect the three separate buildings, several technical and structural interventions were necessary. The corridors of the three buildings were connected to each other, several existing staircases and lifts were relocated, new stairs and lifts were added, and the buildings' foundations were reinforced. The plan fitted well within the local government's policy program for urban regeneration, "Region Bruxelles Capitale," that aimed at developing a dynamic and sustainable urban environment. Sustainability in the terms of energy efficiency was important in this project. Heat pumps with a heat recovery system and solar panels on the roof are part of the measures that were taken.

Australia – Sydney

The Gantry project, at 139–143 Parramatta Road, Camperdown, Sydney, was completed in 2013 and comprises a former motor car works and a pottery warehouse dating from 1922 converted into four apartment buildings and 26 terraces located around a large landscaped courtyard. The developer was City Freehold Projects and the contractor Bates Smart. The project integrates new residential buildings to preserve the sites' industrial history. Over 190 residences are provided, including one and two bedroom single story apartments; two and three story terraces and three bedroom penthouses. New residential buildings were integrated within refurbished historic elements with four new five to six story apartment buildings and extensive landscaping over 18,500 m². Environmental and sustainability features include the landscaping elements to humanize the design and introduce local biodiversity, where no permeable ground surface or landscaping had existed. Landscaping increased from zero to 2,180 m² with the deep soil zone to 1,000m² to increase local biodiversity. These landscaped areas are serviced by rainwater collected from roofs and stored in tanks under a paved laneway. All buildings were orientated to maximize solar access and communal courtyards also benefit from good solar aspect. The amount of new materials and the amount of landfill waste were minimized, for example, by reusing brickwork for landscaping and salvaging steel roof trusses as an aesthetic feature (see Table 18.2).

Conclusions

This chapter has focused on evolving practices of conversion in the Sydney CBD in Australia and has examined the prevailing residential and office markets, before exploring the drivers and barriers to successful conversion, key location and building attributes, through a literature review and a comparison of international cases in the Netherlands, Belgium, and Australia.

It has been shown that conversion is well established in several countries. A variety of physical, economic, environmental, legal/regulatory, social, and political factors influence and impact the degree of conversion. Sustainability is an important driver for conversion, though mostly as an intrinsic value; conversions have the image of being sustainable. Still, few conversions are carried out that adopt specific sustainability measures. The most important drivers are related to the market and location, whereas building costs and estimated risks of conversion are the most important barriers.

Table 18.2 Key criterion for conversion compared

Criteria	Netherlands	Belgium	Australia
Location	Central Rotterdam	Central Brussels	Camperdown, Sydney
Land use type (original)	Office	Office	Industrial (motor car works and pottery warehouse)
Land use type (after conversion)	Residential with ground floor commercial	Residential and hotel	Residential
Size of building	30,000m²	50,000m²	18,500 m²
Number of floors	14	8	5 to 6
Year of construction	1989	1960–1989	1922
Accommodation provided post conversion	600 studios and small apartments (20–55m²) 400 car parking and 700 bicycle parking spaces 2000 m² commercial space	31 apartments Hotel with 405 rooms 1200 m² retail 200 car parking spaces	190 residences (including 1, 2 bed single story apartments, 2 and 3 story terraces and 3 bed penthouses)
Planning issues	Rezoning required for site	None, according to local policy	Heritage
Other regulation issues	Fire safety for residential	Fire safety for hotel and residential	
Cost issues	Extra unforeseen costs associated with regulation and compliance	Lower investment and shorter construction time than demolition and new build	
Aesthetic issues	Out-dated and façade had an "office" look, not necessarily suited to housing	Out-dated office façades on 3 buildings were adapted to one new exterior	Introducing permeable ground surfaces to site to increase biodiversity and amenity
Drivers for conversion	Economic – high vacancy and low demand Regeneration of the location Municipality stimulating housing development in the city centre	Economic – high vacancy, high demand for other functions Regeneration of the location	Economic – accommodating increased population Regeneration of the location Maximize solar access to buildings Communal gardens Reduce new materials use and landfill waste Reused brickwork for landscaping and roof trusses retained
Barriers for conversion	Regulatory, zoning plan, fire safety requirements	Technical, increasing load bearing capacity of the foundations, modifying and adding stairs and lifts	

Source: Authors

The drivers for the Sydney market vary in scope and degree, for example, investment from China is having a major influence on values of residential property at the time of writing. To a lesser extent, planning and regulations are also driving residential conversion. Interestingly, there is a view that some buildings lend themselves easily to conversion and this is a driver. Financial considerations and risks were found to be the biggest barrier, along with aesthetics, planning, contamination, and technical issues. The Sydney CBD is a strong market within Australia, and different degrees of residential conversion, drivers, and barriers exist in other regional cities in New South Wales and other major State capitals within Australia.

The three case studies have similar drivers to those found in the literature review. Looking specifically at sustainability measures, no new information was found. The improvements in operational energy efficiency achieved when converting a building using the standards of the local building codes is seen as a significant improvement. Although these are not recognized as specific sustainability measures as such, conversion is a means to preserve embodied energy. Moreover, the retention of the existing building structure, fabric, and materials inherent in conversion are explained as sustainable. The lessons learned from literature and case studies are applicable to a large degree in the Sydney market, although the level of activity here is much lower and less developed than in some European cities. There are lessons from the European approach to conversions for Australian practice to learn, but also lessons for students and researchers about the applicability of research results from one case study in another context.

Knowing that 87 percent of the buildings we will have in 2050 are already built, conversions should be considered widely as a more important part of real estate development for a sustainable future. Conversions, to a large extent, are comparable to new construction. However, this chapter has shown that some specific aspects need extra attention, as conversion projects always deal with an existing location and building structure. Related to that, cost issues are more complex than for new construction. Furthermore, building law and planning regulations are often directed towards new construction. These aspects, together with the financial and environmental value of conversion, need to be studied more in detail to embed conversion as part of sustainable real estate practice.

References

Barlow, J. and Gann, D. (1993). *Offices into flats*. York: Joseph Rowntree Foundation.

Barlow, J. and Gann, D. (1995). Flexible planning and flexible buildings: Reusing redundant office space. *Journal of Urban Affairs*, *17*(3), 263–276.

Beauregard, R. A. (2005). The textures of property markets: Downtown housing and office conversions in New York City. *Urban Studies*, *42*(13), 2431–2445.

Bullen, P. A. (2007). Adaptive reuse and sustainability of commercial buildings. *Facilities*, *25*(1/2), 20–31.

Bullen, P. A. and Love, P. E. D. (2010). The rhetoric of adaptive reuse or reality of demolition: Views from the field. *Cities*, *27*(4), 215–224.

CBRE (2015a). Viewpoint residential conversion plays: What do they mean for Sydney suburban office markets? Retrieved December 22, 2015 from: www.cbre.com.au/o/sydney/pages/market-reports.aspx

CBRE (2015b). Viewpoint Sydney CBD urban renewal projects. Retrieved December 22, 2015 from: www.cbre.com.au/o/sydney/pages/market-reports.aspx

Coupland, A. and Marsh, C. (1998). The cutting edge 1998: The conversion of redundant office space to residential use. RICS Research: University of Westminster.

City of Sydney (2010). Sydney growth centres strategic assessment program report. Retrieved January 19, 2015 from: www.environment.gov.au/.../sydney-growth-centres-program-report.pdf

EC Harris (2014). Office to residential convert or redevelop? Retrieved January 30, 2015 from: www.echarris.com/pdf/8502_Office%20to%20Residential%20Report_Final.pdf

Geraedts, R. P. and Van der Voordt, D. J. M. (2007) A tool to measure opportunities and risks of converting empty offices into dwellings. In: Boelhouwer, P., Groetelaars, D., Ouwehand, A., and Vogels, E. (Eds) *Conference Proceedings: ENHR International Conference on Sustainable Urban Areas, Rotterdam.*

Heath, T. (2001). Adaptive re-use of offices for residential use: The experiences of London and Toronto. *Cities, 18*(3), 173–184.

Investa (2014). Investa quarterly office report, Q1 2014. Retrieved February 19, 2015 from: www.investa.com.au/quarterly- report-q1-2014-1/

JLL (2014). Sydney CBD office conversions: Manhattan – a template for Sydney? Retrieved February 19, 2015 from: www.jll.com.au/australia/en-au/Research/Pulse%20-%20Sydney%20Office%20Conversions%20May%202014.pdf

Kelly, M. (2008). Britain's building stock: A carbon challenge. Retrieved April 20, 2017 from: www.lcmp.eng.cam.ac.uk/wp-content/uploads/081012_kelly.pdf

Langston, C. (2014). Identifying adaptive reuse potential. In: Wilkinson, S. J., Remøy, H., and Langston, C. (Eds). *Sustainable Building Adaptation.* Oxford: Wiley, pp. 187–206.

Langston, C., Wong, F. K. W., Hui, E., and Shen, L. Y. (2008). Strategic assessment of building adaptive reuse opportunities in Hong Kong. *Building and Environment, 43*(10), 1709–1718.

Laserre, C., Lconte, P., Böhlke, A. and Dooreman, B. (2011). *Gisteren kantoren, vandaag woningen: De conversie van kantoorgebouwen in het Brussels Hoofdstedelijk Gewest.* Brussels: Brussels Hoofdstedelijk Gewest.

Leupen, B. (2006). *Frame and generic space.* Rotterdam: 010 Publishers.

m3property (2015). Commentary, Sydney CBD office market. Retrieved September 30, 2015 from: www.m3property.com.au/wp-content/uploads/2015/10/Sydney-CBD-Office-market-September-2015.pdf

Minami, K. (2007). A study of the urban tissue design for reorganizing urban environments. In *BSA 2007 conference proceedings.* Tokyo: Tokyo Metropolitan University.

Newell, G., MacFarlene, J., and Kok, N. (2011). *Building better returns.* Sydney: Australian Property Institute and Property Funds Association.

Ogawa, H., Kobayashi, K., Sunaga, N., Mitamura, T., Kinoshita, A., Sawada, S., and Matsumoto, S. (2007). A study on the architectural conversion from office to residential facilities through three case studies in Tokyo. In *BSA 2007 conference proceedings.* Tokyo: Tokyo Metropolitan University.

Remøy, H. (2010). *Out of office: a study of the cause of office vacancy and transformation as a means to cope and prevent.* Amsterdam: IOS.

Remøy, H. (2014). Preserving cultural and heritage value. In: Wilkinson, S. J., Remøy, H., and Langston, C. (Eds). *Sustainable Building Adaptation.* Oxford: Wiley, pp. 159–181.

Remøy, H. and de Jonge, H. (2007). Transformation and typology: vacancy, characteristics and conversion-capacity. In *BSA 2007 conference proceedings.* Tokyo: Tokyo Metropolitan University.

Remøy, H. and Van der Voordt, D. J. M. (2007a). Conversion of office buildings: A cross-case analysis. Paper presented at the BSA 2007, Tokyo.

Remøy, H. and Van der Voordt, D. J. M. (2007b). A new life: Conversion of vacant office buildings into housing. *Facilities, 25*(3/4), 88–103.

Remøy, H. and Van der Voordt, D. J. M. (2014). Adaptive reuse of office buildings: Opportunities and risks of conversion into housing. *Building Research and Information, (42)*3, 381–390.

Savills, (2015a). Sydney office quarter times. Retrieved May 24, 2015 from: http://pdf.savills.asia/asia-pacific-research/australian-research/australia-office-/savillsresearch-quarter-times-sydney-office-q1-2015.pdf

Savills, (2015b). Spotlight Sydney residential January 2015. Retrieved January 24, 2015 from: www.savills.com.au/research/australian-research/residential-market.aspx

Tiesdell, S., Oc, T., and Heath, T. (1996). *Revitalizing historic urban quarters.* Oxford: Architectural Press.

United Nations Environment Programme (UNEP) (2009). Buildings and climate change: Summary for decision-makers. Retrieved April 1, 2015 from: www.unep.org/sbci/pdfs/SBCI-BCCSummary.pdf

Wilkinson, S. J. and Remøy, H. (2011). Sustainability and within use office building adaptations: A comparison of Dutch and Australian practices. Paper presented at the Pacific Rim Real Estate Society, Gold Coast.

Wilkinson, S. J. and Remøy, H. (2015). Building resilience in urban settlements through conversion adaptation. In: *RICS Cobra Aubea 2015 Conference Proceedings,* Sydney 8–10 July 2015.

Wilkinson, S. J., James, K., and Reed, R. (2009). Using building adaptation to deliver sustainability in Australia. *Structural Survey*, *27*(1), 46–61.

Wilkinson, S. J., Remøy, H., and Langston, C. (2014). *Sustainable Building Adaptation*. Oxford: Wiley.

19

Sustainability of office buildings

Lessons learned from academic and professional research

Sofia Dermisi

Abstract

Sustainability has been pursued by owners of office buildings with varying frequency since the 1990s, as owners and managers seek improvements in capital and operational costs, return on investment (ROI), retrofit length, and tenant attractiveness/comfort, among a variety of building specific benefits. The millennial generation of workers seems to be more socially and sustainability conscious than generation X and the baby boomers, therefore further solidifying the adoption of sustainable practices in the workplace. A building's sustainability performance can be benchmarked based on recent building codes in certain communities or assessed and certified on a more holistic basis by private rating systems certifying buildings worldwide. Prominent holistic building sustainability assessment systems with worldwide outreach include: Leadership in Energy and Environmental Design (LEED) by the US Green Building Council (USGBC) – based in and dominating the USA market; Building Research Establishment Environmental Assessment Method (BREEAM) – based in and dominating the UK; and Green Globes – based in Canada and the USA.

The adoption of sustainable attributes and certifications among office buildings may be viewed through the lens of quantitative and qualitative outcomes of worldwide research, with a special focus on the construction budget impact. Since 2010, academic and professional research has tried to shed light on the adoption of sustainable practices among key office building performance indicators. The top drivers of green building are client demand, financial returns, environmental regulations, and occupant wellbeing. A five- to eight-year payback period is expected for green property investment with sale premiums ranging from 0% to 30% and rents 0% to 17.3%. Construction cost premiums can vary significantly with a range from −0.4% to 12.5% with the location and rating level pursued influencing the cost. In addition to the rationale for the adoption of sustainable certifications and the evidence of the success of the modifications, there is also a list of issues that the developer should explore before proceeding.

Introduction

The United Nations projects an increase of the current worldwide population from 7.3 billion to 8.5 billion by 2030 and 9.7 billion by 2050.[1] Additionally, urbanization is projected to grow to 66% from the current 54%, by 2050.[2] These trends suggest the need for additional construction and continuous repurposing of existing buildings, to accommodate the growing pressures, especially in urban areas. Even though this chapter focuses on office buildings the projected population growth requires development across all uses (e.g. residential, retail, industrial, etc.). Determining the global building stock is a daunting task, however, Navigant Research estimates its size at 151.8 billion square meters in 2014 with a projected increase of 13% (171.6 sq. m.) in 2024.[3] Although the size of the building stock can be indicative of economic activity, there is also an environmental footprint, which cannot be ignored. The US Environmental Protection Agency (EPA) estimated in a 2009 report that buildings consume 13% of the total water and 39% of total energy in the USA, while generating 38% of the carbon dioxide emissions.[4] Exploring these consumptions by major real estate use, residential buildings consumed larger amounts of water (74%), energy (54%), and produced more CO_2 (21%) compared to the commercial buildings (which experienced 26%, 46%, and 18% respectively). Furthermore, in 2014 alone the commercial real estate sector consumed 20% of US energy consumption, with an increase being predicted to 27% between 2012 and 2040 (CBRE, 2015). These trends underscore the need to address the environmental impact between both uses even though this chapter is focusing on office buildings.

In order to protect and sustain the environment for future generations there is a need to balance the profit of companies with their social (people) and environmental (planet) responsibility. These are the three Ps of sustainability that John Elkington highlighted in his famous quoted phrase of "the triple bottom line" in 1994.[5] Sustainability can be achieved in existing and new structures through a combination of various actions and efficient equipment. The US National Institute of Building Services – Whole Building Design Guide (WBDG)[6] highlights six critical elements in any sustainable design, consisting of the optimization of: (1) the site potential, (2) energy use, (3) water use, (4) building space and material use, (5) operational and maintenance practices, and (6) enhancement of the indoor environmental quality. These elements are also guiding principles for the sustainable US federal buildings, with the exception of the site potential.[7] Beyond the aforementioned elements, a developer and/or owner can view a building before its construction as a four-stage process (life cycle), which includes: materials selection/manufacturing, construction, use/maintenance, and end of life. Bayer et al. (2010) suggest that applying a Life Cycle Assessment (LCA) before construction improves the sustainability of a building and the payback period, because of the improved decision-making process during all four stages of a building. Most recently, LCA was adopted by Leadership in Energy and Environmental Design (LEED) v4, which is one of the prominent sustainability certifications.

Prominent sustainable certifications

Worldwide there are many certifications acknowledging a building's sustainability performance based on a comprehensive assessment of multiple building and area characteristics. The most widely used globally are: Building Research Establishment Environmental Assessment Method (BREEAM), based in the UK; Leadership in Energy and Environmental Design (LEED), based in the USA; and Green Globes, based initially in Canada, which is receiving increasing interest. BREEAM presence is significant with more than 535,967 certificates issued and 2.2 million buildings registered for assessment in 72 countries.[8] BREEAM is the most prominent sustainability certification in Europe, accounting for more than 80% of the sustainable commercial buildings (office, retail, logistics, hotels, etc.) (RICS, 2013). The US Green Building Council, which developed LEED, reports that more than 13.8 billion square feet are LEED certified in the USA, and more than 72,500 certified buildings exist worldwide in more than 150 countries as of August 2015.[9] CBRE's National Green Building Adoption Index trends also show the increasing support for LEED certification among office buildings in the USA,[10] with the adoption of a sustainable certification (Energy Star label or LEED) being more prominent among buildings of more than 250,000 square feet for the ten largest markets versus all other buildings. Finally, Green Globes has nearly 4,000 buildings certified across Canada and the USA.[11]

Although buildings typically obtain one certification, there are cases where buildings pursue multiple. For example, the Crystal in London has attained the highest certification level under LEED (Platinum) and BREEAM (Outstanding).[12] The building is 6,300 square meters and among the innovations adopted is the use of renewable energy (geothermal, photovoltaic, and wind) rather than fossil resources (e.g. oil or gas). It also has a natural ventilation system and triple-glazed windows. Rainwater, graywater, and even black water are purified in the building.

Building Research Establishment Environmental Assessment Method (BREEAM)

BREEAM (Building Research Establishment Environmental Assessment Method) is a building's sustainability assessment tool offered by the Building Research Establishment (BRE), with four assessment schemes, of which three are relevant to office buildings.[13] BREEAM was launched in 1988 with the intention to: mitigate the life cycle impacts of buildings on the environment; enable buildings to be recognized according to their environmental benefits; provide a credible, environmental label for buildings; and stimulate demand for sustainable buildings (Arup, 2014). A survey of construction and other industry professionals (Parker, 2012) has shown that the main driver of BREEAM in the UK has been local planning authority requirements followed by the policies of organizations that might adopt certain requirements in their procurement strategies. The key benefits highlighted in the survey included the social aspect (building recognition), reduction in construction waste and material use as well as improved occupant satisfaction and operation cost savings.

Leadership in Energy and Environmental Design (LEED)

LEED (Leadership in Energy and Environmental Design) was developed by the US Green Building Council (USGBC) in 1993, based on BREEAM. There are currently five ratings, of which three are the most relevant to office buildings.[14] Each of these ratings has a set of prerequisites the project is required to meet to be considered for LEED certification. Both the private and public sectors have recognized LEED as a sustainability standard for buildings, including the US federal government.[15] USGBC argues that LEED properties can differentiate themselves in the marketplace because of the healthier/happier occupants, reduced utility costs and improved occupancies, even though the rents are higher than in non-LEED buildings.[16] Another key driver of LEED is the increasing adoption of various sustainable practices by municipal building codes, including the requirement of an increasing number of communities requiring all the new public buildings to be LEED certified (e.g. San Francisco, Atlanta[17]).

Green Globes

Green Globes was developed in Canada (2000), based on BREEAM as a sustainability assessment tool for buildings.[18] BOMACanada has the Canadian license for Existing Buildings and Green Building Initiative (GBI) for both new and existing buildings in the USA, with GBI being the first organization accredited by the American National Standards Institute (ANSI). Green Globes is increasing its popularity because of the streamlined certification process and the much lower cost than LEED. In 2013, the US federal government determined that either Green Globes or LEED can be used for the certification of government buildings.

Differences between sustainable certifications

A number of differences exist between BREEAM, LEED, and Green Globes with the most notable shown in Table 19.1. Additional differences include:

1   BREEAM's focus on the type of building (use) versus LEED which focuses on status – new or existing (CBRE and EMEA, 2009);
2   BREEAM's passing score for buildings complying with area regulations, even though regulations differ among the various countries. In contrast LEED does not provide such an accommodation;
3   BREEAM's minimum standards, in contrast to LEED, which require buildings to meet certain prerequisites before the certification process begins;
4   BREEAM's location sensitivity, which affects the assessment of a building's environmental and economic conditions more than LEED;
5   LEED's requirement for recyclable materials in contrast to BREEAM which does not make it compulsory;
6   BREEAM's requirement for an assessor to be trained by BRE, in contrast to LEED which adds a credit if a LEED AP is used on a project;
7   BREEAM's ability to compare different buildings in contrast to LEED.[19]

Some major differences between LEED and Green Globes (GG) include:

1   GG's use of an ANSI process in assessing a building versus LEED which does not;
2   GG's use of an interactive program delivery versus a template submission in LEED;
3   GG's lack of prerequisites in contrast to LEED;
4   GG's use of the Life Cycle Assessment (LCA) process, which has just begun in LEED v4;
5   GG's allowance of multiple forest certification versus LEED;
6   GG's certification cost which is significantly less than LEED.[20]

Business case for sustainable buildings

Maximizing the returns from the construction of a building is critical in the success and longevity of any development venture. Therefore, addressing consumption levels (energy and water) and features that can improve the performance of a building are important in raising the value of the delivered asset. Even though residential buildings have a larger environmental footprint compared to commercial, embracing sustainability can significantly improve a building's performance regardless of use. Focusing on office buildings, which represent the majority of commercial buildings, a number of scholars, professionals, and agencies have compared green versus non-green building performance, shedding light on the differences.

Table 19.1 Credential differences

	LEED	BREEAM	Green Globes
Country based	USA	United Kingdom	Canada (initially) USA (since 2004)
Year	1993	1988	2004
Current versions	LEED 2009 and v4	BREEAM International 2013	
Certifications (relevant for office buildings)	Building design and construction Building operations and maintenance Interior design and construction	New construction, Refurbishment and fit-out, In-use international	New construction/significant renovations Commercial interiors Existing buildings
Levels	Certified (40–49 points) Silver (50–59) Gold (60–79) Platinum (80–110)	Unclassified Pass (30–44%) Good (45–54%) Very good (55–69%) Excellent (70–84%) Outstanding (>84%)	1 Globe 2 Globe 3 Globe 4 Globe
Categories	Energy and atmosphere Indoor environmental quality Water efficiency Materials & resources Sustainable sites Regional priority Integrative process Location and transportation Innovation[a] and design process	Energy Health and wellbeing Water Materials Land use and ecology Pollution Management Transport Waste Total Innovation credits[b]	Energy Indoor environment Water Materials & resources Site Emissions Project management

Notes: [a]: LEED innovation credits are awarded under one of three options offered: www.usgbc.org/node/2613903?return=/credits/new-construction/v4/innovation
[b]: BREEAM innovation credits are awarded when: a) a building has exemplary performance within existing guidelines (all assessment areas have exemplary criteria), and b) an application is made by the project's assessor for a technology applied, feature, design/construction method or process to be recognized as innovative.

Performance of green versus non-green buildings

Although scholars, certification organizations (USGBC, BREEAM, etc.), and government agencies (e.g. US Environmental Protection Agency, US General Services Administration (GSA), etc.) increasingly provide evidence of the benefits achieved from the embracing of sustainable practices in the built environment, delays in adoption of such practices exist. Three reasons for these delays stand out in several notable research studies.

First, cost and complexity of the process are cited by many managers and owners as their primary reason for not beginning the certification. Marker, Mason, and Morrow (2014) emphasize the top reasons for such delays, which include real and perceived costs, as well as process logistics and paperwork. Depending on the owner's objectives in owning the building, the large investment in retrofits and the extended payback period may influence the decision. The paperwork and the documentation required for certification are hugely time-consuming and many buildings simply do not have the number of employees necessary to undertake the process, which often requires many months to complete.

Second, as is usually the case in the commercial building marketplace, the location of sustainable buildings within an area often has a significant impact on the decision to seek sustainability certification. For example, Braun and Bienert (2015) argue that commercial green buildings are more likely to be found in prime versus non-prime locations within a Metropolitan Statistical Area (MSA). Dermisi (2014) also found that LEED buildings tend to be located closer to each other by 21%, with differentiation existing between the ratings.

Third, easy and close access to mass transit stations encourages owners and managers to initiate the certification process. Transportation is a key attraction because of its importance in a building's scoring. In addition to the desirability to tenants and residents of buildings near public transportation, sustainability and transportation are a marketing duo that decreases vacancies.

It is important to note as Fuerst et al. (2012) suggest, that sustainable class A office space is becoming less unique, since it accounted for almost half of the US transactions from 2007 through 2012. The increasing adoption of sustainable practices and certifications is highlighted in a recent survey of more than 1000 participants from 69 countries (Dodge Data, 2016). This survey suggested that commercial construction, which has been the premier adoption sector of green practices, is now closely followed by institutional construction.

The two top drivers of green building worldwide are client demand (increasing to 40% from 35% in 2012), environmental regulations (increasing to 35% from 23% in 2012), while market demand followed (although slightly decreasing to 30% from 33% in 2012). The top social reason for building green was the encouragement of sustainable business practices especially in the US (74%) and UK (72%). Another survey of US executives (Turner, 2015) suggested that the most important reasons for the adoption of green features are financial (energy efficiency, asking rents, ongoing operations and maintenance costs, and occupancy rates) as well as non-financial (occupant wellbeing, indoor air quality, employee productivity) with a five-year or more payback period being acceptable by the majority. The Dodge (2016) survey estimated a payback period for green investments of eight years with a 14% decrease in operating costs over five years and an 8% increase in building value over non-green projects.

Dermisi's (2013) survey of class A office buildings achieving LEED certification showed an average retrofit cost of $0.21/sf and an average 1.8-year payback period. Aggregation of relevant research on sustainable versus non-sustainable buildings by the World Green Building Council (2013) suggested that the rewards of a sustainable building (asset value, operating efficiencies, workplace productivity, and health as well as risk mitigation) outperform the assumed costs (including the cost of construction). Key academic studies comparing LEED and non-LEED office buildings identified significant sales and rental premiums, as well as lower operating expenses for LEED buildings. For example, Eichholtz et al. (2013) found a sale premium of 13% and Fuerst and McAllister (2011) 25% for LEED buildings. The World Green Building Council (2013) reports sale price premiums across the various sustainable designations ranging from 0% to 30%, and rent premiums ranging from 0% to 17.3%. Kok et al. (2012) found a rental premium of 7.1%, and Fuerst and McAllister (2011) a 5% premium. Wiley et al. (2010) found a much higher premium of 15–17% and Dermisi (2013) a 30% premium among certified class A buildings, with differentiations existing among certification levels (e.g. Gold 17% and Silver 16.7%). Similar to the results of Reichardt et al. (2012), Devine and Kok's (2015) analysis of US leases suggested that sustainable buildings achieve 3–4% higher rents with LEED buildings achieving occupancy rates 4% higher compared to non-LEED. Their results also suggest that lease renewal is significantly higher among LEED buildings with rent concessions leading to a smaller reduction of average rent. A more focused study by Dermisi (2013) on class A buildings found that certified buildings achieve 18.8% lower vacancy.

Finally, looking at the differences in operating expenses between sustainable and non-sustainable buildings, Reichardt (2014) finds that LEED buildings command lower operating expenses in contrast to buildings with only the Energy Star label, which experience higher operating expenses. A survey of US General Services Administration (GSA) buildings found a 19% decrease in operating costs for LEED vs non-LEED buildings (GSA, 2011). Dermisi (2013) found that operating expenses decreased on average by 8.09%, while energy costs decreased by 7.02% among class A sustainable buildings. A survey of GSA buildings (GSA, 2011) found energy savings of 25% for LEED buildings, while the World Green Building Council (2013) reports that energy use among LEED buildings can decrease by 25–30%. Energy savings have a direct effect on a building's budget allowing for the increase of the Net Operating Income (NOI) when significant savings are present. IMT et al. (2015) highlight the effect cost savings have on a building's value (Table 19.2).

Beyond the direct monetary effects of sustainable buildings, the World Green Building Council (2013) acknowledges the productivity and health benefits to the occupiers, with shorter hospital stays (8.5%), improved mental function and memory (10% to 25%), increased sales (15% to 40%), and higher productivity (18%). Miller et al. (2009) provide some more in-depth evidence of the improved productivity, especially for those who moved to sustainable buildings (either LEED or Energy Star). A survey of focus groups in green office buildings of four major US cities identifies the importance of green attributes over the economic and environmental impact from these attributes (Simons et al., 2014). Additionally, a CBRE survey

Table 19.2 Effect of energy savings on building value per square foot

Energy Saved	Savings per square foot	Value per Square foot
Baseline	–	–
5%	$0.13	$2.38
10%	$0.25	$4.76
15%	$0.38	$7.14
25%	$0.63	$11.90

Notes: Assumptions: Income capitalization approach to value is used and the owner is responsible for utilities. The energy baseline is $3.50/sf/year and the cap rate 5.25%.

found higher employee satisfaction among tenant managers of sustainable buildings, with executives finding it easier to recruit in sustainable versus non-green buildings (CBRE, 2011).

Sustainable construction costs of green vs non-green

A critical component of any development is the project's cost and a key question developers must ask is: Is there a construction premium for sustainable versus non-sustainable labeled buildings? There are a variety of responses to this question. For example, Morris and Matthiessen's (2007) comparison of green vs non-green construction does not find any significant differences. Aggregated research from various worldwide sustainable designations from 2000-2012 finds cost premiums to be within the −0.4% to 12.5% range, with the highest values experienced for LEED Platinum, BREEAM Excellent or zero carbon projects (World Green Building Council (2013)).[21] The Green Building Council (2013) estimates the cost differentiation between sustainable and non-sustainable construction to the lower end of the previous bracket, between 0–4%, with developers being able to mitigate it even further (e.g. early budget development, construction team with sustainability experience, use of Integrated Design Process (IDP) and economies of scale which have led to product cost decreases caused by both increase of sustainable buildings and embracing of sustainability by building codes). Another study of LEED and BREEAM buildings by CBRE and EMEA (2009) revealed a 2–3% construction premium for a basic certification and building certified at higher levels experiencing a 5–7.5% construction cost increase. A survey of construction and other industry professionals on BREEAM adoption premiums in the UK found that for less than half of the group surveyed the costs incurred were significant, however these costs were viewed by the majority as recoverable through more efficient operation of the building in the long term. The group experienced costs premiums from 1–20% with the median being 7.5% (Parker, 2012). In contrast, Yetunde et al. (2014) suggested that developers invest only a maximum of 2% to achieve a higher BREEAM rating, while lower ratings could be achieved with no or minimal additional cost. The same study finds that even though developers might need to spend more upfront for a higher rating their payback period is 2–5 years based on their energy and water costs.

Justification of a developer's pursuit of sustainability

Reasons a developer should adopt sustainable practices and/or certification

The World Green Building Council (2013) provides a visual argument behind the reasons developers and owners pursue sustainable features and certifications for their buildings. Although each group has multiple reasons for such pursuits common reasons among all three are (WGBC, 2013): lower refurbishment costs, corporate image, compliance with legislation and CSR requirements and lower transaction fees. Combining the perspectives of WGBC (2014), Arup (2014), USGBC,[22] the author's experience, and the opinions of numerous managers and owners, a more comprehensive assessment can be offered on the reasons developers would be interested in adopting sustainability:

- **Legislative requirements**: an increasing number of communities/cities are adopting and mandating higher sustainability standards through their building codes, which creates an alignment with sustainable certifications options.
- **Investor, owner, tenant mission/vision alignment and corporate social responsibility**: In recent years, sustainability has been embraced by real estate and other

corporations as part of their mission statement with a number of them as well as government agencies mandating sustainable practices.[23] These actions create a prerequisite for the space they occupy and owners/developers pursuing these tenants will need to adopt sustainable practices in their buildings to maintain their appeal. Corporations are also increasingly focusing on corporate social responsibility, with one of the many aspects being their employee experience in the buildings they occupy in the form of occupant health and wellbeing benefits. Corporate responsibility is also focusing efforts to decrease pollution through a sustainable footprint and help in protecting the environment.

- **Building economics**: Adoption of sustainable systems improves the energy and water consumption of a building, allowing improved performance through the commissioning process. The use of efficient equipment and an environmentally friendly design allows for an increase in the equipment life and the continuous building system monitoring coupled with an expedited identification of failures, leaks, etc.
- **Market dynamics**: Real estate trends highlighted at length earlier suggest that green properties attain improved vacancies, rents, and sales prices over their comparable non-sustainable buildings.
- **Financial, permitting, tax and other incentives**: Certain lenders offer loans with more favorable terms for sustainable construction and renovation due to permitting and other efficiencies. Correspondingly, an increasing number of communities offer incentives in the form of zoning exemptions, expedited city approval process for the adoption of sustainable practices. Additional incentives and rebates are also offered by utility companies/US states on energy and water efficient equipment or fixtures, in addition to tax benefits.
- **Risk management**: Sustainability ratings require building features that enhance commissioning requirements, establish handover documentation requirements for proper equipment use and monitoring, decrease reliance on conventional energy use, and incorporate passive design elements.
- **Effect of millennials**: The millennial generation, which just surpassed the baby boomers in number, puts a premium on sustainability in their workplace and beyond and is actively seeking companies which share the same values.[24] For a tenant to attract such a workforce in a development, they need to be in a sustainable building, which offers open-collaborative spaces, walkable and appealing internal and external surroundings, close proximity to mass transit, and even wellness programs.

Issues a developer should explore before deciding to build a sustainable structure

Arup (2014) provides a framework for the identification of a primary and secondary sustainability certification systems and ratings. A project may determine which ones to adopt by answering five questions in advance:

1 Where is the project to be located?
2 What legislative requirements exist in this location?
3 What are the local market dynamics/expectations for new construction?
4 Who are the prospective tenants (names or types) and what requirements have they identified for sustainability certification?
5 Who is investing in this building?

Depending on the answers to the questions, the primary system can be then selected by exploring further answers such as the level of legislative sustainability requirements (e.g. building codes, etc.), the actual expectations of the owner, tenants and the tangible benefits by the adoption of sustainable practices at a higher versus a lower level of certification. If the assessment of a secondary certification system is required the expectations need to be determined similar to the primary system with an additional emphasis on the cost–benefit analysis of this system.

Beyond the questions offered in the Arup study there are some additional issues a developer needs to explore in advance of making the decision to build a sustainable project:

1   Existence of financial incentives and grants:
  1.1   Tax incentives: tax inducements can take multiple forms such as income tax (e.g. Green Building Tax Credit,[25] Sustainable Building Tax Credit[26]), property tax[27] (e.g. Alternative Energy Improvements[28]), corporate tax (e.g. Energy-efficient commercial business deduction – U.S. Tax code 179D, Business energy investment tax credit – U.S. Tax code 48, Qualified reuse and recycling property – U.S. Tax code 168m, Renewable electricity production credit – U.S. Tax code 45), city tax exemptions.
  1.2   Lower permit fees: communities are increasingly offering incentives such as lower fees (e.g. City of Chicago,[29] Charlotte – Mecklenburg County, North Carolina[30]).
  1.3   Access to financing with better terms (Green community initiative – Seattle).
  1.4   Utility rate reduction: websites such as the Database of State Incentives for Renewables and Efficiency offer information on the various monetary initiatives developers can utilize for a sustainable project.[31]
  1.5   Other monetary initiatives: such as certification fee reimbursement by city are already offered and will in all likelihood expand in the future.[32]
2   Regulatory environment:
  2.1   Building codes: adoption of ASHRAE 90.1 and 189.1 have a significant effect on a project. Cities are increasingly requiring all new public or publicly funded constructions to be LEED certified based on specific square feet or investment size (e.g. Atlanta – ordinance #03-0-1693 and Chicago).
  2.2   Renewable energy permits: certain areas require such installations (e.g. the US Department of Interior – Bureau of Land Management offers a list of resources for Geothermal, Solar and Wind for multiple US states).
  2.3   Bonus density: some communities allow for an increase of the Floor to Area Ratio (FAR) if the building is sustainable (e.g. Chicago building code: 17-4-1015, Seattle[33]).
  2.4   Expedited permitting: certain communities offer a more streamline and expedited issuing of permits if the project is sustainable (e.g. Dallas – Ordinance 27131).
3   True market value of a sustainable project in a specific market:
  3.1   Premium offered in a sustainable project: identification of the value (higher occupancy, rents, etc.) assigned by a market (tenants, buyers) for sustainable versus non-sustainable buildings and assessment of the potential construction cost versus the reward (premium).
  3.2   Green leases: explore if green leases are offered in the area and the ways area tenants participate in the building's performance and savings (e.g. energy, water use) returns.
  3.3   Insurance costs: explore the difference in insurance costs between sustainable and non-sustainable buildings in the area.
4.   Developer's objective/plans for the property:
  4.1   Development's objective: how green should the project be? How low should the energy costs, GHG emissions and water use be? Is the developer interested in a building's life cycle analysis? Is the developer targeting specific users and should the project cater to their needs?

4.2 Holding period: the application of certain sustainable attributes to a building requires longer payback periods and if a developer is interested in a short holding period the costs might overshadow the short-term profit.

4.3 Site and scale: is the developer looking for a site promoting sustainability or the site is already selected and the building needs to be designed sustainably? What is the scale of the building (one or multiple-campus) and current status (new or existing).

Conclusions

Academic and professional research on the reasons behind the adoption of sustainable practices and certifications among office buildings show significant financial (increased rents, lower vacancies, etc.) and non-financial (occupant wellbeing, employee productivity, etc.) benefits for certificated buildings. Construction costs for sustainably certified buildings, however, may have a sizable premium, of up to 12.5%, requiring the developer to carefully assess if a sustainable building will be cost effective. If the answer is "yes" then a decision will need to be made on the type of certification and rating. Developers should be aware of the evidence that sustainability might not be valued (increased rent, occupancy, etc.) everywhere (downtown area versus suburban and other areas) in the same way.

Beyond presenting current information on the benefits of sustainable office buildings, this chapter highlights certain elements a developer needs to assess to determine if such a construction can be cost effective and profitable at a specific site. These elements include the existence of financial incentives, regulatory environment, true market value of a sustainable project in an area, and the developer's objectives.

Notes

1 UN, World population projected to reach 9.7 billion by 2050, July 29, 2015, www.un.org/development/desa/en/news/population/2015-report.html
2 UN, "World Urbanization Prospects," 2014 revision http://esa.un.org/unpd/wup/Highlights/WUP2014-Highlights.pdf
3 Navigant, Global Building Stock Database, www.navigantresearch.com/newsroom/the-global-building-stock-is-expected-to-increase-to-171-6-billion-square-meters-by-2024
4 US EPA, "Buildings and their Impact on the Environment: A Statistical Summary," 2009 http://archive.epa.gov/greenbuilding/web/html/whybuild.html
5 John Elkington, www.sustainability.com/history
6 The National Institute of Building Services: www.wbdg.org/design/sustainable.php
7 Guiding Principles for Sustainable Federal Buildings: http://energy.gov/eere/femp/guiding-principles-sustainable-federal-buildings
8 BREEAM market penetration statistics: www.breeam.com/projects/explore/, www.breeam.com/filelibrary/Briefing%20Papers/BREEAM-Annual-Digest-August-2014.pdf
9 USGBC, Green Building facts: www.usgbc.org/articles/green-building-facts
10 CBRE's National Green Building Adoption Index (CBRE, 2015), analyzes Energy Star label and LEED certification adoption trends among office markets in the 30 largest US cities.
11 Green Globes statistics: www.greenglobes.com/about.asp#history (Canada) and www.thegbi.org/green-globes-certification/ (US)
12 www.siemens.com/press/en/pressrelease/?press=/en/pressrelease/2013/infrastructure-cities/building-technologies/icbt201311093.htm&content[]=ICBT&content[]=BT
13 BREEAM schemes: New Construction, Communities, Refurbishment and Fit-out, In-Use International, www.breeam.com/resources
14 LEED ratings: Building Design & Construction, Interior Design & Construction, Building Operations & Maintenance, Neighborhood Development and Homes (www.usgbc.org/leed#rating)
15 US Federal government sustainable requirements: www.gsa.gov/portal/category/25999

16   LEED benefits: http://leed.usgbc.org/leed.html
17   San Francisco Building code adoption of LEED certification: http://sfenvironment.org/sites/default/files/policy/sfe_zw_cd_envcode_ch7.pdf. Atlanta's ordinance: http://carbonn.org/uploads/tx_carbonndata/Atlanta_%20GA_LEED%20Adopted%20Ordinance_2003.pdf
18   Green Globes website: www.greenglobes.com/about.asp. A list of Green Globes case studies can be found at: www.greenglobes.com/casestudies.asp
19   Differences of BREEAM & LEED: www.bsria.co.uk/news/article/breeam-or-leed-strengths-and-weaknesses-of-the-two-main-environmental-assessment-methods/. BREEAM www.breeam.com/
20   Differences between LEED and Green Globes: www.greenglobes.com/about.asp#ncvsleed
21   Assessment of construction costs were based on BREEAM (UK based), LEED/Energy Star (USA based), Green Mark (Singapore based), Green Star (Australia based) and Standard 5281 (Israel based).
22   www.usgbc.org/articles/green-building-facts
23   US General Services Administration (GSA) requires all new federal buildings to be LEED Gold from 2013 and leased space to be at least LEED Silver (www.gsa.gov/portal/category/25999)
24   US Census, "Millennials Outnumber Baby Boomers and Are Far More Diverse," June 25, 2015, www.census.gov/newsroom/press-releases/2015/cb15-113.html
25   Maryland and other states offer income tax reduction for businesses in buildings which save on energy: http://taxes.marylandtaxes.com/Business_Taxes/General_Information/Business_Tax_Credits/Green_Building_Tax_Credit.shtml
26   Sustainable Building Tax Credit Program-NM: www.emnrd.state.nm.us/ECMD/CleanEnergyTaxIncentives/SBTC.html
27   http://en.openei.org/wiki/Property_Tax_Incentive
28   Real Property Tax Exemption for Alternative Energy Improvements (Hawaii): https://energy.gov/savings/city-and-county-honolulu-real-property-tax-exemption-alternative-energy-improvements
29   Green Permit Program – City of Chicago: www.cityofchicago.org/city/en/depts/bldgs/provdrs/green_permit.html   http://en.openei.org/wiki/City_of_Chicago_-_Green_Permit_Program_(Illinois)
30   Permit fees for Charlotte – Mecklenburg County: http://charmeck.org/mecklenburg/county/LUESA/CodeEnforcement/Documents/fees.pdf
31   The site for the Database of State Incentives for Renewables & Efficiency (www.dsireusa.org/) offers a "comprehensive source of information on incentives and policies that support renewables and energy efficiency in the USA."
32   Cities are promoting the sustainable certification of buildings by offering certification fee reimbursments (e.g. City of Columbus, OH http://energy.gov/savings/city-columbus-green-columbus-fund; City of Chandler, AZ www.chandleraz.gov/Content/GB_Application.pdf; City of Encinitas CA www.encinitasca.gov/modules/showdocument.aspx?documentid=226 etc.)
33   City of Seattle ordinance: http://clerk.ci.seattle.wa.us/~scripts/nph-brs.exe?s1=LEED&s2=&s3=&s4=&s5=&Sect4=AND&l=20&Sect1=IMAGE&Sect2=THESON&Sect3=PLURON&Sect5=CBOR1&Sect6=HITOFF&d=CBOR&p=1&u=/~public/cbor1.htm&r=1&f=G

References

Arup & Partners (2014) "International Sustainability Systems Comparison – Key International Sustainability Systems: Energy and Water Conservation Requirements," prepared for CoreNet Global, March.

Bayer C., M. Gamble, R. Gentry, and S. Joshi (2010) "AIA Guide to Building Life Cycle Assessment in Practice," AIA. http://aiad8.prod.acquia-sites.com/sites/default/files/2016-04/Building-Life-Cycle-Assessment-Guide.pdf

Braun T. and S. Bienert (2015) "Is Green (still) a Matter of Prime? Stylized Facts about the Location of Commercial Green Buildings," *Journal of Sustainable Real Estate*, 7:1, 160–182.

CBRE (2011) "Building Performance and Occupier Satisfaction Produce Improved Return on Green Building Investments," Press Release, October 6. www.cbre.com/EN/aboutus/MediaCentre/2011/Pages/10062011.aspx

CBRE (2015) National Green Building Adoption Index 2015. www.cbre.com/~/media/files/corporate%20responsibility/green-building-adoption-index-2015.pdf?la=en

CBRE and EMEA Research (2009) "Who Pays for Green? The Economics of Sustainable Buildings." http://portal.cbre.eu/uk_en/imgs_styles/emea_economics_of_sustainable_building2009.pdf

Dermisi, S. (2013) "Performance of Downtown Chicago's Office Buildings Before and After their LEED Existing Buildings' Certification," *Real Estate Finance*, 29:5, 37–50.

Dermisi, S. (2014) "A Study of LEED vs Non-LEED Office Buildings: Spatial & Mass Transit Proximity in Downtown Chicago," *Journal of Sustainable Real Estate*, 6:1, 115–142.

Devine A. and N. Kok (2015) "Green Certification and Building Performance: Implications for Tangibles and Intangibles," *The Journal of Portfolio Management*, Special Real Estate Issue, 7th edition. www.iinews.com/site/pdfs/JPM_RE_2015_Kok.pdf

Dodge Data & Analytics (2016) "World Green Building Trends 2016: Developing Markets Accelerate Global Green Growth." http://naturalleader.com/wp-content/themes/natlead/pdf/World%20Green%20Trends%20ExecSummary_1110%20(1).pdf

Eichholtz, P., N. Kok, and J. Quiqley (2013) "The Economics of Green Building," *The Review of Economics and Statistics*, 95:1, 50–63.

Fuerst, F. and P. McAllister (2011) "Green Noise or Green Value? Measuring the Effects of Environmental Certification on Office Values," *Real Estate Economics*, 39:1, 45–69.

Fuerst, F., T. Gabrieli, and P. M. McAllister (2012) "A Green Winner's Curse? Investor Behavior in the Market for Eco-Certified Office Buildings." http://papers.ssrn.com/sol3/papers.cfm?abstract_id=2114528

General Services Administration (GSA) (2011) "Green Building Performance: A Post-Occupancy Evaluation of 22 GSA Buildings." www.gsa.gov/graphics/pbs/Green_Building_Performance.pdf

Institute for Market Transformation, Appraisal Institute, and Department of Energy & Environment (2015) "High-Performance Buildings and Property Value." www.imt.org/uploads/resources/files/LenderGuide_FINAL.pdf

Kok, N., N. G. Miller, and P. Morris (2012) "The Economics of Green Retrofits," *Journal of Sustainable Real Estate*, 4:1, 4–47.

Marker, A. W., S. G. Mason, and P. Morrow (2014) "Change Factors Influencing the Diffusion and Adoption of Green Building Practices," *Performance Improvement Quarterly*, 26:4, 5–24.

Miller, N. G., D. Pogue, Q. D. Gough, and S. M. Davis (2009) "Green Buildings and Productivity," *Journal of Sustainable Real Estate*, 1:1, 65–89.

Morris, P. and L. F. Matthiessen (2007) "Cost of Green Revisited: Reexamining the Feasibility and Cost Impact of Sustainable Design in the Light of Increased Market Adoption," Davis Langdon. www3.cec.org/islandora-gb/en/islandora/object/islandora%3A948/datastream/OBJ-EN/view

Parker, J. (2012) "The Value of BREEAM." http://breeam.es/images/recursos/inf/informe_schneider_electric_the_value_of_breeam.pdf

Reichardt, A. (2014) "Operating Expenses and the Rent Premium of ENERGY STAR and LEED Certified Buildings in the Central and Eastern US," *Journal of Real Estate Finance and Economics*, 49:3, 413–433.

Reichardt A., F. Fuerst, N. B. Rottke, and J. Zietz (2012) "Sustainable Building Certification and the Rent Premium: A Panel Data Approach," *Journal of Real Estate Research*, 34:1, 99–126.

RICS (2013) "Going for Green: Sustainable Building Certification Statistics Europe," RICS. www.buildinggreen.net/assets/cms/File/NEW/Sustainable%20rating-2011.pdf

Simons R. A., S. Robinson, and E. Lee (2014) "Green Office Buildings: A Qualitative Exploration of Green Office Building Attributes," *Journal of Sustainable Real Estate*, 6:1, 211–232.

Turner (2015) "2014 Green Building Market Barometer." www.turnerconstruction.com/download-document/turner2014greenbuildingmarketbarometer.pdf

Wiley, J., J. Benefield, and K. Johnson (2010) "Green Design and the Market for Commercial Office Space," *Journal of Real Estate Finance and Economics*, 41:2, 228–243.

World Green Building Council (WGBC) (2013) "The Business Case for Green Building: A Review of the Costs and Benefits for Developers, Investors and Occupants." www.worldgbc.org/files/1513/6608/0674/Business_Case_For_Green_Building_Report_WEB_2013-04-11.pdf

World Green Building Council (WGBC) (2014) "Health, Wellbeing & Productivity in Offices: The Next Chapter for Green Building." www.worldgbc.org/files/6314/1152/0821/WorldGBC__Health_Wellbeing__productivity_Full_Report.pdf

Yetunde A., R. Quartermaine, and D. Sutton (2014) *Delivering Sustainable Buildings: Saving and Payback*, IHS BRE Press.

20

Post-disaster recovery for real estate development

An analysis of multi-family investment from the perspective of a low income housing tax credit (LIHTC) project

Elaine Worzala and V. Lynn Hammett

Abstract

The 2004 hurricane season in Florida and resulting development of low income housing tax credit (LIHTC) projects in the years following the disaster are used as a case study to examine how public policy can influence redevelopment during recovery. The extent of housing losses in Florida after four hurricanes hit within six weeks, costing billions of dollars in property damages with hundreds of thousands of low-income housing units damaged or destroyed, clearly demonstrates the risks associated with coastal development for this vulnerable population. To identify the influence of public policy on development patterns and assess the risk associated with government directives, a content analysis of Florida's Qualified Action Plans for the LIHTC program from 2004 to 2010 was undertaken and results revealed that in the first year counties in the direct path of the hurricanes were given development preference. In 2006, the second year after the disaster event, preferences were expanded to more heavily populated counties, such as Miami-Dade. The location of new LIHTC developments built between 2005 and 2010 were then analyzed to determine how many had been approved and built in locations that would be impacted by storm surges from Category 3 or 5 hurricanes. More than two-thirds of new LIHTC projects post-2004 were constructed within storm surge boundaries, suggesting the need for proactive site planning from public and private stakeholders during the pre-construction phase of LIHTC development, or a shift in public policy to give preference to locations that are not as vulnerable to flooding.

Introduction

During disaster recovery, homeowners and businesses get the bulk of subsidies, grants, and low-interest loans to expedite recovery efforts. Owners of low-income multi-family housing face barriers that inhibit gaining access to grants or low-interest loans to help them fully participate in rebuilding their investments. Current policies surrounding subsidies, grants, and loans do little to address the additional risks faced by owners of multi-family housing, particularly for low-income housing projects in areas at risk for natural disasters. The low income housing tax credit program (LIHTC) is the primary subsidy available to help promote development of affordable multi-family housing in the USA. It is used across the country, set by federal law but administered at the state level, to help provide affordable housing. It is useful to use this program to better understand the impact that existing policy can have on promoting real estate development and to examine the potential unintended consequences that might arise for post-disaster redevelopment for the LIHTC multi-family housing sector.

Gaining a better understanding of the historical and economic factors that have created the current conditions requires an examination of past policies and implementation for both disaster relief and housing. To that end, a history of these policies and practices in the USA is included in this chapter. A short evaluation of how an LIHTC project is typically structured is presented along with discussion of the risks and rewards from the perspective of a real estate developer. The development patterns of LIHTC projects and the effect of government policy preferences on the location of these projects will be examined with a content analysis of the Qualified Action Plans that were set in Florida in 2005 and 2006, after a particularly bad hurricane season in 2004. Risks associated with development in general and LIHTC projects in particular are then highlighted and the chapter ends with some suggestions for mitigating these risks for future LIHTC development in areas that are prone to natural disasters.

The evolution of disaster policy and affordable housing recovery

1950–1979: Disaster Relief Act and implementation of FEMA and HUD

In 1950, the Disaster Relief Act was signed into law, which put into effect a permanent disaster relief program for the USA that included general disaster relief. In 1965, Hurricane Betsy hit Florida with a six-foot storm surge that flooded Miami and Ft. Lauderdale. A few days later, it turned and hit the Gulf Coast (National Hurricane Center, 1965). The National Flood Insurance Protection Act followed in 1968 establishing the National Flood Insurance Program, which is still a primary source of funding for disaster recovery today.

In the meantime, the first authorization for production of public housing occurred in 1955. While 810,000 units were authorized for construction, only 125,000 were ultimately built. At the time, urban renewal and slum clearance projects were a priority, which resulted in the destruction of more low-income housing units than were replaced (Thomas, 1997; Teaford, 2000). This would be repeated more than 50 years later during the recovery phase of Hurricane Katrina in 2005, when 4,500 public housing units, marginally damaged by the storm, were demolished and sold to private investors for redevelopment with LIHTC and Hope VI programs. One-to-one replacement of low-income housing was not a priority, resulting in displacement of renter households (PolicyLink, 2007; Unity of Greater New Orleans, 2010).

The 1960s ushered in housing legislation that adopted affordable housing production for the elderly and the disabled as well as the support of non-profit organizations as development partners. In 1963, when Lyndon Johnson took office, his vision of a Great Society to eliminate poverty and racial injustices resulted in establishment of the Department of Housing and Urban Development (HUD) and the underfunded Model Cities Program.

Disaster recovery and housing policy merged in 1973 when responsibility for post-disaster relief and reconstruction was assigned to the Department of Housing and Urban Development

(HUD) under the Office of Emergency Preparedness (Lindell et al., 2006, p. 17). The 1970s also ushered in federal legislation establishing executive offices of technological hazards programs including Dam Safety Coordination, Earthquake Hazard Reduction Program, and the Warning and Emergency Broadcast system, among others (p. 18). The National Governor's Association also got engaged in setting policies and procedures focusing efforts on the adoption of a comprehensive emergency management plan, particularly mitigation and recovery (Drabek, 1991, p. 18). In 1979, all federal disaster agencies were consolidated under the Federal Emergency Management Act (FEMA) including agencies under the Department of Defense, HUD, the National Weather Service (NWS), and the Executive Office of the President.

In 1974, the Community Development Block Grant (CDBG) program was created to provide state controlled subsidies for housing and urban development. By the 1980s, the private sector was heavily involved in low-income housing production with direct payments made to private developers and non-profit organizations (Schwartz, 2014) while HUD coordinated both disaster and non-disaster related housing resources for the public and private sector.

1980–2001: the growth of FEMA and HUD

Disaster legislation in the 1980s began with a response to the partial nuclear meltdown at Three Mile Island in late 1979. Legislation encompassed nuclear contingency planning, the Superfund law, and coordination at every level of government for emergency contingency plans. In 1988, the Robert T. Stafford Disaster Relief and Emergency Assistance Act was signed into law. This established federal cost sharing for planning and public assistance while assisting local and state authorities in the development of emergency management plans. Over the following three years, Hurricane Hugo in the south-east, the Loma Prieta earthquake, and Hurricane Andrew in Florida were all major disasters causing tremendous damage. In all three cases, FEMA was heavily criticized for failing to respond adequately to these events. As noted by Tierney et al., "the public expects government to respond swiftly and effectively in emergencies and has little tolerance when those expectations are not met" (Tierney et al., 2001, p. 152). During the 1990s, disaster programs began to focus on mitigation. The 1993 Hazard Mitigation and Relocation Act was the first proactive legislation to reduce flood hazards through relocation and acquisition of floodplains.

Major changes were also taking place in housing policy during the 1980s. Budget cuts forced state housing agencies to cut back on maintenance, resulting in the deterioration of low-income housing projects that had been built 20 years earlier. The low income housing tax credit (LIHTC) program was established under the Tax Reform Act of 1986 to balance the need for low-income rental housing production against the increase in homeownership incentives that were also included in the Tax Reform Act. The goal of LIHTC was to encourage private investment in multi-family housing for low-income households by increasing equity contributions from investors and reducing debt burdens on multi-family developments so that some of the units could be leased at below market rents. The program was innovative because it moved the production of low-income housing from direct government funding to the Internal Revenue Service. Instead of direct payments, investors purchase tax credits to offset taxable income that are used to provide more equity for the low-income housing projects. Since it began, LIHTC has been used to produce over half of the multi-family rental housing constructed around the USA (Khadduri et al., 2012) with more than 2.6 million housing units being built between 2006 and 2013 according to HUD. The program is currently the single largest subsidy for low-income rental housing production (Schwartz, 2014, p. 135). A detailed explanation of the LIHTC program follows.

2001–current: low-income housing and disaster recovery

The 9/11 terrorist attacks on the World Trade Center led to the creation of the Department of Homeland Security (DHS). More than 22 agencies, including FEMA, were folded into DHS and operated under a single cabinet agency. In 2005, all emergency preparedness activities under FEMA were moved to the Office of Preparedness and FEMA was left to focus on response and recovery. When Hurricane Katrina struck in 2005, FEMA was put to the test under this new agency. In spite of multiple evacuation exercises prior to Katrina, funding shortfalls prevented FEMA from making improvements to address deficiencies.

Subsidies were available for housing recovery in the Gulf Coast states of Florida, Mississippi and Louisiana in 2005 as a response to the devastation caused by Hurricane Katrina. Businesses could apply for economic development assistance with New Market Tax Credits, CDBG, and HOME Investment Partnership subsidies through the state, and disaster loans and grants through HUD (see Table 20.1). CDBG subsidies have historically been used for housing assistance, business assistance, and reconstruction of infrastructure after disaster. Both CBDG and HOME funds have been used in conjunction with LIHTC to address low-income housing needs.

Income provisions tied to CDBG are often waived during disaster recovery allowing states to allocate additional funds for needed development. Income provisions typically require at least 70 percent of CDBG funds to benefit low- to moderate-income households but these targets have been reduced during disasters. For instance, income provisions were reduced to 50 percent after the Florida hurricanes of 2004 (Boyd, 2011).

Disaster programs and policies that favor homeowners and commercial interests during recovery do not always extend to multi-family housing developers. Owners and developers of affordable and low-income multi-family housing often struggle post-disaster because recovery programs do not meet their financial needs (Wu and Lindell, 2003). Priorities for subsidies and low-interest loans are directed toward infrastructure, economic development, and homeowners (Comerio et al., 1994; Mueller et al., 2011). A lack of incentives for investors results in housing disparity and shortages in affordable housing for several years after the initial disaster. McCarthy and Hanson (2008) found permits for multi-family housing were issued less often than for single-family units. After Katrina, even single-family housing units with less severe damage were more likely to be issued permits quicker than multi-family units with extensive damage contributing to the overall supply shortages of available low-income rental housing during recovery (McCarthy and Hanson, 2008; Unity of Greater New Orleans, 2010).

The LIHTC program

LIHTC is the biggest supply-side program for low-income multi-family housing production in the United States. LIHTC developers can combine government subsidies, such as Community Development Block Grants (CDBG) or HOME Investments Partnership Program (HOME), with the federal LIHTC program to reduce reliance on debt for housing production. In return for tax credits, developers agree to maintain a percentage of housing units at affordable rents calculated according to area median income. Tax credits are awarded annually through a competitive process. Each state uses the same basic structure but is given flexibility to identify housing preferences according to local needs.

Table 20.1 Housing subsidies and loans for disaster recovery

Program	Purpose
HOME Investments Partnerships Program	Provides grants to states and localities that are often used in partnership with local non-profits. Eligible activities include constructing, buying, and rehabilitating affordable housing. Funds can be targeted for rentals or home ownership. Low income households may qualify for direct rental assistance. HOME is the largest block grant program available for affordable housing.
Community Development Block Grant (CDBG)	Provides resources to communities for a wide range of community development needs. Annual grants are allocated to larger cities and urban counties for housing and expansion of economic opportunities. The primary beneficiaries are principally low- and moderate-income households. CDBG has multiple programs for a wide variety of activities. Disaster Recovery Assistance is a flexible grant dispersed under the program and is subject to availability.
Supplemental LIHTC	Supplemental LIHTC were granted by Congress after Hurricane Katrina. In some cases, credits were advanced from traditional LIHTC that were to be awarded in future years.
Federal Disaster Loans	Individual assistance in the form of housing, grants for personal use, low interest loans, counselling and other assistance. Public assistance is available for communities. Low interest loans for renters and homeowners may be available. Issued through the Small Business Administration (SBA).
Hazard Mitigation Grants (HMG)	Applicants come from state, local government, Indian tribes, and private non-profit organizations. According to FEMA, homeowners and businesses must apply through one of these applicants.
Disaster Bonds	Tax exempt debt instruments issued by Congress and administered by states to direct private investment dollars to disaster recovery.
New Market Tax Credits	Targeted tax credits to low income markets. Credits are used to encourage investment in economic development and jobs creation in low income communities.
Physical Disaster Loans	Administered under the SBA, Physical Disaster Loans can be used to repair or replace real and personal property. Businesses of any size are eligible. Interest rates are capped at 4% if no other financing is available or 8% if credit can be obtained elsewhere.
Small Business Administration (SBA) Loans	Other types of disaster loans include Home and Personal Property Loans, Economic Injury Disaster Loans, and Military Reservists Economic Injury Loans.

The process

Tax credits are awarded annually on a competitive basis according to the priorities identified by the state housing finance agency (HFA) in each state (Figure 20.1). Each year, the Internal Revenue Service (IRS) awards tax credits to each state on a per capita basis. Tax credits are managed and awarded through the HFAs who publish criteria with guidelines and preference systems in a Qualified Action Plan (QAP). QAPs are published each year with the input of various stakeholders, including developers. When responding to disaster, priorities might include location preferences that coincide with housing losses. This was the case in Florida after the 2004 hurricanes caused devastating damage to the housing supply, which will be analyzed in more detail later in this chapter.

Developers use the QAP criteria to prepare proposals for potential LIHTC projects that are then awarded in a competitive bid process. As with most government programs, there are limited subsidies available compared to housing needs and developers have strict eligibility guidelines that must be met in order to participate. In most states, there are more proposals than funds available so the awards are very competitive. Typically, state housing agencies choose the winning proposals using either a point system or a subjective process depending on state rules.

Once a project is accepted by the HFA, tax credits are awarded to the developer who will either keep the credits for their own income taxes or find investors that can use the tax credits, often through a third-party organization called a Syndicator. The Syndicator puts together an investment fund for investors. The equity raised through the sale of tax credits typically covers development costs and reduces the debt incurred by a project. This reduction in costs is what allows the developer to charge below market rent for some of the units in the project. The goal of the tax credit program is to maximize investor participation so that debt can be minimized and rent can be kept at a below market rate for a portion of the units. Maximum tax credits are achieved when projects designate all units as affordable units.

KEY STEPS AND ENTITIES IN THE LIHTC PROCESS

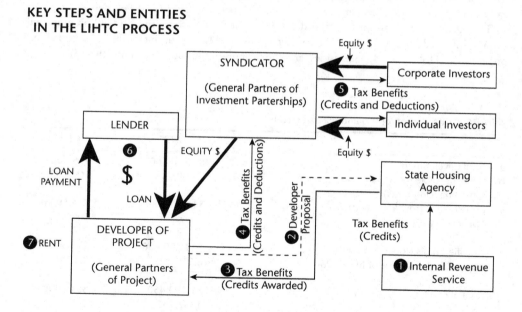

Figure 20.1 Key steps and entities in the LIHTC process

Source: Modified from a diagram prepared for Congressional testimony, Danter Company, 2015.

The value and risk of tax credits

Tax credits give investors the ability to offset future income, dollar for dollar, over a ten-year period. In return, developers agree to operate under LIHTC guidelines for 15 years. Two types of tax credits can be claimed under LIHTC, either for new construction or to renovate existing buildings (see Table 20.2). The 9 percent LIHTC covers 70 percent of new construction that does not include other subsidies. The 4 percent LIHTC funds 30 percent of new construction that can include other subsidies. It also can be used to fund the acquisition costs of existing buildings. Land, cash reserves, and some financing costs cannot be covered under LIHTC. The annual credit is derived from the development costs (qualified basis) multiplied by the applicable federal rate published annually by the IRS. Basis boosts are also set in the policies and give the HFAs flexibility to address local priorities. Boost provisions can be awarded for difficult development areas (DDAs), development in qualified census tracts (QCTs), elderly housing, disaster areas, or households with a targeted income. The Gulf Opportunity Act of 2005 (Go Zone) legislation, signed into law on December 15, 2005 after Hurricanes Katrina and Rita, designated DDAs in Louisiana, Mississippi, and Florida making them eligible for a 130 percent basis boost. The impact of the basis boosts on eligible tax credits is illustrated in Table 20.2.

Qualified action plans

A qualified action plan (QAP) is the tool used to communicate LIHTC priorities to the real estate development community. Under federal law, each state must develop a QAP annually. The QAP specifies the criteria used to select LIHTC proposals and preferred characteristics that will be given greater weight between competitive projects. The QAP process includes early input from stakeholders and advocacy groups, including LIHTC developers.

The QAP serves as an information document that explains how the LIHTC program will be administered in the context of local housing needs within established set-asides and preferences (Hollar, 2014). Set-asides are the minimum number of units that must meet

Table 20.2 Comparison of 9% and 4% tax credits

		9% LIHTC	4% LIHTC
a.	Total development costs	$15,000,000	$15,000,000
b.	LESS: ineligible costs (land, cash reserves, some financing costs)	$(500,000)	$(500,000)
c.	Eligible basis (row a – row b)	$14,500,000	$14,500,000
d.	Applicable fraction	100%	100%
e.	Qualified basis (row c × row d)	$14,500,000	$14,500,000
f.	Applicable % (AFR) as of July 2006	8.21%	3.52%
g.	Annual credit (row e × row f)	$1,190,450	$510,400
h.	Basis boost (130%)	$357,135	$ –
i.	**Annual credit after basis boost**	**$1,547,585**	**$510,400**
j.	**Total credit over 10 years (i × 10)**	**$15,475,850**	**$5,104,000**

established criteria, for instance the number of very-low-income units, or numbers of units set aside for families, to be eligible to be considered for the tax credits. Preferences allow states to determine specific criteria that include characteristics or amenities that are preferred by the state. Some common characteristic preferences might include housing targeted to serve rural communities, single parent families, or seniors. Preferences for amenities might be the inclusion of a business center, community room, or activity center. Federal statutes require certain set-asides and preferences to be included in QAPs, however states may incorporate additional criteria, such as set-asides for additional units for extremely low income households or specific location requirements. States can also reserve flexibility for project preferences through public policy statements and basis boosts (Gustafson and Walker, 2002). A basis boost is an additional credit bonus of up to 30 percent for projects meeting special criteria, such as building in a difficult to develop area (DDA). Basis boosts are only available to 9 percent tax credit projects, and since the Housing and Economic Recovery Act of 2008, can be awarded whenever they are needed for project viability.

Set-asides

LIHTC projects must meet minimum set-asides for the total number of low-income units in a project. Developers can choose to comply with either a 20–50 rule or a 40–60 rule. Under the 20–50 rule, at least 20 percent of units must be set aside as affordable to households who have incomes of 50 percent or less of area median income (AMI). The 40–60 rule requires at least 40 percent of units be set aside as affordable to households with incomes of 60 percent or less of AMI. These minimum set-asides are irrevocable and cannot be revised for the duration of a 15-year compliance period. The minimum set-aside must be met within 12 months from when the development is placed into service. Failure to maintain these set-asides during the compliance period results in the recapture of tax credits claimed by investors.

Syndicators and investors

Nearly all tax credits are sold to syndicators who act as an intermediary between investors and developers. Syndicators will package tax credits into funds that can include one or more developments. Several LIHTC projects can be included in a single equity fund that is sold to multiple investors. The investment funds are the source of equity dollars a developer will supplement with conventional loans. Some developers acquire additional subsidies from bonds, block grants, CBDG, HOME funds, and low interest loans. Equity raised through tax credits and other subsidies reduces the debt burden on the LIHTC property, making lower rents financially feasible. Investors benefit with a dollar-for-dollar tax credit against their future income for a ten-year period. That is, if an investor buys $10,000,000 in tax credits they will have a $1,000,000 credit against their tax liability each year for 10 years. The greatest risk to investors is the risk of recapture on previously filed tax returns if there is a period of non-compliance. In other words, if management fails to maintain the ratio of low income units for a 15-year period, investors will be required to refund any tax credits claimed in their tax filings.

Hurricane disaster and affordable housing recovery

When communities engage in post-disaster recovery, the initial focus is on economic development. A limited pool of resources is primarily dedicated to infrastructure and commercial development to try to expedite a swift economic recovery (Comerio et al., 1994;

Mueller et al., 2011). Studies of previous hurricane events have demonstrated that multi-family housing recovery often occurs at a slower pace than that of the single-family housing market (Comerio et al., 1994; GAO, 2010; Wu and Lindell, 2003). Homeowners have access to federal disaster assistance and low-interest loans as well as insurance coverage to carry out recovery activities. In addition, rental housing shortages are not uncommon as affected homeowners temporarily occupy available rental units resulting in an imbalance between supply and demand. This, of course, causes rental rates to increase. As a result, rising rents disenfranchise low-income families who cannot compete in the rental market.

To make matters worse, after any presidentially declared disaster in the USA, IRS policies provide some relief for investors and owners of LIHTC properties by relaxing rules for transient households. Often income limits are lifted so LIHTC housing can accommodate local property owners who need temporary housing while they repair damage to their homes. This provision amplifies the reduction in supply of housing for the low-income populations in a market.

Thus, recovery in the single-family housing sector gets underway while owners of multi-family housing face barriers that slow the process (Comerio et al., 1994). Owners of multi-family housing, particularly LIHTC, are less likely to qualify for meaningful disaster assistance so reconstruction of the damaged units is delayed. Increased risk is also a factor since the available disaster relief programs compel owners to take on additional debt in spite of constraints and limitations for increasing rental income (Wu and Lindell, 2003). This is particularly true for LIHTC properties because there are few disaster relief programs specific to multi-family housing. Of the disaster programs that are available for multi-family owners, such as disaster loans or low interest loans from the small business administration (SBA), nearly all of them involve taking on additional debt. LIHTC properties constructed in communities subject to ongoing hazards, such as in coastal counties, present additional risks to taxpayers and LIHTC owners when subsidies and tax credits fund these projects on high hazard sites.

Natural disaster risk for LIHTC properties in coastal counties: the case of Florida

Urban coastal environments are particularly subject to the challenges of climate change. According to the latest Intergovernmental Panel on Climate Change (IPCC) report, Florida and Texas are the most vulnerable states in the USA and are likely to see extreme storm effects and sea level rise as a result (IPCC, 2014).

Florida has experienced exponential growth over the last 40 years. Between 1970 and 2010, Florida's population grew 195 percent (Florida Department of Health, 2012). The effects of population growth and global warming in coastal areas intensify damage caused by extreme tropical storms and hurricanes with associated flooding and storm surges.

It is well understood that hurricanes are the most prevalent natural disaster risk for Florida. In 2004, in a span of six weeks, the state was hit with four hurricanes, costing billions of dollars in housing losses alone. Each of the four hurricanes contributed to one of the worst single years of hurricane disasters in Florida history, particularly with the devastating number of housing units damaged or destroyed as a result of storm surges and wind damage. The first to make landfall was Hurricane Charley when it hit Port Charlotte as a Category 4 hurricane on August 13. Storm surges were relatively small, not exceeding 7 feet, and reached only 6 to 7 miles from the center (Pasch, Brown, and Blake, 2011). In spite of the small storm surges, Charley caused damage estimated at $15 billion, making it the second costliest hurricane in US history. Property damage in Florida was $5.4 billion. Nearly 95 percent of buildings in downtown

Arcadia in DeSoto County were damaged. Hardee County saw nearly 5,000 houses damaged or destroyed not including 23,000 buildings damaged near Lake Wales.

Charley was followed by Hurricane Frances which hit on September 4, in both Palm Beach and Martin counties. Frances was a Category 2 storm with winds of 105 miles per hour and rains that caused a portion of Interstate 95 to collapse. The storm surge was 6 feet along the east coast where Frances hit with total damage at an estimated at $9.5 billion, including more severe losses in the housing sector. In 2011, Frances ranked as the eighth costliest hurricane in the USA (Beven, 2014, p. 4).

When Hurricane Ivan hit the panhandle as a Category 3 hurricane on September 16, an additional $18.82 billion in damage was added to the already devastating losses. A portion of Interstate 10 collapsed as a result of the storm surge and wave action. Baldwin, Escambia, and Santa Rosa counties lost thousands of homes. Ivan was the third costliest hurricane in the USA (Stewart, 2011, p. 6).

The final hurricane made landfall just ten days later, following the same path as Hurricane Frances. Jeanne was a Category 3 storm with storm surges from 3.5 feet to 6 feet along the east coast. Total damage was estimated at $7.66 billion (Lawrence and Cobb, 2014). Initial damage from all four storms was estimated at $51 billion, but as response and recovery efforts unfolded, it became obvious that these initial figures barely scratched the surface of losses in infrastructure, housing, and business assets.

According to the Housing Working Group, convened by Governor Jeb Bush to analyze storm damage, all of Florida's 67 counties felt the impact of the hurricanes. The housing sector was expected to sustain losses of more than $213 billion with more than 700,000 homes damaged or destroyed (HWG, 2005). Households occupying 52,916 multi-family housing units registered with the Federal Emergency Management Agency (FEMA) for housing assistance. FEMA provided rental assistance to 148,803 households; Structural Housing Assistance was provided to 116,000 households. An estimated 250,000 households were in need of rental units as a result of the hurricanes. Many of these households were placed in available LIHTC units temporarily, regardless of income, creating a shortage in supply of affordable units for low-income households. In response, the Florida State Housing Finance Corporation (HFC) looked to the LIHTC program to direct housing recovery in areas that experienced significant housing losses.

After the 2004 hurricane season in Florida, HFC included additional location preferences in the 2005 and 2006 QAPs. Normally, Florida QAPs limited location preferences to small, medium and large counties in general. Specific counties were not identified in Florida QAPs in any other year from 2004 through 2010. A point system was used in Florida to rank preferences, but it was limited to the project application and unit amenities without regard for location. Instead of assigning point values to location characteristics within the QAP, Florida maintained flexibility in project selection by including specific keywords to identify preferences. Preference keywords included the phrases "targeted" and "gives preference." Searching for these keywords in QAPs between 2004 and 2010 indicated that priorities for LIHTC were given for counties directly impacted by hurricanes Charlie, Frances, Ivan, and Jeanne in the 2005 and 2006 QAP. In 2006, some of the directly hit counties were replaced with counties that did not have severe housing losses, but did have large populations, such as in Miami-Dade. After 2006, specific county preferences were no longer identified for LIHTC development through the 2010 QAP.

Policy preferences were also expressed in statements from other agencies. After the 2004 hurricanes, Florida Governor Jeb Bush assembled the HWG to identify the extent of housing losses throughout the state. The HWG then issued policy statements that essentially supported

HFA preferences as defined in the 2005 and 2006 QAPs (HWG, 2005). In 2005, targeted statements within the QAP supported assistance for counties with the most severe housing losses, primarily coastal counties. When HWG completed their work, these were the counties identified with the most severe housing damage. In 2006, coastal counties with large populations and some rural counties were targeted in the QAP even though these counties had a smaller percentage of housing losses. The HWG report singled out all but two of these counties for significant to severe housing damage (HWG, 2005). In spite of the potential economic risk resulting from future storms, HFA and HWG preferences were primarily in coastal counties. In other words, LIHTC were earmarked for the highest risk counties for natural disasters.

Hurricane risk for LIHTC projects in Florida

Given the risk of storm surge as a result of hurricanes, Florida LIHTC properties are subject to significant potential financial losses when a hurricane hits. According to the Florida Finance Housing Corporation, over $201 million in housing credits have been allocated for over 53,000 affordable units in Florida since 1987. In just the two years following the 2004 hurricane season, over $65 million was awarded in tax credits for LIHTC development in Florida.

To analyze this potential loss, Hammett (2015) identified almost 45,000 LIHTC units that were constructed in Florida between 2004 and 2010 and found 74 percent of them were located in coastal counties. Using Geographic Information Systems (GIS) and a Sea, Lake, and Overland Surge from Hurricanes (SLOSH) model, Hammett modeled two possible storm hazards: an average Category 3 hurricane at median tide and a worst case scenario with a Category 5 hurricane hitting the coast at high tide. Results indicate that slightly over 26,000 LIHTC units along the Florida coast would be subject to storm surges between 5.2 and 23.3 feet in the first scenario (Category 3 hurricane) with estimated damage of close to $1 billion ($988 million). The county with the highest financial risk was Miami-Dade with estimated damage of slightly over $300 million while Monroe County had the highest percentage of its LIHTC units at risk (98 percent).

In the second scenario (Category 5 hurricane), over 48,000 LIHTC units would be subject to storm surges between 4.2 and 37.9 feet with estimated damage of over $1.8 billion. More than half of all LIHTC units in Miami-Dade would be at risk (12,833) with estimated damage of more than $485 million.

Additional development risk post-disaster

In addition to typical real estate development risk, LIHTC is vulnerable to recapture risk because of complicated management and reporting requirements (Roberts, 2009). Recapture risk after disaster occurs when the required numbers of low-income units are not back online within a specified grace period. Management plays a significant role in maintaining the required number of LIHTC units, therefore management and recapture risk are closely tied when disaster strikes, especially if redevelopment subsidies are unavailable or inadequate to repair damaged units.

The risk of foreclosure increases after disaster. Typically, foreclosure risk among LIHTC is less than 1 percent even with tight cash flow margins (Reznick, 2011). However, disaster-related costs have a negative effect on performance of an LIHTC investment if rehabilitation costs fall above the break-even point for operations. Considering the low profit margins typically achieved by LIHTC, damage to units after disaster could increase the incidence of foreclosure if they cannot be brought back online in a reasonable amount of time. Risk of recapture due to a brief noncompliance period from downed units is reduced because the IRS typically waives compliance requirements, at least temporarily, during the disaster response and restoration period. When foreclosure does occur, IRS rules provide for an extended compliance period and recapture is again mitigated under the assumption that a subsequent owner will continue affordability status. Regardless, the risk of recapture is one of the greatest concerns for tax credit investors.

Strategies for managing private sector risk have received some attention. Harrington (2006) uses economic theory to propose catastrophic risk insurance while Kunreuther (2006) uses risk decision theory to argue for a comprehensive natural disaster insurance program. Both of these ideas increase developer costs, creating a barrier for participation because of the tight profitability margins associated with LIHTC. In a post-disaster community, rising insurance costs put negative pressure on profitability.

Conclusions

Policies that drive LIHTC housing production also drive LIHTC developers into high hazard areas creating additional risk to investors. As detailed by Hammett (2015), over 340 LIHTC developments in Florida are located within storm surge boundaries with potential flooding between 4.2 to 37.9 feet, putting more than 48,000 LIHTC units at risk from storm surge. Given that Florida averages 2.6 persons per household, this means that approximately 125,315 people live in LIHTC properties that are susceptible to significant damage in the wake of storm surges from a Category 5 storm. Millions of dollars in physical damage and financial losses are at risk, both to owners and taxpayers who subsidize these developments. Ultimately, government policy and proactive developers need to take steps to mitigate risk in hazardous storm surge areas.

As the coastal population continues to increase, policy and market forces are likely to continue to support LIHTC growth in hazardous coastal communities. Potential losses for at-risk development suggest that an assessment should be undertaken to identify solutions to mitigate risks for both public and private interests. As more is understood about the impact of climate change on coastal systems, it has become obvious that a concerted effort should be made to identify risks and adopt policies for adaptation and risk reduction techniques before development approval is given and LIHTC credits are awarded. Public and private participants should proactively analyze potential development sites and assess the storm surge risk during the pre-development stage and policies should be implemented to not allow development in low-lying areas with significant storm surge risk.

Storm surge analysis can also be used to identify existing LIHTC sites that are in storm hazard locations and would allow existing owners to take a proactive approach to address potential capital expenditures should a storm surge occur and cause major damage to existing projects. These existing LIHTC projects could then be supported by redevelopment subsidies that allow for adaptation and mitigation techniques to reduce potential damage such as the construction of artificial barriers and physical defenses to storm surge hazards. In addition, non-structural solutions could be implemented as well, such as the use of natural vegetation, protecting coastal wetlands, and creating a wide buffer with natural canopies as an aesthetic amenity.

Technology already exists that will allow local governments and LIHTC developers to identify storm surge boundaries and to mitigate potential damage by improving site analysis and building design. Additional hazard insurance can be obtained to reduce risk once existing inventory at risk is identified. Disaster grants, mitigation grants, CDBG, HOME funds, and low-cost loans for multi-family development can be combined with LIHTC awards to encourage development in non-hazardous areas. In addition, grant programs could be developed to encourage developers toward adaptation measures for resilience in existing LIHTC developments. Cooperative efforts of public and private interests can reduce risk for LIHTC development in areas like Florida that are prone to hurricane disaster. In short, climate change is inevitable and storm surges are relatively predictable so all levels of government as well as the private sector should work to ensure that our most vulnerable citizens are not the ones put in harm's way in the event of a hurricane.

Acknowledgments

We would like to thank Adam Swalley, Center Coordinator for the Carter Real Estate Center, for his assistance in putting the final manuscript together.

References

Beven, J. L. 2014. Tropical Cycle Report: Hurricane Frances. National Hurricane Center, www.nhc.noaa.gov/data/tcr/AL062004_Frances.pdf (retrieved December 30, 2014).

Boyd, E. 2011. Community Development Block Grant funds in disaster relief and recovery. Congressional Research Service.

Comerio, M. C., Landis, J. D., and Rofe, Y. 1994. Post-Disaster Residential Rebuilding, Working Paper 608, Institute of Urban and Regional Development, University of California, Berkeley, CA.

Danter Company. 2015. Follow the Money: How the LIHTC Program Works, www.danter.com/taxcredit/lihtccht.htm (retrieved March 21, 2015).

Drabek, T. 1991. The Evolution of Emergency Management. In *Emergency Management: Principles and Practice for Local Government*, Thomas E. Drabek and Gerard J. Hoetmer (eds), pp. 3–29. Washington, DC: International City Managers Association.

Florida Department of Health. 2012. Florida Population Atlas, www.floridacharts.com/Charts/atlas/population/PopAtlas2012/narrative.pdf (retrieved February 5, 2015).

GAO (US Government Accountability Office). 2010. *Disaster Assistance: Federal Assistance for Permanent Housing Primarily Benefited Homeowners; Opportunities Exist to Better Target Rental Housing Needs* (Publication Number GAO-10-17). Washington, DC: US Government Accountability Office.

Gustafson, J. and Walker, J. C. 2002. Analysis of State Qualified Action Plans for the Low Income Housing Tax Credit Program. Submitted to the US Department of Housing and Urban Development. Washington, DC: The Urban Institute, Metropolitan Housing and Communities Policy Institute.

Hammett, V. L. 2015. Risk analysis and disaster recovery: A Florida LIHTC case study. PhD dissertation, Clemson University.

Harrington, S. E. 2006. Rethinking Disaster Policy after Hurricane Katrina. In *On Risk and Disaster: Lessons from Hurricane Katrina*, R. J. Daniels, D. F. Kettl, and H. Kunreuther (eds), pp. 203–221. Philadephia, PA: University of Pennsylvania Press.

Hollar, M. K. 2014. *Understanding Whom the LIHTC Program Serves: Tenants in LIHTC Units as of December 31, 2012*. Washington, DC: US Department of Housing and Urban Development, Office of Policy and Research.

Hurricane Housing Working Group (HWG). 2005. Recommendations to Assist in Florida's Long Term Housing Recovery Efforts. Report and Recommendations of the Hurricane Housing Working Group created by Executive Order 04-240, February 2015.

IPCC. 2014. Summary for policymakers. In *Climate Change 2014: Impacts, Adaptation, and Vulnerability. Part A: Global and Sectoral Aspects. Contribution of Working Group II to the Fifth Assessment Report of the Intergovernmental Panel on Climate Change*, C. B. Field, V. R. Barros, D. J. Dokken, K. J. Mach, M. D. Mastrandrea, T. E. Bilir, M. Chatterjee, K. L. Ebi, Y. O. Estrada, R. C. Genova, B. Girma, E. S. Kissel, A. N. Levy, S. MacCracken, P. R. Mastrandrea, and L. L. White (eds), pp. 1–32. Cambridge: Cambridge University Press.

Khadduri, J., Climaco, C., Burnett, K., Gould, L., and Elving, L. 2012. *What Happens to Low-Income Housing*

Tax Credit Properties at Year 15 and Beyond? Washington, DC: US Department of Housing and Urban Development, Office of Policy and Development Research.

Kunreuther, H. 2006. Has the Time Come for Comprehensive Natural Disaster Insurance? In *On Risk and Disaster: Lessons from Hurricane Katrina*, R. J. Daniels, D. F. Kettl, and H. Kunreuther (eds), pp. 176–192. Philadephia, PA: University of Pennsylvania Press.

Lawrence, M. B. and Cobb, H. D. 2014. Tropical Cycle Report: Hurricane Jeanne. National Hurricane Center, www.nhc.noaa.gov/data/tcr/AL112004_Jeanne.pdf (retrieved December 30, 2014).

Lindell, M. K., Perry, R. W., Prater, C., and Nicholson, W. C., 2006. *Fundamentals of Emergency Management.* Washington, DC: FEMA.

McCarthy, K. F. and Hanson, M. 2008. *Post-Katrina Recovery of the Housing Market along the Mississippi Gulf Coast. A Technical Report.* Santa Monica, CA: Rand Gulf States Policy Institute.

Mueller, E. J., Bell, H., Chang, B. B., and Henneberger, J. 2011. Looking for Home after Katrina: Postdisaster Housing Policy and Low-Income Survivors. *Journal of Planning Education and Research*, 31(3), 291–307.

National Hurricane Center. 1965. Preliminary Report on Hurricane Betsy, Page 1 (GIF). Tropical Cyclone Report (Report) (United States National Oceanic and Atmospheric Administration), p. 6. https://docs.lib.noaa.gov/rescue/hurricanes/QC9452B48H81965.pdf (retrieved July 12, 2014).

Pasch, R. J., Brown, D. P. and Blake, E. S. 2011. Tropical Cycle Report: Hurricane Charley. National Hurricane Center, www.nhc.noaa.gov/data/tcr/AL032004_Charley.pdf (retrieved December 30, 2014).

PolicyLink. 2007. Fewer homes for Katrina's poorest victims: An analysis of subsidized homes in post Katrina New Orleans, www.habitants.org/content/download/11398/111431/-le (retrieved April 17, 2017).

Reznick Group. 2011. The Low-Income Housing Tax Credit Program at Year 25: A Current Look at Its Performance. www. reznickgroup.com/sites/reznickgroup.com/files/papers/reznickgroup_lihtc_survey_2011.pdf (retrieved December 15, 2014).

Roberts, B. 2009. Strengthening the Low Income Housing Tax Credit Investment Market. *Community Investments*, 21(3), 11–13.

Schwartz, A. F. 2014. *Housing Policy in the United States*, 3rd Edition. New York: Routledge.

Stewart, S. R. 2011. Tropical Cycle Report: Hurricane Charley. National Hurricane Center, www.nhc.noaa.gov/data/tcr/AL092004_Ivan.pdf (retrieved December 30, 2014).

Teaford, J. C. 2000. Urban renewal and its aftermath. *Housing Policy Debate*, 11(2), 443–465.

Thomas, J. M. 1997. *Redevelopment and Race*. Baltimore: Johns Hopkins University Press.

Tierney, K. J., Lindell, M. K., Perry, R. W. (eds) 2001. *Facing the Unexpected—Disaster Preparedness and Response in the United States*. Washington, DC: Joseph Henry Press.

Unity of Greater New Orleans. 2010. Search and Rescue Five Years Later: Saving People Still Trapped in Katrina's Ruin. A report of the Abandoned Buildings Outreach Team, August 2010. http://unitygno.org/wp-content/uploads/2010/08/UNITY_AB-Report_August2010.pdf (retrieved April 17, 2017).

Wu, J. and Lindell, M. K. 2003. *Housing Reconstruction after Two Major Earthquakes: The 1994 Northridge Earthquake in the United States and the 1999 Chi-Chi Earthquake in Taiwan*. College Station, TX: Hazard Reduction & Recovery Center, Texas A&M University.

21

Drivers and opportunities of sustainability in real estate development

Sven Bienert and Rogerio Santovito

Abstract

Our interpretation of sustainable development is that the balance of the Triple Bottom Line principles (ecological, economic, and socio-cultural dimensions) is a prerequisite for real estate, in order for it to contribute to intergenerational justice and mitigate global climate change, whilst at the same time ensuring long-term business success in the sector. For too long, real estate companies only felt obliged to generate profits. An awareness of the finite nature of natural resources, progressive climate change, as well as social injustice, and spectacular business failures, have all increased awareness of the need for a clear sustainability agenda at all levels of the sector.

This increasing pressure from society and tighter regulatory framework have placed the real estate industry's contribution to sustainable development at the forefront of discussions. The industry share amounts to 30–40 percent of all resource utilization and GHG emissions, which emphasizes that the real estate sector is in the limelight of political debate on tackling climate change.

Sustainable real estate development covers all aspects of mitigation. Reducing GHG emission can be achieved, for example, with recycled building materials, higher density and energetic retrofitting. Property certification and a clear focus on life-cycle-costing (LCC) are key elements. Furthermore, adaption of the property stock becomes more and more relevant, as climate change and extreme weather events become more frequent over time. Sustainability must be implemented at the company level, so that every decision is made in accordance with the long-term benefits for all stakeholders.

Introduction

Extreme weather events, which have been reported with increasing frequency by several countries (NOAA, 2016), are a very tangible consequence of climate change. The average global temperatures across land and ocean surfaces have been increasing consistently over the last 100 years, and are now the highest they have been over the past five centuries. In January 2016, the record was broken once more, with the highest level registered for a January in the 137-year period of measurements (NOAA, 2016; IPCC, 2013).

Proof of considerable global warming: Record high temperatures are regularly being reported

Climate change is already causing massive changes in ecosystems. Moreover, experts say, the ability of ecosystems to adapt to relatively rapidly changing conditions will be exceeded if global temperatures increase by more than 2°C and with disastrous consequences for the plant and animal biosphere, and hence for humans and the built environment.

Until the 1990s, there was no consensus among the international scientific community regarding the prevalence of anthropogenic global warming. However, this situation has changed with the continuing publication of the IPCC Reports and other scientific studies. Today, there is very substantial agreement that climate change is caused by human activities. The scientific community has put considerable effort into studying the emissions of greenhouse gases from anthropogenic sources, and it is now considered a clear phenomenon and key driver for ongoing climate change.

In 2013, the concentration in the atmosphere of carbon dioxide (CO_2), one of the most important greenhouse gases, exceeded what is considered the critical level of 400 parts per million (ppm), compared to 280ppm during the pre-industrial era (IPCC, 2007; IPCC, 2013). In 2014, emissions from fossil-fuel combustion and from industrial processes (production of cement clinker, metals, and chemicals) totaled over 36 billion tons of CO_2 (Olivier et al., 2015).

Climate change is manmade: COP21 put pressure on industries to act more sustainably

Without additional actions and merely maintaining policies and regulations currently in place, the US's Energy Information Administration (EIA, 2013) forecasts further growth of 46% in global energy-related CO_2 emissions by 2040. Historically, there have been many observed variations in CO_2 emissions in certain regions, and this can be explained partially by events such as the end of coal subsidies in the United Kingdom, the reunification of Germany, or more recently, changes in the Japanese energy matrix due to the Fukushima accident. However, such events only explain short-term variations in CO_2 emissions and not meaningful changes in emissions trends. It is a sad fact that to date, on a global scale, there has not been any real success in effectively reducing the overall amount of GHG emissions.

The reduction of GHG emissions is, however, of central importance to counteract dramatic climate change and unpredictable tipping points in the world's ecosystem. The international community is well aware of its responsibilities. On December 12, 2015, at the 11th meeting of the Parties to the Kyoto Protocol (CMP11) and the 21st United Nations Climate Change Conference in Paris (COP21), the Assembly adopted the new climate agreement. Those meetings reinforced the main climate objectives – to keep the rise in global average temperature below 2°C. In the framework of the so-called Lima Call for Climate Action, the participating states were urged to submit their national climate action programs (INDCs, Intended Nationally Determined Contributions) in the run-up to the conference. Topics such as the Green Climate Fund or CO_2-pricing were on the agenda, in order to achieve the "2-degree goal" (the limit of global warming to 2 degrees compared to pre-industrial levels) (UN-FCCC, 2016; UNDP, 2016). Of course, there are many regional differences to be considered when analyzing those commitments and reduction targets, especially when evaluating the changing economies of China, India, and other large developing countries.

The role of sustainable cities

Urban emissions from cities of emerging economies are already converging with those of developed cities (World Bank, 2014). Although there is no evidence of a consensual percentage of this contribution, mostly due to compatibility issues with existing GHG emissions-accounting procedures, it is clear that cities worldwide play a major role in achieving carbon-reduction targets. Several studies make an extensive comparison of GHG emissions from several large cities (Kennedy et al., 2009; Dhakal, 2010; Hoornweg et al., 2011). Beijing, Shanghai, and Tianjin, for example, have per capita emissions comparable to those of large European and some North American cities (Sugar et al., 2012).

Due to the fact that cities are a major emitter, several initiatives by city and local authorities were initiated, focusing on decarbonizing, or achieving low-carbon levels. A new report by the New Climate Economy has found that investing in public and low-emission transport, waste management, and building efficiency in cities could generate savings of over €15m by the year 2050 (Gouldson et al., 2015). Aside from these economic benefits, greenhouse gas emissions could be reduced by 3.7 gigatonnes carbon dioxide equivalent per year by 2030, exceeding the current annual emissions of India. This shows the relevance of low-carbon initiatives in cities all over the world.

However, creating sustainable cities is not easy. For example, the goal of creating healthy cities with a lot of green space is to some extent contradictory with aiming at high density in order to reduce emissions from transportation. For developers this is of major importance, as they can moderate and steer the necessary change.

Real estate industry with high potential for mitigation and need for adaptation

The construction industry is a major global employer and can be of significant relevance to the GNP of countries (Murray and Cotgrave, 2007), therefore playing a central role in climate protection, due to its high resource consumption (Nelson et al., 2010; ZIA, 2015). That puts the sector in a position to take meaningful sustainable development actions, for example, contributing to climate change mitigation by fostering sustainable buildings and adopting sustainable management practices.

Aside from its impact on the level of emissions arising from individual buildings ("property level"), there are also opportunities for action on other levels, such as intense collaboration with local governments and urban planners. Given the rapid urbanization trend in cities, the re-urbanization of old central areas in cities, or decontamination of old industrial sites for new developments (brownfield developments), are practical examples of how the real estate industry can act collaboratively to mitigate climate change.

Furthermore, it should be noted that the real estate industry can not only contribute to mitigating climate change and fostering sustainable urban development, but is at the same time highly affected by climate change. In that sense, adaptation measures are also needed, in order to create a less vulnerable and more resilient property stock.

Drivers of and opportunities for sustainability in real estate development

Key challenges (changes in ecological, socio-cultural, and economic conditions)

The overall social responsibility for sustainable action and enhanced environmental protection will shift from voluntary to compulsory, when states anchor the appropriate conditions in their legal systems. In the context of many fields such as energy consumption, pollutant emissions in the production of construction materials or the future disposal costs of problematic materials, this has not yet been done in all countries on a large scale. Against this background, many policy initiatives are currently in place (and planned), aimed at limiting the consumption of resources by and emissions of the property and construction sector (IIGCC, 2013; Guyatt et al., 2011; Leurig and Dlugolecki 2011).

Regulatory framework for climate protection will be tighter

Due to the EU directive on the energy performance of buildings, all new buildings in Europe from 2021 onwards must comply with the standard of "nearly zero-energy buildings" (European Council, 2010, 2012, 2013; RICS, 2013a). It is likely that this holding "trend" for greater regulation of the legal framework will be expanded in the wake of COP21. Nevertheless, it should be noted that tightening the legislative framework regarding emissions and conserving resources is occurring not only in Europe – which is often cited as a pioneer – but also worldwide. As an example of those enforcements, mandatory "energy disclosure laws" for real estate portfolio holders are already in place in many cities in the USA (Makower 2013).

It is important to note that the introduction of such laws, regulations, and directives also understandably indirectly affects the behavior of all market participants (for example real estate investors or tenants) (Dent et al., 2012). Consequently, no individual elements can be considered in isolation in terms of achieving sustainable development in the real estate industry. As an example, the changing laws and increasing transparency are changing the behavior of investors, which in turn has a direct effect on their willingness to pay, thus influencing the economic sustainability.

In addition to extensive legislative initiatives at the object level, sustainability is increasingly regulated by law even at the enterprise level. In this context, for example, the European Parliament and the Commission have strengthened regulations on CSR (Howitt, 2013) and commitment to sustainability reporting for large companies (Wensen et al., 2011). Real estate developers should proactively address these aspects, and, for example, carry out reporting (such as Global Reporting Initiative), voluntarily, in order to be ahead of upcoming regulations and ensure a "future-proof" business model.

Demographic change, urbanization, digitalization – challenges beyond climate change

It is not only the ongoing climate change and regulation related to climate protection, but also the demographic change in developed countries, and the continuing global population increase, urbanization, or the increasing scarcity of natural resources, that will change the built environment in this century.

The world is becoming more urbanized. Estimates made by the United Nations (UN-DESA 2014, 2015) show a trend of migration from rural to urban areas. As of 2014, 52 percent of the global population live in urban areas, and this percentage is likely to increase up to 66.4 percent, with the expectation that the global population will reach 9.55 billion by 2050. Some cities in Europe have experienced a population decline in recent years. Nevertheless, Europe has 73 percent of its population living in urban areas, and this is expected to exceed 80 percemt by 2050. This trend is relevant, since most GHG emissions that contribute to global climate change come from urban areas (UN-Habitat, 2011).

Urban areas are associated with around 70 percent of global energy consumption and energy-related greenhouse gas emissions. The IPCC estimates that in 2010, urban areas accounted for 67–76 percent of global energy use and 71–76 percent of global CO_2 emissions from final energy use (Seto et al., 2014; IEA, 2015).

In the wake of the financial and economic crisis of 2008/2009, the restoration and permanent support of reputation is a major corporate challenge. Through using third-party recognition schemes, sustainable corporate governance can significantly enhance the reputation of the company (Schleich, 2012).

Another challenge is the progressive transformation of values. This is taking place in the entire economy and hence applies equally to the real estate industry. Companies' human resources are confronted in this context with issues of work–life balance, the demand for meaningful corporate objectives, and an overall high degree of flexibility. Sustainable corporate strategies in this case can have a supportive effect (e.g. through flexible working hours, health management, open office structures to improve the communication culture) and thereby increase staff motivation and identification with one's own company (e.g. corporate image, contributions to the common good under the banner "Corporate Citizenship" high standards of environmental protection, etc.) (Rabe von Pappenheim, 2012). The so-called "Generation Y" – that is, people born between 1980 and 2000 – are radically changing the current working and living structures. This generation wants to live, work, and consume in a way that can be clearly distinguished from previously conventional structures (Schleich, 2012).

Besides the economic implications that emerge from the subject area of environmental or social sustainability, there are also a number of other economic challenges that the sector must face. For example, the sovereign debt crisis in Europe and the subsequent low interest rates aimed at supporting the public sector in selected European countries, have already caused strong inflows to real estate investments, leading partially to overheated real estate markets. For real estate developers, this increasing volatility and potentially abrupt shifts are a result of its broad integration with capital markets.

Short-term vs long-term sustainability agenda

The challenges these so-called megatrends pose on the economy and society are already clearly noticeable. As examples, commodity prices are exposed to ongoing fluctuations (Brown, 2008), whole regions are suffering under changing climatic conditions, and extreme weather events are causing massive damage to infrastructure and real estate assets (Messervy et al., 2014; Bienert, 2014). Responsible acting corporate decision-makers must be able to anticipate fundamental structural changes in time and translate them into viable, value-creating business models with reasonable returns for each investor.

Impact on profitability and business cases in the real estate industry

For capital-market oriented companies, maximizing the return on equity is still the central financial target of the so-called shareholder-value-oriented management approach – often associated with short-term thinking (ZIA, 2015; Laasch and Conaway, 2014). At first glance, this ultimate goal seems incompatible with a strong sustainability agenda that focuses on longer periods and might cause significant costs in the present. A more accurate analysis, however, reveals that thinking to secure one's own competitiveness is important, especially in times of scarce resources, increasing environmental pollution, a critical and networked society, and changing customer needs. More and more decision-makers in companies are aware of their responsibilities and make efforts to avoid a negative impact of their actions on humans and the environment, and attempt to implement sustainable business models.

How climate change and increasing extreme weather events are affecting property values

Especially for developers, location (location, location!) is essential for a viable and successful development. Taking this fundamental aspect into account, it might be interesting to think for a minute about the way climate change influences property values.

Property values are based on the highest and best use of a given location. Income-generation potential is directly linked to aspects such as climatic conditions, particularly illumination, wind, emissions (noise, smoke, dust), and rainfall, as well as soil conditions such as surface formation, natural cover, bearing capacity, groundwater conditions, mudslide areas, and exposure to natural hazards (flooding, avalanches, storms, hurricanes).

Location is fixed by definition – increasing extreme weather events imply only downside risks

Good protection against natural hazards is therefore essential for both commercial and residential real estate, and also for infrastructure assets. Negative changes can already be noticed all over the world. Reduced real estate values have negative consequences for the economy as a whole (Bienert, 2014). In developed economies, the value of property amounts to an average of 3.5 times a country's gross domestic product (GDP) (Brandes and LeBlanc, 2013). Thus, even marginal changes in value can lead to massive monetary damage. In addition, in real estate valuation, present value is regularly considered in assessing future potential benefits. Thus, even relatively moderate reductions in value can trigger substantial losses in expected returns. Table 21.1 summarizes possible negative impacts.

In essence, negative impacts of climate change are likely to result in significant adjustments in the real estate industry, until a new balance has been reached. However, some regions or market participants might even derive benefits out of climate change.

Developers must now evaluate "location, location, location" in a different sense

Although still of great relevance, the locations of projects no longer rely solely on demographics and market data, as there are now many other aspects to be considered. For example, agriculture will now be feasible in regions that were formerly too cold, allowing vineyards to move further north and opening new opportunities for real estate developments. For developers, it will be of increasing importance to analyze these (positive or in most cases negative) aspects in advance, also considering new technologies for land decontamination (for brownfield developments), exposure to natural hazards, alternative and sustainable sources of energy and water, etc.

The essence of pay-off studies: isolating the "green value"

Enhanced sustainability at all levels of a real estate company is often associated with higher costs (Epstein and Buhovac, 2010). In market-driven companies, higher expenses, if not connected with the fulfilment of legal requirements, must be justified. Put another way, no investments in sustainable action programs will be made, unless their added value, i.e. the positive effect on profits and shareholder value, can be clearly worked out in advance. The *Stern Report* (Stern, 2006) stressed, for the first time, that the monetarization of manmade climate change is important. Numerous studies attempting to quantify the profitability of sustainable products in different sectors and regions were created as a result of that report (for a review of relevant studies, see Clark et al., 2014; Fulton et al., 2012). For a developer, the crucial question might be whether a LEED-platinum certified building will merely trigger higher construction costs or whether the additional investment will pay off.

Core question – does sustainability pay off?

While the costs of sustainable actions are relatively easy to quantify, it is often difficult to isolate

Table 21.1 Effects of climate change on property values

Climate aspect	Commercial and residential real estate	Forestry	Agriculture	Infrastructure
Rise in temperature	Reduced ground rent (lower potential revenue, in the case of regional population changes; also, increased need for cooling, and thus higher operating costs)	Reduced ground rent (in the case of increase in forest fires, pest infestation, extinction of species)	Reduced ground rent (in the case of increasing drought, pest infestation)	Increased wear on installations; unstable ground
Water scarcity	Decline in attractiveness of a region/decline in ground rent; higher costs for water supply and treatment	Reduced revenues from forestry/ increased danger of forest fires	Reduced harvests; increased costs for irrigation	Decline in bearing capacity of soil
Rising sea level	Reduced settlement area in coastal regions	—	Reduced agricultural land area/loss of potential revenues	Danger to port facilities
Increase in extreme weather events	1. Direct loss (e.g., hail damage to buildings) 2. Indirect loss (e.g., through gaps in production or rent after hurricanes) 3. Consequential loss (e.g., declining number of tourists in flood areas, rising insurance premiums)	1. Direct loss 2. Consequential loss 3. Depreciation of natural capital (permanent damage to ecosystems, extinction of species)	1. Direct loss 2. Consequential loss 3. Depreciation of natural capital	1. Direct loss 2. Indirect loss (infrastructure damages due to extremes in temperature, precipitation/ flooding/overload of urban drainage systems/storm surges, which can lead to damage to roads, rail, airports, and ports; electricity transmission infrastructure is also vulnerable)
Increased regulation	Higher construction costs and running costs; higher costs, particularly in the case of carbon taxation	—	—	Higher construction costs and running costs

Table 21.1 continued

Climate aspect	Commercial and residential real estate	Forestry	Agriculture	Infrastructure
Increased adaptation costs due to climate change	Higher adaptation costs to protect properties and to make buildings energy- and resource-efficient	Higher adaptation costs	Higher adaptation costs	Higher adaptation costs

Source: Authors, based on Bienert, 2014

the resultant financial benefits and ultimately define the added value of a more sustainable property, compared to a non-green peer group. The willingness to pay can be, in some studies, isolated by using multiple regressions (hedonic pricing) (Dent et al., 2012). Here, the basic idea is that the willingness to pay for sustainable real estate properties cannot derive from purely theoretical aspects, but rather directly from the market. The price of a good will be, for example, a property that is "dismantled" by means of statistical methods into its individual components. To obtain meaningful results, correspondingly large data sets need to be used, as a single purchase price cannot be evaluated (Auer, 2005; Maier and Herath, 2015).

Table 21.2 represents only a sample excerpt from reports generated in different markets around the world. It is noteworthy that the majority of these evaluations were able to determine, from real market data, a significant positive relationship between sustainable properties and rents or values.

For developers it is essential to inform end-investors or potential buyers about the upsides that more sustainable properties achieve in the markets. This will speed up demand for these products and clearly also stimulate higher willingness to pay.

Altruism vs business case: what's the industry's logical response?

Some people now find it hard to approach the "sustainability" topic with the necessary neutrality. Too much has been said about the need for action, and the subject has been intensely exploited for marketing purposes, without actually leading to appropriate operational implementation steps.

We clearly see the challenges facing the industry because of climate change, but also note in this respect other socio-cultural and economic issues. In particular, risks arise that could lead to increasing values for more sustainable buildings in the case of opportunities, but also potential dangers in terms of objects exposed to natural hazards. Sustainable thinking and economic analysis should therefore be based on the real estate business decision-making processes of developers (Brounen and Kok, 2011).

Serious commitment to sustainability is not necessarily altruistic, but a consequence of the fact that an understanding of the changing context for purely economic reasons for the individual company is no longer an option. Real estate developers must therefore perceive securing their long-term sustainability as an economically and strategically decision-relevant element and implement this accordingly in all areas and levels of the company.

Table 21.2 International studies on the added value at the property level

Authors	Title	Research details	Research subject	Sustainability measure	Main findings
Eichholtz, Kok, Quigley (2010)	Doing well by doing good? Green office buildings	Around 10,000 Office properties. Period: 2004–2007 (USA)	Impact of sustainability certification on the tenant/transaction prices	LEED/ Energy Star	Energy Star-certified buildings have 3.3% higher rents. LEED-certified buildings have 5.2% higher rents. A 10% reduction in energy consumption causes an increase in the net present value by 1% and 16.8% to 18.8% higher sales prices.
Fürst and McAllister (2011a)	Green Noise or Green Value? Measuring the effect of environmental certification on office values	24,479 Office properties. Period: 1999–2008 (USA)	Influence of double standards on lease/ transaction prices	LEED/ Energy Star	LEED-certified buildings have 5% higher rents and 25% higher transaction prices. Energy Star-certified buildings have 4% higher rents and 26% higher transaction prices.
Fürst and McAllister (2011b)	Eco-labelling in commercial office markets: Do LEED and Energy Star offices obtain multiple premiums?	2,688 LEED/Energy Star Office properties. Period: 1999–2009 (USA)	Influence of sustainability certificates on rental/ sale prices and occupancy	LEED/ Energy Star	LEED-certified buildings have 4–5% higher rents and 25% higher transaction prices. Energy Star-certified buildings have 3–4% higher rents and 18% higher transaction prices. Double certified buildings have 28–29% higher transaction prices. Energy Star buildings have better occupancy.
Miller, Spivey, Florance (2009)	Does Green Pay off?	1,200 LEED/Energy Star Office properties. Period: 2004–2008 (USA)	Impact of sustainability certification on transaction prices	LEED/ Energy Star	Energy Star-certified properties have 6% higher transaction prices. LEED-certified properties have 12% higher transaction prices.

Table 21.2 continued

Authors	Title	Research details	Research subject	Sustainability measure	Main findings
Wiley, Benefield, Johnson (2010)	Green Design and the market for commercial office space	7,308 Office properties. Period: 2008 (USA)	Impact of energy-efficient design on rental/transaction prices	LEED/ Energy Star	In comparison to traditional buildings Energy Star-certified buildings have Rent: + 7.3%–8.9% occupancy: + 10–11% LEED-certified buildings have: higher rent: + 15.2%–17.3% higher occupancy: + 16–18%.
Reichardt, Fürst, Rottke, Zietz (2012)	Sustainable Building Certification and the rent premium: A panel data approach	1,768 LEED/Energy Star Büroimmobilien Zeitraum: 2000–2010 (USA)	Impact of sustainability certification on rents	LEED/ Energy Star	Energy Star-certified office buildings have 2.5% higher rents. LEED-certified office buildings have 2.9% higher rents. Positive correlation between Energy Star certification and building occupancy is confirmed. Rent Premium varies over time.
Eichholtz, Kok, Quigley (2013)	The economics of Green Building	21,000 office buildings; 6,000 transactions Period: 2007, 2009 (USA)	Influence of Green Buildings in effective rents, rates, occupancy	LEED/ Energy Star	Increase the number of green buildings between 2007–2009. Crisis has little impact on rents. Price premium for green buildings has dropped slightly, but still high occupancy/rent. Age of the label has a negative impact on the premium; Energy Star-certification premium decreases with 0.4% per year (base: Technological progress of the building). LEED or Energy Star-certified buildings have 3% higher rents; effective rents (consideration of utilization) are about 8% higher (due to the high utilization of certified buildings). Certified buildings have 13% higher transaction prices.

Source: Authors

283

Implementation of sustainable value creation at the company level

Redefine normative and strategic framework

One of the largest initiatives for sustainable investments, the UN Principles for Responsible Investment Initiative, has identified short-term thinking among economic leaders as one of the main barriers to more sustainable initiatives within the business environment (Engshuber, 2013). Therefore, a more long-term view of opportunities and threats will be crucial to enabling future-proof business models to evolve.

Sustainability pays off – maturity as a key challenge

When discussing the pay-off for sustainable actions, the issue is frequently not the lack of profitability of "investment in the future" per se, but the challenge of dealing meaningfully with the maturity and calculability of seemingly intangible benefits. In the long term, it becomes meaningful for sustainability investments to optimize profits by avoiding future – and probably higher – adjustment costs, thus developing new markets, creating competitive advantage, and improving the corporate image (Seidel and Menn, 1988; Bleis, 1995). This positive relationship between success and sustainability was described by Porter and van der Linde in 1995, in their internationally acclaimed research work "Green and Competitive: Ending the Stalemate." The results were impressively confirmed in a variety of current real estate related studies (see Table 21.2) (Porter and van der Linde, 1995).

Triple-bottom-line guidance for normative positioning and goal setting

The sustainable corporate governance of real estate developers is characterized by the simultaneous and equal implementation of environmental, economic, and social dimensions of project, portfolio, and company-level objectives. Some specific examples of the various dimensions of sustainability in the real estate industry are:

- Ecological component: Reduction of resource consumption in the life cycle of developed projects and avoidance of emissions (in particular direct, and possibly indirect GHG emissions) at all levels of the company. Introduction of appropriate control instruments and indicators (Key Performance Indicators, KPIs), sustainability reporting, and benchmarking. Certification of the portfolio and general focus on "Green buildings" (green building or expanding "sustainable building," i.e. sustainable buildings) (Sayce et al., 2007).
- Socio-cultural component: Maximizing comfort, health, satisfaction, and well-being of future occupiers/integration of projects in the direct local environment/consideration of interests of all stakeholders, especially employees.
- Economic component: Reducing the life cycle costs of buildings and maximizing revenues. Ensuring high use capacity and functional flexibility of the property portfolio in order to support stable and sustainable financial performance. Foster innovations for an economical use of resources and energy. Ensuring the long-term competitiveness of the company.

An essential prerequisite for meaningful adjustments to the business model is initially the corresponding normative positioning. In this respect, the developer formulates a corporate vision, corporate mission, and its fundamental values, in order to define the main aspects of corporate sustainability and endeavor, in an attempt to derive overall corporate goals.

After establishing the normative frame of reference, the strategic direction of the project development company is based on overarching company targets, in which the specific sustainability goals and strategies, and the performance potential of the company, are determined, so that the strategic planning achieves the defined objectives.

Resolve trade-offs and making complementarities of transparent goals

For the real estate industry, the strong interactions and trade-offs between the abovementioned dimensions are particularly valid. It is important in this context, to ensure complementarity and balance between objectives. While short-term trade-offs between profit maximization and an intensification of the environmental benefits are often obvious, the long-term trade-offs and complementarity are not normally readily visible. However, the achievement of complementary targets is possible, which can contribute through a corresponding ecological work environment for achieving better productivity together with a socio-cultural improvement of employee comfort, health, and satisfaction, which will ultimately reflect positively in monetary results.

However, even these simple examples are sometimes beset by long payback periods, interdependencies, uncertainty and not entirely transparent effect chains, all of which represent significant challenges for sustainable management in the real estate industry.

Implementation of CSR agenda in property development companies

Sustainable corporate management within the CSR-understanding used here, is nowadays directed to the entire enterprise as an economic entity. This includes the executives, employees, suppliers, customers, and thus in effect all the company's relations with its immediate and broader environment (Schwerk, 2012). A consistent implementation of CSR is thus manifested in the real estate industry in addition to the management functions in the value chain.

Implementation of sustainable management follows a clear and stringent logic

Achieving sustainability-oriented corporate management requires a structured integration of all the above-mentioned CSR elements in the values and objectives catalog, planning the appropriate programs of action in various corporate functions and operational processes, and in the subsequent monitoring of any implementation failure within the company. In this context, it is important to differentiate and integrate the sustainable corporate governance at the normative, strategic, and operational levels (Baumgartner, 2010). Figure 21.1 illustrates the implementation steps and the individual consideration levels, which are explained in more detail in the next section.

Contextual factors as the basis for individualized approach

When undertaking the first implementation, it should be noted that it is either not possible or does not make sense to proceed without serious consideration of the specific business situation. In other words, there is no single solution. Rather, the individual, so-called contextual factors of each real estate development company are of particular importance. Mainly the demands of internal and external stakeholders, as well as the various factors in the general business and industry environment, should be considered in the context of sustainability.

Critical aspects of sustainable property development

Sustainability orientation needs to be fully integrated into value-added features of the real estate company. It is impossible to cover details of all of these aspects in a short chapter. Therefore, we will briefly discuss a couple of critical aspects that should be considered when discussing the development of sustainable real estate.

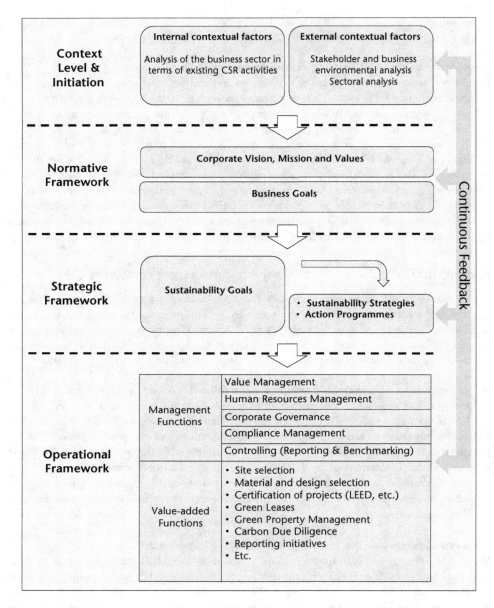

Figure 21.1 Implementation of sustainable management in real estate development companies

Source: Authors based on Schwerk (2012) and Baumgartner (2010)

Green building as the starting point

Of course, the crucial starting point for a sustainable real estate development is the project itself. Terms like "green building" or "sustainable property" and "responsible property investment" are widespread in the real estate sector (Roberts et al., 2007), although a really satisfactory "green building" definition is always difficult to formulate, since it is a controversial topic. The United States Green Building Council defines it as:

> the planning, design, construction, and operation of buildings with several central, foremost considerations: energy use, water use, indoor environmental quality, material section and the building's effects on its site.
>
> *(US Green Building Council, 2016)*

Figure 21.2 shows the elements that can be cited as relevant features of green buildings.

Already a market standard: property certification and labels

Looking to the increased attention paid to real estate market stakeholders regarding sustainability issues, certification labels have been tried worldwide, in an attempt to create standards for rewarding particularly sustainable buildings, so-called "green buildings." These certification systems work together with planning instruments and with consideration for aspects of sustainable construction, such as energy efficiency, life cycle assessment, and life cycle costs. The certification scheme is designed to provide tenants and investors, property developers, and project developers with verifiable proof of the sustainable quality of the building after its completion. However, certification schemes rely not only on ecological aspects, but also on offering a number of advantages for a sustainability-oriented enterprise:

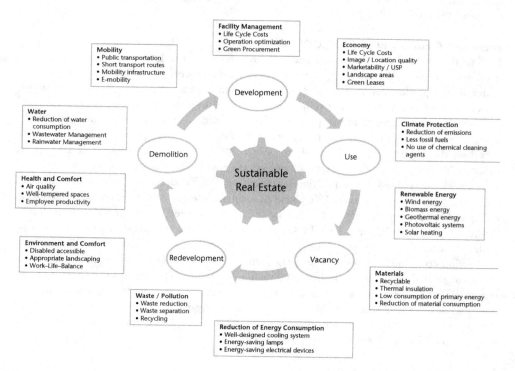

Figure 21.2 Green building features in a building's life cycle

Source: Authors based on ZIA, 2015

- reduction and control of pollution of buildings;
- improved national and international comparability of the quality of real estate;
- improved competitiveness over the entire life cycle;
- reduced life cycle costs and insurance amounts;
- simplified financing or favorable lending;
- minimizing the risk of vacancies due to increasing demand for certified real estate (especially in government agencies and multinationals);
- improved marketability as a result of existing excess demand in the investment market.

Nowadays many different national certificates exist, which have been developed and awarded by the Green Building Councils in the various countries. On an international level, there are the stablished English Building Research Establishment Environmental Assessment Method (BREEAM) and the American Leadership in Energy and Environmental Design (LEED).

In Germany, the Deutschen Gesellschaft für Nachhaltiges Bauen (DGNB) also plays an important role (Ebert et al., 2010; RICS, 2013b). The Green Star (Australia), HQE (France), and the CASBEE (Japan) also have high regional significance. These systems share similar criteria in the fields of economy, ecology, and social issues, but with different priorities, and the standards for achieving these goals are determined in comparison with national building standards. Especially in the field of large-volume projects, certifications are now designated as the market standard and thus provide less of a unique selling proposition for a project, but in many markets are already a sine qua non. The most widely used certification system globally, LEED, has certified more than 90,000 and 140,000 registered buildings (US Green Building Council, 2014; UNEP FI, 2012; RICS, 2013a).

Increased focus on LCA and LCC

Due to the wide range of built-up materials in stocks and the longevity of properties, it is becoming increasingly important to obtain reliable data concerning the LCA of construction materials and production inputs, and also to check and monitor this data. This is of particular importance when it comes to the subject area of CO_2 footprints.

There is general agreement that greater inclusion of natural cycles on economic value chains is of increasing importance. By focusing on biodegradable raw materials and increasing recycling, this circuit concept is actively supported. While ecological systems generally consist of closed material cycles, this is certainly not yet the case in the real estate industry.

It is therefore the task of product development to enable the development of such innovative solutions. The stronger focus on Life Cycle Assessment (life cycle analysis, often referred to as LCA) and concomitantly, Life Cycle Costing (LCC) are targeted steps in this direction (RICS, 2013b). The term LCC, or the total costs of a product – here the property itself – should be understood as for its entire lifetime. Generally, in this case, the initial cost (investment, production costs) and follow-up costs (usage, renewal, maintenance or demolition costs) are to be distinguished from one another (Wübbenhorst, 1984). The term LCA describes the systematic analysis of environmental aspects and potential environmental impacts during the life cycle of a product, that is, in relation to the widest possible application "from the cradle to the grave." Environmental influences are shown and analyzed across the entire life cycle of a property (Feifel et al., 2009).

Innovative technical solutions

Innovation management is one of the three most important aspects of a business strategy (Capgemini, 2010). This is also true for the real estate industry, especially in terms of technology, quality construction, materials, and employed processes. Because the real estate sector, as mentioned, has a very high proportion of global resource consumption, it plays a central role in generating innovations for greater dematerialization (i.e. the reduction of materials used) and substitution (i.e., the use of ecologically offered products). Integral planning and thus the early inclusion of a broad variety of disciplines in the development of the projects is a useful tool to achieve progress in this area.

Truly innovative in this context is the research area of bionic buildings, which take nature as a role model for building components and try to implement them architecturally, while increasingly changing the working environments of the users of buildings (Rabe von Pappenheim, 2012). The continuous monitoring of innovative breakthroughs is needed, not only to meet the quality requirements of users in the future, but also the quantitative development of space requirements. In that sense, the real estate development company plays a major role to overcome the apparent contrast between increasing density and healthy, livable neighborhoods with innovative real estate projects. For that, the involvement of different stakeholders in the planning phase is essential, especially for companies that are involved in neighborhood and large project developments.

Retrofitting the property stock

A major part of the existing property stock in many countries requires some degree of retrofit in order to improve energy efficiency. Refurbishing the existing property stock is key to reducing carbon emissions from heating and other energy-related demands. In addition to improving the energy consumption in the use phase of the properties, the real estate industry should also address the mitigation of indirect emissions, incurred in the course of the construction. It is interesting, that the advantages of a new building compared to the refurbishment of the stock through a so-called "whole life carbon approach" can relativize the decision-making (RICS, 2013a), since the already-emitted CO_2 (so-called "sunk cost") may no longer influence the decision.

Brownfield vs greenfield development

In many cities, the saturated urban infrastructure and scarcity of available areas, combined with the high value of land, makes it virtually impossible to carry out large real estate projects in consolidated central areas. Given the lack of options in the urban core, urban expansion ultimately occurs horizontally, in an "urban sprawl" manner. However, the urban sprawl in undeveloped areas consumes large amounts of agricultural land, eliminates forests, seizes habitats and natural resources, requires considerable public investments in infrastructure, increases the demand for energy, and promotes the intensive use of private vehicles. Thus, this pattern of unlimited urban expansion is no longer viable, and in fact, is regarded as an outdated, highly unsustainable urban model. The search for solutions for growth within the urban core must restore urban environmental quality, from the perspective of compactness and diversity, promoting diversity of use in a dense and compact urban fabric. In this sense, the redevelopment of urban areas (brownfield) is a suitable alternative to be explored, as it takes advantage of existing resources, and reconnects the urban fabric of underutilized areas of the city.

Brownfields are potentially sustainable development opportunities for neighborhoods in which these kinds of sites exist, and can provide a range of benefits (Glumac et al., 2014). For example, the physical improvements of a brownfield result in better site characteristics, such as lower contamination levels, better site accessibility, and ultimately improve the image of a neighborhood. Brownfield sites present particular challenges to national and regional policy makers in terms of bringing the land back into beneficial use and in terms of cleaning up contaminated land and groundwater. While contaminated land management aims at the management or elimination of risks, the primary objective of brownfield redevelopment is the reuse of land and reintegration of properties into the economic cycle. Brownfield redevelopments entail much broader aspects that need to be considered and converted into action to achieve effective solutions. Not only the environmental aspect, but also the political, social and economic context of brownfields becomes clearer through an analysis of related aspects in these areas (Schädler et al., 2011).

On the other hand, greenfield developments (business as usual) lead to a less complicated ownership structure and might result in a faster approval process by local governments. Nevertheless, the physical improvements of land quality and legal endeavors often lead to financial benefits, such as value capturing or improved real estate value. However, the need to deal simultaneously with physical improvements, legal frameworks, financing, and the associated trade-offs may explain why brownfield problems are not easily resolved. Dealing with brownfields in this broader sense also means dealing with the different interests of a variety of stakeholders, including regulators, investors, landowners, developers, consultants, academics, community groups, technology providers, and the financial sector.

Conclusion

Sustainable real estate development can deliver healthy and energy-efficient property, with benefits for all stakeholders, being the users, investors, or on a broad scale, the city. Every property type can be built or refurbished with sustainable attributes, to a highly variable degree. Mitigation and adaptation options for real estate development companies demand a full range of interdisciplinary actions, at property, company, and city level.

Therefore, a coherent strategy for sustainable development demands transparency and true commitment. Communication, planning, and coordination are the key to establishing strategies that harmonize the requirements and priorities of distinct stakeholders, aligned with national and regional climate policies, regulations and programs. A systemic view, accounting for causal relations between every stakeholder, is needed to integrate all sources of information relevant to the implementation process and, in that sense, the development of a sustainable development strategy and implementation plan establishes a common base, enabling stakeholders to identify both adaptation and mitigation options, and find common ground.

It is clear that the real estate sector plays an important role in achieving the commitments made by many countries with the Paris Agreement. For many years, "to raise awareness" was a dominant mantra when discussing the subject of sustainable development. Now, there is already a critical mass of knowledge, expertise, and technology available, which allows the real estate sector to lead the way and take action towards a more sustainable real estate development.

References

Auer, L. von. 2005. *Ökonometrie. Eine Einführung*, 2nd edition. Berlin: Springer.

Baumgartner, R. J. 2010. *Nachhaltigkeitsorientierte Unternehmensführung*. Munich: Mering.

Bienert, S. 2014. *Extreme Weather Events and Property Values: Assessing New Investment Frameworks for the Decades Ahead*. London: Urban Land Institute.

Bleis, T. 1995. *Öko-Controlling*. Frankfurt am Main: P. Lang.

Brandes, U. and LeBlanc, A. 2013. *Risk and Resilience in Coastal Regions*. Washington, DC: Urban Land Institute.

Brounen, D. and Kok, N. 2011. On the Economics of Energy Labelling in the Housing Market, *Journal of Environmental Economics and Management*, Vol. 62, 166–179.

Brown, O. 2008. *From Feast to Famine: After Seven Good Years What Now for Commodity Producers in the Developing World*. Manitoba: International Institute for Sustainable Development.

Capgemini Consulting. 2010. *Global Innovation Survey*. London.

Clark, G. L., Feiner, A., and Viehs, M., 2014. *From the Stockholder to the Stakeholder: How Sustainability Can Drive Financial Outperformance*. Oxford: University of Oxford and Arabesque Partners.

Dent, P., Patrick, M., and Xu, Y. 2012. *Property Markets and Sustainable Behavior*. Abingdon: Routledge.

Dhakal, S. 2010. GHG Emissions from Urbanization and Opportunities for Urban Carbon Mitigation, *Current Opinion in Environmental Sustainability*, Vol. 2, No. 4, 227–283.

Ebert, T., Eßig, N., and Hauser, G., 2010. *Zertifizierungssysteme für Gebäude: Nachhaltigkeit bewerten – Internationaler Systemvergleich – Zertifizierung und Ökonomie*. Munich: Detail.

Energy Information Administration (EIA). 2013. *International Energy Outlook 2013*. Washington, DC: US Department of Energy.

Engshuber, W. 2013. Transparency, Engagement, Bringing Down Barriers, in: *PRI Annual Report 2013*. London: PRI Association.

Epstein, M. J., and Buhovac, A. R. 2010. Solving the Sustainability Implementation Challenge, *Organizational Dynamics*, Vol. 39, No. 4, 306–315.

European Council. 2010. Directive 2010/31/EU of the European Parliament and of the Council of 19 May 2010 on the energy performance of buildings (recast), L 153/13, Brussels.

European Council. 2012. Directive 2012/27/EU of the European Parliament and the Council of 25 October 2012 on energy efficiency, amending Directives 2009/125/EC and repealing Directives 2004/8/EC and 2006/32/EC, L315/1, Brussels.

European Council. 2013. Directive 2002/91/EC of the European Parliament and of the Council of 16 December 2002 on the energy performance of buildings, L 1/65, Brussels.

Feifel, S., Walk, W., Wursthorn, S., and Schebek, L. (eds). 2009. *Ökobilanzierung 2009 – Ansätze und Weiterentwicklungen zur Operationalisierung von Nachhaltigkeit*. Karlsruhe: Karlsruher Institut für Technologie.

Fulton, M., Kahn, B. M., and Sharples, C. 2012. *Sustainable Investing: Establishing Long-Term Value and Performance*. Frankfurt am Main: DB Climate Change Advisors, Deutsche Bank Group.

Glumac, B., Han, Q., and Schaefer, W. M. 2014. Actors' Preferences in the Redevelopment of Brownfield: Latent Class Model, *Journal of Urban Planning and Development*, Vol. 141, No. 2.

Gouldson, A., Colenbrander, S., Sudmant, A., Godfrey, N., Millward-Hopkins, J., Fang, W., and Zhao, X., 2015. Accelerating Low-Carbon Development in the World's Cities. Contributing paper for Seizing the Global Opportunity: Partnerships for Better Growth and a Better Climate. New Climate Economy, London and Washington, DC. http://newclimateeconomy.report/misc/working-papers

Guyatt, D. et al. 2011. Climate Change Scenarios: Implications for Strategic Asset Allocation, Mercer LLC.

Hoornweg, D., Sugar, L., and Gomez, C. L. T. 2011. Cities and greenhouse Gas Emissions: Moving Forward, *Environment & Urbanization*, Vol. 23, No. 1, 207–227.

Howitt, R. 2013. Report on Corporate Social Responsibility: promoting society's interests and a route to sustainable and inclusive recovery (2012/2097(INI)) A7-0023/2013, European Parliament, Committee on Employment and Social Affairs, draft parliamentary resolution – Plenary sitting, Brussels, January 29.

IEA (International Energy Agency). 2015. *World Energy Outlook 2015*. Paris: International Energy Agency.

IIGCC. 2013. Protecting Value in Real Estate: Managing investment risks from climate change. London: Institutional Investors Group on Climate Change (IIGCC). www.iigcc.org/files/publication-files/IIGCC_Protecting_Value_in_Real_Estate.pdf

IPCC, 2007. Summary for Policymakers. In: *Climate Change 2007: The Physical Science Basis. Contribution of Working Group I to the Fourth Assessment Report of the Intergovernmental Panel on Climate Change*, Solomon, S., D. Qin, M. Manning, Z. Chen, M. Marquis, K.B. Averyt, M. Tignor and H.L. Miller (eds). Cambridge: Cambridge University Press, pp. 1–18.

IPCC, 2013. *Intergovernmental Panel on Climate Change. Climate Change 2013: The Physical Science Basis. Contribution of Working Group I to the Fifth Assessment Report of the IPCC*. Cambridge: Cambridge University Press.

Kennedy, C.A., Ramaswami, A., Carney, S., and Dhakal, S. 2009. Greenhouse Gas Emission Baselines for Global Cities and Metropolitan Regions, paper presented at the World Bank's Fifth Urban Research Symposium, Marseille, France, June 28–30.

Laasch, O. and Conaway, R. N. 2014. *Principles of Responsible Management: Global Sustainability, Responsibility, and Ethics*. Stamford, CA: Cengage Learning.

Leurig, S. and Dlugolecki, A. 2011. *Insurer Climate Risk Disclosure Survey: 2012 Findings & Recommendations.* Boston: CERES.

Maier, G. and Herath, S. 2015. *Immobilienbewertung mit hedonischen Preismodellen – Theoretische Grundlagen und praktische Anwendung.* Wiesbaden: Springer.

Makower, J. 2013. *State of Green Business Report 2013.* Oakland, CA: GreenBiz Group.

Messervy, H., McHale, C., and Spivey, R. 2014. *Insurer Climate Risk Disclosure Survey, Report & Scorecard, 2014: Findings & Recommendations.* Boston: CERES.

Murray, P.E. and Cotgrave, A. J. 2007. Sustainability Literacy: The Future Paradigm for Construction Education?, *Structural Survey*, Vol. 25, No. 1, 7–23.

Nelson, A., Rakau, O., and Doerrenberg, P. 2010. *Green Buildings – A Niche Becomes Mainstream.* Frankfurt: Deutsche Bank Research.

NOAA (National Oceanic and Atmospheric Administration). 2016. State of the Climate: Global Analysis for February 2016. Asheville, NC: National Climatic Data Center.

Olivier, J. G. J., Janssens-Maenhout, G., Muntean, M., and Peters, J. A. H. W. 2015. *Trends in Global CO$_2$ Emissions: 2015 Report.* The Hague: PBL Netherlands Environmental Assessment Agency; Brussels: Joint Research Centre.

Porter, M. and van der Linde, C., 1995. Green and Competitive: Ending the Stalemate, *Harvard Business Review*, Vol. 73, No. 5, 120–133.

Rabe von Pappenheim, J. 2012. Schaffe neue Arbeitswelten, in: *Grundsätze nachhaltiger Unternehmensführung*, Günther, E. and Ruter, R. X. (eds). Berlin: Erich Schmidt, pp. 103–119.

RICS. 2013a. *Sustainable Construction: Realising the Opportunities for Built Environment Professionals.* London: RICS Europe Sustainability Task Force.

RICS. 2013b. *Going for Green: Sustainable Building Certification Statistics Europe.* Frankfurt: RICS Deutschland.

Roberts, C., Rapson, D., and Shiers, D. 2007. Social Responsibility: Key Terms and their Uses in Property Investments, *Journal of Property Investment & Finance*, Vol. 25, 388–400.

Sayce, S., Ellison, L., and Parnell, P. 2007. Understanding Investment Drivers for UK Sustainable Property, *Building Research & Information*, Vol. 35, No. 6, 103–119.

Schädler, S., Morio, M., Bartke, S., Rohr-Zänker, R., and Finkel, M. 2011. Designing Sustainable and Economically Attractive Brownfield Revitalization Options Using an Integrated Assessment Model, *Journal of Environmental Management*, Vol. 92, No. 3, 827–837.

Schleich, H., 2012. *Sustainable Property Portfolio Management – With Special Consideration of Energy Efficiency Improvements in the Property Portfolio Stock.* Cologne: Immobilienmanager-Verlag.

Schwerk, A. 2012. Strategische Einbettung von CSR in das Unternehmen, in: *Corporate Social Responsibility – Verantwortungsvolle Unternehmensführung in Theorie und Praxis*, Schneider, A. and Schmidpeter, R. (eds). Heidelberg: Springer, pp. 331–356.

Seidel, E. and Menn, H. 1988. *Ökologisch orientierte Betriebswirtschaft.* Stuttgart: W. Kohlhammer.

Seto, K. C., Dhakal, S., Bigio, A., Blanco, H., Delgado, G. C., Dewar, D., Huang, L., Inaba, A., Kansal, A., Lwasa, S., McMahon, J. E., Müller, D. B., Murakami, J., Nagendra, H., and Ramaswami, A. 2014. Human Settlements, Infrastructure and Spatial Planning, in: *Climate Change 2014: Mitigation of Climate Change. Contribution of Working Group III to the Fifth Assessment Report of the Intergovernmental Panel on Climate Change*, Edenhofer, O., Pichs-Madruga, R., Sokona, Y., Farahani, E., Kadner, S., Seyboth, K., Adler, A., Baum, I., Brunner, S., Eickemeier, P., Kriemann, B., Savolainen, J., Schlömer, S., Stechow, C. von, Zwickel, T., and Minx, J. C. (eds). Cambridge: Cambridge University Press, pp. 1–1454.

Stern, N. 2006. *The Economics of Climate Change: The Stern Review.* Cambridge: Cambridge University Press.

Sugar, L., Kennedy, C. and Leman, E. 2012. Greenhouse Gas Emissions from Chinese Cities, *Journal of Industrial Ecology*, Vol. 16, No. 4, 552–563.

UN-FCCC. 2016. United Nations Framework Convention on Climate Change. COP21 Report of the Conference of the Parties on its twenty-first session, FCCC/CP/2015/10/Add.1).

UN-DESA. 2014. World Urbanisation Prospects. Population Division of the United Nations Department of Economic and Social Affairs. https://esa.un.org/unpd/wup/Publications/Files/WUP2014-Highlights.pdf.

UN-DESA. 2015. World Population Prospects. Population Division of the United Nations Department of Economic and Social Affairs. https://esa.un.org/unpd/wpp/publications/files/key_findings_wpp_2015.pdf.

UN DP. 2016. United Nations Development Programme. Sustainable Development Goals. Goal 13: Take urgent action to combat climate change and its impacts.

UNEP FI. 2012. Property Working Group of the United Nations Environment Programme Finance Initiative (Editors), 2012. Responsible Property Investment – What the leaders are doing, 2nd edition.

UN-Habitat. 2011. *Cities and Climate Change: Global Report on Human Settlements, 2011.* United Nations Human Settlements Programme.

US Green Building Council. 2014. Infographic: LEED in the World. www.usgbc.org/articles/infographic-leed-world.

Wensen, K., Broer, W., Klein, J., and Knopf, J. 2011. *The State of Play in Sustainability Reporting in the EU.* Brussels: European Commission.

World Bank, 2014. *Urban China: Toward Efficient, Inclusive, and Sustainable Urbanization.* Washington, DC: World Bank.

Wübbenhorst, K. 1984. Konzept der Lebenszykluskosten; Grundlagen, Problemstellungen und technologische Zusammenhänge (Diss.), Techn. Hochschule Darmstadt.

ZIA (Zentraler Immobilien Ausschuss) 2015. *Nachhaltige Unternehmensführung in der Immobilienwirtschaft.* Cologne: Zentraler Immobilien Ausschuss.

22

The foundations of sustainability in real estate markets

Richard Reed

Abstract

The chapter examines the concept of sustainability in the built environment and why sustainability has now been fully embedded into the built environment with reference to new and existing buildings, rather than viewing sustainable attributes as "optional" or a sustainable building as "different." In today's real estate market it is rare for a new building to not incorporate some form of sustainability, however this relatively high level of acceptance has taken decades to be embraced in the mainstream built environment. Initially, in the 1990s sustainable buildings were referred to as "green buildings," which then evolved to "sustainable buildings" in the twenty-first century. The chapter commences by examining the commonly accepted definition of sustainability from the Brundtland Report (1987) which introduces the framework for achieving a balance between economic and environmental competing interests. The next section introduces the concept of sustainability into the built environment with reference to how to undertake sustainable development. Here it examines the concept of the "circle of blame" regarding sustainability and how various stakeholders have different levels of responsibility for undertaking sustainable development. The discussion is focused on examples of global sustainable building rating systems, ranging from "design" rating systems to "in use" rating systems. The chapter acknowledges that although most stakeholders in the real estate market are typically profit-seeking with the underlying aim of "maximizing shareholder returns," sustainability can now contribute both directly and indirectly to bottom-line profit rather than being an added financial cost.

Introduction

The chapter examines the concept of sustainable real estate development in the built environment. It considers the impact of and relevance to the term "sustainability" which now is fully embedded in today's mainstream economy and society, as well as when undertaking the development process in the property and real estate discipline (Jailani et al., 2015). The chapter commences with a review of the drivers behind the introduction of sustainability into the built environment. Attention is then placed on the challenges initially faced when the concept of sustainability was introduced, including reference to the "circle of blame" and other barriers such as a lack of understanding about sustainable benefits, uncertainty about how sustainability can be incorporated into a sustainable building, and the indirect economic benefits available from incorporating sustainability such as potentially lower employee absenteeism.

When a new real estate development or a redevelopment of an existing site in today's market is undertaken it is now most likely to incorporate some levels of sustainable attributes (Andelin et al., 2015), however this wasn't always the scenario. This decision may be linked to market pressure from other competing buildings that are perceived to be "sustainable," or at times may be due to demand from tenants or owners seeking a minimum level of sustainability. Alternatively, this decision could be based on a legal requirement affecting real estate development; an analogy can be made with vehicle manufacturers in some countries who are legally restricted to producing vehicles that use only unleaded fuel in order to reduce exhaust emissions. Previously there were very few buildings incorporating sustainable features. However, in the real estate market these buildings offered perceived benefits for investors and tenants and an above-market premium was observed in some instances, although the additional construction cost did not exceed the added benefit in most cases (Cajias and Piazolo, 2013). Today this relationship has changed, where the number of new sustainable buildings (i.e. a building with some degree of sustainability) is now in the majority in some markets. Therefore, buildings with no sustainable attributes are discounted from the market rate rather than a premium paid for sustainable buildings (Mandell and Wilhelmsson, 2011). Therefore, every benchmark real estate development now includes some level of sustainability, either directly or indirectly.

The evolution of sustainability into mainstream acceptance also acknowledges the successful contribution of green building organizations in raising the profile of sustainable buildings in the marketplace, especially to investors and occupiers. Most stakeholders have become familiar with previously little used sustainable terms which are now in mainstream use including photovoltaic cells, green buildings, natural lighting, and hot-desking. In addition, the uptake of sustainability has been observed on a global basis in both developed and developing countries across different continents including the UK, USA, China (Zhai et al., 2014), India (Reed, 2015a) and New Zealand (Warren-Myers and Reed, 2010). This chapter is presented as follows: the next section discusses global change and how sustainability has evolved, and is followed by a discussion on the incorporation of sustainability into real estate development. This is followed by an examination of the barriers to the uptake of sustainability.

Global change and the evolution of sustainability

The early focus on sustainability in the built environment was linked to the increased profile of global change or climate change with increasing concern about the substantial contribution of buildings to CO_2 emission levels. There is a substantial body of knowledge about climate change and varying views on the level of change that is occurring, however there appears to be a general consensus that climate change is "real" and the human race's effect on the planet needs to be considered in a long-term perspective. Here, a starting point for sustainable real estate development is to consider the different types of change in the built environment that collectively contribute to global change, as listed in Table 22.1. While most of the focus is commonly placed only on environmental change, other critical aspects of change include social change, political change, and economic change. This also highlights the debate about "sustainable real estate development" being an oxymoron; for example, "how can one develop a parcel of land or a property and view it as sustainable?" The factors in Table 22.1 highlight

Table 22.1 Factors affecting changes in real estate development

Aspects	Phenomenon
Social change	• demographic changes • population growth • aging • urbanization • migration trends • gentrification • spatial segregation in cities
Ecological change	• global environmental changes • global warming • deforestation • acidification of the seas • loss of biodiversity
Political change	• Kyoto Protocol – signed in 1997 • legislation on energy savings and sustainability, resource savings and renewable energy • certificate trading • Sustainable Innovation Forum, Paris (2015)
Economic change	• changes in the job market in the construction industry • new environmental technologies in the construction sector • programs for environmental and energy-efficient buildings and urban redevelopment • consideration of the whole life cycle of buildings (construction and operating costs)

Source: Ebert et al., 2010; Dolezala and Spitzbart-Glasla, 2015

the different dimensions of changes affecting the real estate market where all improvements (i.e. buildings) begin to age and become obsolete from when they are on the architect's drawing board. So clearly all land needs to be improved to some extent, however it is the sustainable nature of this development which requires careful consideration to address the factors affecting changes in real estate development.

From an international perspective it accepted that the evolution of sustainable buildings commenced with the 1987 publication of *Our Common Future* which is most commonly referred to as the *Brundtland Report* (WCED, 1987) which arose from a World Commission meeting. In this publication the most commonly used definition of sustainability was stated as:

> Development that meets the needs of the present without compromising the ability of future generations to meet their own needs.
>
> *(WCED, 1987, p. 40)*

The benchmark definition has stood the test of time and is still commonly referred to, even though three decades have now passed since 1987. Note the acknowledgement in the definition regarding the intergenerational effect of sustainable behavior and the responsibility of one generation to provide sufficient resources for the following generations to meet their own needs. Since then many different definitions of sustainability have been promoted. For example a later definition of sustainability was stated as:

> Sustainable development involves the simultaneous pursuit of economic prosperity, environmental quality and social equity. Companies aiming for sustainability need to perform not against a single, financial bottom line but against the triple bottom line.
>
> *(Elkington, 1998, p. 397)*

With climate change and sustainability gaining a higher profile there was an accepted emphasis on "treading lightly" on the planet's finite resources. However one of the barriers to the acceptance of sustainability in society and the economy was linked to defining what "sustainability" actually is. Following the release of the *Stern Review* in 2006 it was evident that stakeholders in the built environment needed to take a stronger stance by incorporating sustainable attributes into new and existing buildings (Stern et al., 2006). There are similarities between the earlier definition provided by the *Bruntland Report* (WCED, 1987) and the later WBCSD definition where both refer directly or indirectly to social, economic and environmental aspects of society. In the WBCSD definition there is direct reference to the triple bottom line approach (Elkington, 1994) which has since been widely recognized when seeking to incorporate sustainability. Figure 22.1 highlights the need to consider all three factors (i.e. economic, social, and environmental) in a sustainable context rather than being limited only to environmental factors, for example.

Building upon the three pillar model in Figure 22.1, a fourth pillar was then added that highlighted the importance of incorporating a cultural perspective (Figure 22.2). This provided balance where the four pillars then collectively contributed to a "sustainable community" which typifies the changing view of many neighborhoods, towns, and cities towards embracing sustainability.

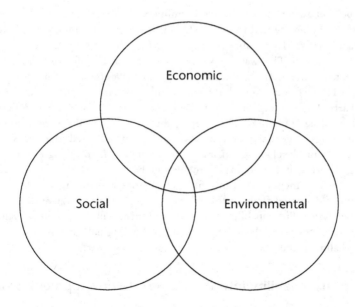

Figure 22.1   "Three Pillars" model of sustainable development based on triple bottom line accounting

Source: Elkington, 1994

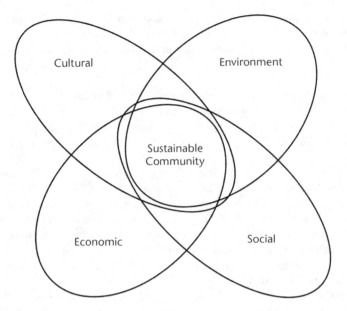

Figure 22.2 "Four Pillars" model of sustainability

Source: Hawkes, 2001

The next important aspect is to consider the timing of changes in sustainability affecting real estate development. As highlighted in Figure 22.3, this can be traced back to the early 1980s, where the climate change debate and the argument that the planet was becoming "hotter" became a mainstream topic. This was followed by the increasing profile of climate change and the broader acceptance in society towards the end of the twentieth century, partly driven by the uptake of social media, coupled with the rise of political parties with a sustainable platform, e.g. the Green Party. The ensuing step was to define sustainability, which was followed by the increased profile of "green buildings." However, by the early twenty-first century the use of the term "green buildings" decreased and it has now been accepted that most real estate developments have some level of sustainability incorporated in the design and construction phases. Looking forward, it is anticipated there will be further technological advances in sustainability that will be incorporated into future real estate developments. This may include, for example, improved storage facilities for electrical energy and innovative design features to maximize heating and cooling capability. It is envisaged there will be ongoing developments in sustainability, however they can all be traced back to the original catalyst being the need to address climate change in the 1980s.

Incorporating sustainability into the real estate development process

It is generally accepted that sustainability, to varying levels, has been adopted in some form or another by all stakeholders who are undertaking real estate development, and this has become "business as usual" (Reed and Sims, 2015). In today's market it would be more unusual for a real estate development *not to include* at least some sustainable attributes; however, it was

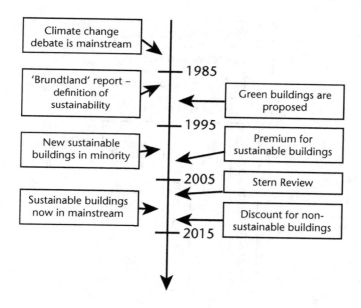

Figure 22.3 Evolution of sustainability in the built environment
Source: Reed and Krajinovic-Bilos, 2013

previously more unusual to include any sustainable attributes. For example it is now commonplace for a development to include some form of rainwater collection, usage of natural light, and on-site generation of electricity, e.g. photo-voltaic cells. Traditionally, the value of a building depreciates by incurring a loss in value, which is primarily due to the negative effect of one or multiple forms of obsolescence. The three primary forms of obsolescence that affect buildings are functional, economic, and physical (Australian Property Institute, 2015). Functional obsolescence refers to the ability of the improvement to operate to maximum productivity, economic obsolescence refers to the actual cost of maintaining the asset, and physical obsolescence refers to the wearing out of the improvements.

One of the challenges with incorporating sustainability in real estate development is identifying and understanding what sustainability actually is, and this requires more discussion. Being a relatively broad discipline (API, 2015), sustainability can be divided into different sub-types (Reed and Sims, 2015), as shown in Figure 22.4. The sub-types are constantly changing and can be summarized as follows.

- Climate change is the overarching catalyst that has led to the higher profile of sustainability.
- Embodied sustainability is the energy used to manufacture a good. For example, there is more embodied energy used in the manufacture of a concrete column as opposed to manufacturing a timber column.
- In-use sustainability refers to the use of energy in the operation of the building. This is the most commonly perceived type of sustainability but is only a minor part of the problem.

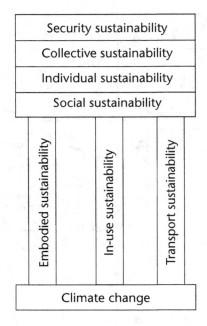

Figure 22.4   Different types of sustainability

Source: Reed and Sims, 2015

- Transport sustainability refers to the use of energy in the transport phase to access the development. For example, it is not productive to build a sustainable building in a remote location which would use a lot of energy to access.
- Evolving types of sustainability include social sustainability, individual sustainability, collective sustainability, and security sustainability, e.g. in response to terrorism.

When examining the process of real estate development in a sustainable context, the potential key sustainability issues change in each phase, as shown in Table 22.2. In addition to the impact on the environment, it is important to consider the impact of each change in more detail.

(a) Land for development

Potentially this is the phase of the development that has the most visible effect on the environment, as well as being the most obvious stage of the development to the broader community. A sustainable development should "tread lightly" on the earth's surface and will pay particular attention to minimizing the long-term effect on the environment from the development (Dolezala and Spitzbart-Glasla, 2015). Often this will incur additional costs for the developer, however, these costs are unavoidable in order to satisfy the requirements of planning authorities and relevant legislation.

(b) Development finance

Increasing focus has been on "green finance" which involves sourcing funds from environmentally responsible and conscious financiers.

(c) Planning

The additional emphasis in the planning phase aligns with the definition of sustainability. Consideration is given to the overall development life cycle, including the construction of real estate, operation of the building, and also the use of the site after the existing use is no longer relevant.

(d) Design and construction

Often underrated with many previous developments, substantial emphasis has been placed on ensuring the design phase optimizes the sustainable aspects of the building. With the construction phase there are opportunities to incorporate sustainable practices. including minimizing wastage and reducing the amount of embodied energy.

(e) Market research

Substantial effort is required to understand the changing perception of tenants and investors towards sustainability. For example, tenants may be willing to compromise on the use of air-conditioning if natural air is a viable substitute in the development.

(f) Promotion and selling

The demands of tenants and buyers have changed on an intergenerational basis. The awareness of current generations about environmental issues is substantially higher than previous generations. This needs to be fully considered in promotional and sales strategies.

Barriers to the uptake of sustainability in real estate development

The initial barriers to the uptake of sustainability included the lack of knowledge about sustainable alternatives in real estate development, the absence of a premium for a sustainable development (or a discount for a non-sustainable development), and the relatively few sustainable buildings in the marketplace. Although these barriers have decreased during the twenty-first century, other barriers to the uptake of sustainable buildings have appeared (Zhai et al., 2014). These include uncertainty about who is responsible for increasing the level of sustainability in real estate development, confusion due to the large number of sustainability rating tools, the requirement for a valid cost–benefit analysis, and the low level of communication in the post-development phase.

Responsibility for the uptake of sustainability

Although it was widely accepted that global change and climate change has been occurring and the built environment is one of the largest emitters of CO_2 emissions, there has been a level of uncertainty about who should take responsibility for increasing sustainability in real estate

Table 22.2 Real estate development stages and potential key sustainability issues

Real estate development stage	Potential sustainability issue
(a) Land for development	• Loss of habitat • Loss of biodiversity • Contamination (naturally occurring or due to previous use)
(b) Development finance	• Consider eco-financing
(c) Planning	• Transport • Ecology and site issues • Zoning and land use issues
(d) Design and construction	• Selection of contractors – including their CSR performance • Reducing carbon dioxide emissions • Minimizing pollution • Use of life-cycle costing or whole life costing techniques • Using resources efficiently • Waste management on site • Re-use of materials and recycling materials • Specification and selection of materials – health and embodied energy issues • Water • Health and wellbeing for users • Environmental assessment ratings
(e) Market research	• Awareness of changing social and cultural perceptions towards sustainability
(f) Promotion and selling	• Awareness of changing social and cultural perceptions towards sustainability

Source: Reed and Sims, 2015

developments (Cadman, 2000). As shown in Figure 22.5, there were four stakeholders, which arguably created a "circle of blame" where this process slowed down the initial uptake of sustainable practices in real estate development. This circular relationship highlighted the motives behind investors, occupiers, constructors, and developers. At times it is also referred to as the "vicious circle of blame."

An updated "circle of blame" was later produced (Warren-Myers and Reed, 2010) that incorporated the underlying aim of the four stakeholders and most organizations, which is to maximize shareholders' wealth and involves an evaluation of financial considerations. In Figure 22.6 the valuation considerations were central to the "circle of blame." It was argued that unless the financial argument was supported and there were increased value benefits for any or all stakeholders, then the level of sustainability would be restricted.

The relationship between stakeholders in every real estate development is crucial to the success of the project, however, this has an added dimension when considering sustainability. In contrast to the earlier resistance to sustainability in Figure 22.5 and Figure 22.6, the existing attitude towards sustainability is positive and driven by six stakeholders as shown in Figure 22.7. In more recent times, each of the stakeholders has acknowledged the positive contribution that sustainability can make to achieving their individual goals.

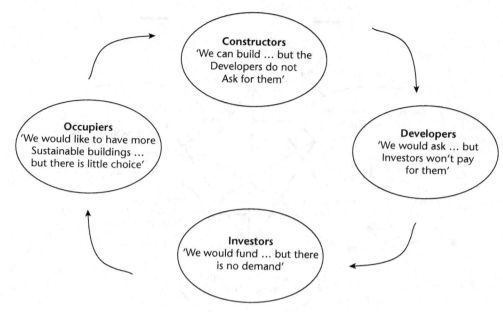

Figure 22.5 Circle of blame

Source: Cadman, 2000

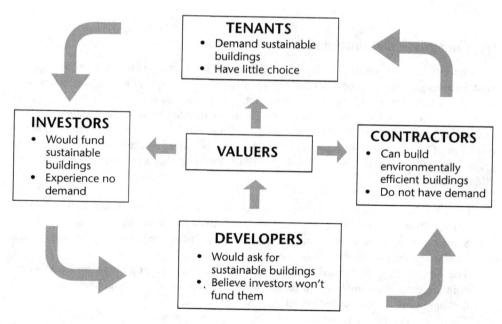

Figure 22.6 Modified circle of blame

Source: Warren-Myers and Reed, 2010

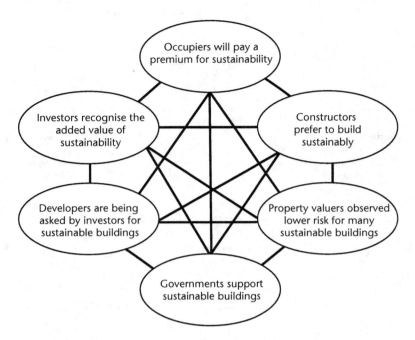

Figure 22.7   Views of stakeholders in sustainable real estate development

Source: Developed from Cadman, 2000

Large number of different sustainability rating tools

Over time there has been the development of sustainability rating tools, which are designed to assist in assessing the level of sustainability in particular buildings. Many organizations own buildings that are located in different countries, however, and this has created confusion in the real estate market due the large number of different sustainability rating tools which are unique to each region or country (Reed and Krajinovic-Bilos, 2013). As shown in Figure 22.8, the evolution of rating tools from the first generation has now moved to the more complex second generation rating tools that are currently available. The complications associated with many different rating tools include:

- Each rating tool measures different aspects of buildings and has been designed to be different from other rating tools. Therefore, the attributes measured also differ.
- Some rating tools examine a building in the "design" phase only, where other rating tools are focused on the "in-use" phase only.
- The outputs from individual rating tools vary. For example, one rating tool may rate a building on a numbered system (e.g. 1–6 stars) where another rating system may rate a building using a color-coded system (e.g. platinum, gold, silver).
- There is a diverse range of sustainability rating tools, which assess different sustainable aspects of buildings to varying levels. However, this has created a level of confusion in the marketplace for stakeholders, including investors, owners, and tenants. Whilst it is unlikely

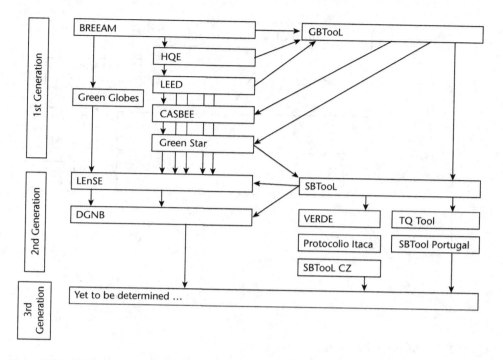

Figure 22.8 Evolution of sustainability rating tools
Source: Ebert et al., 2010

that sustainability rating tools will be standardized and aggregated into fewer tools, the onus rests with the end-user to understand the positive and negative attributes of each rating tool.

Cost versus benefit analysis for sustainability

One of the largest barriers to the uptake of sustainability in real estate development has been the lack of added value. When examining the relationship between market value and the construction cost of undertaking sustainable real estate development as shown in Figure 22.9, there was a substantial difference between: (a) the current market value when completed, and (c) the initial construction cost. However, over time, the financial gap (b) has reduced substantially since the innovative phase when sustainable buildings were first constructed (Cajias and Piazolo, 2013). The main reasons for this reduced gap include a reduced cost to incorporate sustainable features due to technological advances, wider uptake of sustainability attributes, which has reduced the initial cost, and the recognition by investors that sustainable buildings often have a higher long-term market value.

Lack of communication between building stakeholders

A major barrier in the uptake of sustainable real estate developments has been the relatively poor communication between stakeholders (Jailani et al., 2015). If a development benefits from

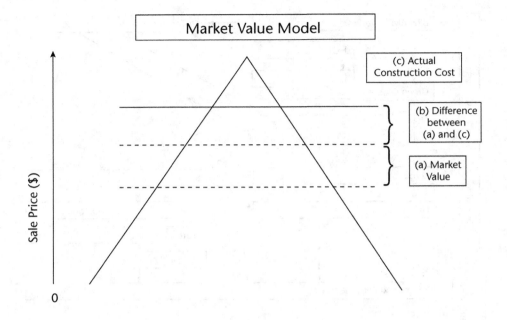

Figure 22.9   Market value versus construction cost

Source: Reed and Sims, 2015

incorporating sustainable attributes, this information is not always openly shared throughout the real estate industry and is referred to as a "knowledge gap" as shown in Figure 22.10. It has been shown that there are communication breakdowns between stakeholders who are involved in the different phases of a building's life cycle (Brown and Cole, 2009). This scenario is also evident between architects and occupants who rarely share knowledge about the real estate development from the design phase to the post-occupancy phase. In addition, this often restricts the operation of future sustainable buildings from taking full advantage of sophisticated design and advanced building operation systems. In other words, an architect focuses on the design phase of a real estate development and does not usually contact the tenants in the post-occupancy phase. Furthermore, the tenants or occupiers of the sustainable building are often unfamiliar with the characteristics of the sustainable building and how to maximize its efficiency level. There are no building operating manuals issued to new occupiers or tenants.

Conclusion

This chapter examined the concept of sustainability in the built environment, with relevance to real estate development. It examined global change and the evolution of sustainability from the latter part of the twentieth century and examined the factors affecting changes in real estate development, identifying the driving influences. The definition of sustainability provides the underlying foundation for how stakeholders in the real estate market need to incorporate sustainability into real estate development: the original "triple bottom line" evolved to a "four pillars" model of sustainability that incorporated cultural sustainability.

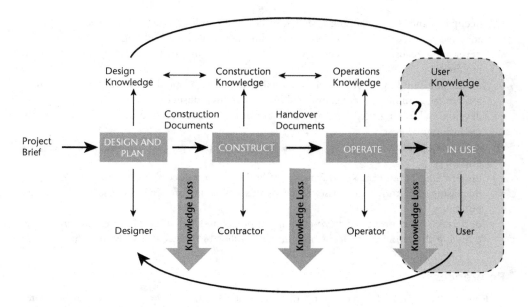

Figure 22.10 The knowledge cycle: Communication and knowledge sharing

Source: Jailani et al., 2015

The main barriers to the uptake of sustainability in real estate development were identified as:

- the "circle of blame" and allocating responsibility for the uptake of sustainability;
- the inclusion of value in a "modified circle of blame";
- the large number of different sustainability rating tools, which has created confusion in the market;
- cost versus benefit analysis when undertaking sustainability;
- lack of communication and knowledge sharing between building stakeholders about the operation of sustainable attributes in real estate developments.

Incorporating at least a basic level of sustainability when undertaking real estate development is now accepted as a minimum standard. An uptake of sustainability in real estate development will benefit this and future generations, however, this must be balanced with the needs of stakeholders who also have their own objectives, which often include a profit-seeking motive. In summary, a compromise between sustainable and financial agendas is required by all stakeholders to produce long-term sustainable solutions in real estate development.

Future research into sustainable real estate (development) will continue to evolve to meet the needs of stakeholders as these needs and demands change. For example, the initial catalyst behind sustainable buildings can be linked back to concerns about climate change and the warming of the planet, however there are many other related issues that have since arisen and others that are yet to surface. Examples include issues relating to:

- occupants and their changing needs in sustainable buildings;
- adaptive re-use of existing buildings including heritage-listed buildings;
- developing re-usable or "Lego block" type of buildings that can be easily adapted as required;
- promoting buildings that have a base shell that can be adapted to suit different uses (e.g. residential, commercial, education) without being demolished;
- developing buildings that have a small environmental footprint and can be replaced at low environmental cost;
- decreasing the amount of wastage in the construction process for new and renovated buildings;
- undertaking off-site mass construction of buildings to reduce the construction time;
- increasing the knowledge base of occupants about maximizing the use of sustainable buildings;
- undertaking a coordinated town planning approach to a sustainable built environment, e.g. reducing the number of vacant buildings, regardless of their level of sustainability.

References

Andelin, M., Sarasoja, A.-L., Ventovuori, T., and Junnila, S. 2015. Breaking the circle of blame for sustainable buildings: Evidence from Nordic countries. *Journal of Corporate Real Estate*, 17(1), 26–45.

API (Australian Property Institute) 2015. *The Valuation of Real Estate*, R. Reed (ed.), 2nd edition. Canberra: Australian Property Institute.

Brown, Z. and Cole, R. J. 2009. Influence of occupants' knowledge on comfort expectations and behaviour. *Building Research & Information*, 37(3), 227–245.

Cadman, D. 2000. The vicious circle of blame. In: Keeping, M. (ed.) *What About Demand? Do Investors Want Sustainable Buildings?* London: RICS Research Foundation.

Cajias, M. and Piazolo, D. 2013. Green performs better: Energy efficiency and financial return on buildings. *Journal of Corporate Real Estate*, 15(1), 53–72.

Dolezala, F. and Spitzbart-Glasla, C. 2015. Relevance of acoustic performance in green building labels and social sustainability ratings. *Energy Procedia*, 78, 1629–1634.

Ebert, T., Essi, N., and Hauser, G. 2010. *Zertifizierungssysteme für Gebäude*. Berlin: Green Books.

Elkington, J. 1994. Towards the sustainable corporation: win-win-win business strategies for sustainable development. *California Management Review*, 36(2), 90–100.

Elkington, J, 1998. *Cannibals with Forks: The Triple Bottom Line of 21st Century Business*. British Columbia: New Society Publishers.

Hawkes, J. 2001. *The Fourth Pillar of Sustainability: Culture's Essential Role in Public Planning*. Melbourne: Common Ground/Cultural Development Network.

Jailani, J., Reed, R., and James, K. 2015. Examining the perception of tenants in sustainable office buildings. *Property Management*, 33(4), 386–404.

Mandell, S. and Wilhelmsson, M. 2011. Willingness to pay for sustainable housing. *Journal of Housing Research*, 20(1), 35–51.

Reed, R. 2015a. India: Real estate development in the fastest growing free market democracy. In Squires, G. and Heurkens, E. (eds) *International Approaches to Real Estate Development*. Abingdon: Taylor & Francis, pp. 150–166.

Reed, R. and Krajinovic-Bilos, A. 2013. An examination of international sustainability rating tools: An update. 19th Annual Pacific Rim Real Estate Society Conference, Melbourne, January 13–16. Pacific Rim Real Estate Society. www.prres.net/

Reed, R. G. and Sims, S. 2015. *Property Development*. Abingdon: Routledge.

Stern, N. S. et al. 2006. *Stern Review: The Economics of Climate Change*. London: HM Treasury.

Warren-Myers, G. and Reed, R. 2010. The challenges of identifying and examining links between sustainability and value: Evidence from Australia and New Zealand. *The Journal of Sustainable Real Estate*, 2(1), 201–220.

WCED (World Commission on Environment and Development). 1987. *Our Common Future*. Oxford: Oxford University Press.

Zhai, X., Reed, R., and Mills, A. 2014. Addressing sustainable challenges in China: The case for increasing the uptake of off-site industrialisation. *Smart and Sustainable Built Environment*, 3, 261–274.

Part VI

Design in real estate development

23

Actor perceptions of good design for real estate development

Peter Hendee Brown

Abstract

Good design increases the economic value of a real estate development project and its intrinsic value to the community, but good design is in the eye of the beholder and there are often many eyes on a development project. As a result, the design process is not a solo act but rather an ensemble performance by a large cast of actors including neighbors, community groups, politicians, investors, lenders, buyers, and architects, led by the real estate developer. These actors represent a broad range of competing and sometimes conflicting interests and they exert varying degrees of influence on the design process and the physical form of the building. They also have very different ideas of what good design means, seeing a project differently based on whether they are more interested in their experience of the exterior or the interior of the building. The real estate developer must listen carefully and balance these interests to ensure a successful project – one that attracts capital, community support, city approvals, and interested buyers. This chapter begins with a cast list of the actors and their interests and then considers good design from three perspectives: high vs functional design, the monetary vs intrinsic value of property, and what users of real estate really want. The case of a high-rise condominium project in Chicago illustrates how one real estate developer successfully reconciled these competing interests and the chapter concludes by arguing that savvy developers increase their chances of success by using the design process to manage risk.

Introduction

Real estate developers combine a set of resources – land, improvements, and professional services – to increase the value of an existing piece of "real property" beyond the costs of capital and operating improvements. The difference between costs and value is the developer's profit.

But beyond the numbers, the result of the real estate development process is a tangible asset with physical attributes that make it valuable and worth paying a price to own or use, for a potential tenant, user, or buyer. Many of these physical attributes are the result of a design process. Good design produces more valuable real estate, whether a basic warehouse, a strip retail center, an apartment, an office tower, or a luxury hotel. Design must take into account a broad range of building systems, features, and characteristics including: location, site geometry, and soil conditions; size, mix, and layout of units or spaces; structural, mechanical, electrical, and plumbing systems; and exterior and interior building materials and finishes. A good design maximizes efficiency and reduces capital and operating costs while providing a functional and aesthetically pleasing building that will attract buyers, enhance the built environment, and contribute to the community.

Many people think that the design for a building is the product of an architect's creative vision. But it is the architect's client – the developer – who actually orchestrates the design process. The developer also has his or her own vision for a project, one that extends well beyond the building design. In addition to earning significant profits, the developer's motives include everything from the thrill of deal-making to city-building on a large scale, enhancing his or her corporate or personal image, creating a memorable place and experience for users of the building, or leaving a legacy in one's community. But the developer does not have free rein to implement his or her grand vision either, because there are a whole host of other actors with varied and often conflicting preferences who play direct and indirect roles in the design process. The developer, therefore, must do everything possible to identify and address the interests of these actors – often expressed through the language of design – to maximize his or her chances of success.

Introducing the actors and their interests

A real estate development project is a private sector business venture that is acted out on a public stage on a large scale. This section will serve as a cast list, introducing each of the typical actors on that stage and their usual interests and concerns.

The architect: What is my creative opportunity on this project?

The architect's objective is typically twofold. First, the architect seeks to exercise his or her creative abilities in the design of a unique building they can be proud of and that will potentially win awards and elevate the architect in the eyes of his or her peers. Second, they seek to earn a fee, satisfy the client, and generate repeat business.

The contractor: How can I earn my fee and generate repeat business?

The developer typically engages a contractor during the design phase to provide initial cost estimates based on the architect's preliminary designs. Through a process called "value engineering," the contractor will also make suggestions for changes to the design with the intent of preserving the architect's basic vision while simplifying construction and reducing costs.

The city planner: Does it comply with policies, plans, and codes?

The job of the city planner is to review a proposed project and any requests for variances and make a recommendation to city leadership to approve, modify, or reject the project. Planners consider site plan, massing, density, materials, uses, parking arrangements, traffic impacts, and other aspects of the design when determining whether or not the project conforms to a city's development policies, comprehensive plans, and codes.

The politician: Does it maximize growth without alienating my voting base?

Elected officials typically support development because new investment increases the tax base and creates jobs but they must also balance these benefits with the concerns of community members and nearby neighbors, if they hope to be re-elected. Politicians view the development process through a self-interested lens because they care most about whether their constituents think they are working on their behalf. They can help smooth the path for a project, kill it, or cause delays through indecisiveness.

Community members: Is it good for the community?

Community members want to ensure that a project fits into their community. Community organizations often promote appropriate development while conditioning their support by requiring the developer to make certain changes that reflect their local knowledge and preferences.

Nearby neighbors: How will it affect my property and me?

The people most affected by a development project are the nearby neighbors. First, they must suffer the temporary inconveniences of construction. More important, once the building is complete it may block views and sunlight, change surrounding traffic patterns, and attract new uses – and new neighbors. Existing neighbors typically care primarily about their own self-interests and they often oppose projects that the larger community supports. But just a handful of neighbors can still be very vocal, which is why community groups, city planners, and elected officials are careful to listen to their concerns.

Other special interest groups: Does it address our specific concern?

Special interest groups promote positions that go beyond the individual project, such as historic preservation, sustainability, public safety, transit, environmental protection, habitat preservation and restoration, and water and storm water use and management.

Investors: What is my rate of return and how risky is this deal?

The developer and sometimes a few investors fund the early "pursuit phase" of a development project with their own cash. Funds are spent to obtain site control and to finance due diligence, design, legal work, entitlements, and market studies. During the pursuit phase the investors' dollars are 100 percent at risk and if the project is not approved or does not sell, the developer will be unable to secure a construction loan, in which case the project is dead and most or all of the investment will have been lost.

Lenders: Will the developer be able to fully repay the loan on schedule?

Once the project has received city approvals and commitments from buyers or tenants, the developer will close on a construction loan, permanent loan, or mortgage from a bank or other source of construction debt. Because entitlement risk and market risk have been significantly reduced by this point, a lender will be willing to loan as much as 70 percent to 80 percent of the total project cost. The bank is lending its depositor's funds, which will be recouped from mortgage payments made to the bank and from sales or refinancing proceeds, therefore the bank is motivated to make a safe loan.

Buyers: Does this product suit my needs for a price I am willing to pay?

While community members are interested in a building's exterior design, the actual buyers and users of real estate care as much or more about what happens on the inside, so they focus on interior layout, views, daylight, finishes, building amenities, and parking arrangements. Buyers want a product that is in a good location and that satisfies their needs for a price they are willing and able to pay. Astute developers sometimes engage potential buyers early in the design phase to vet proposed design features.

The developer: Will this project earn a profit that justifies the risks?

The developer's primary purpose is to earn a large profit – one that is commensurate with the significant risks inherent in real estate development. Throughout the design process the developer will focus intently on reducing risk and uncertainty. They will listen to all of the other actors and use the design process to reconcile and accommodate their various opinions and suggestions, integrating this feedback into the building design as much as is practical. In addition to risking his or her pursuit phase cash and personal reputation, the developer often is required to personally guarantee repayment of all loans, effectively putting all of his or her personal assets at risk. Once the project is complete, the developer must account for future risks and costs related to construction defects and warranties.

When considering how each of these different actors views a proposed development project, it is helpful to remember that, "Where you stand depends on where you sit." This public policy concept, known as "Miles's Law," is named for a Truman-era bureaucrat named Rufus E. Miles, who observed that actors pursue policies that benefit their own interests over collective interests. Miles's Law applies in real estate development too, where architects care about design, contractors care about costs, politicians want growth, communities want investment, neighbors resist change, investors expect to earn a profit, lenders want to make a safe loan, and developers must somehow satisfy them all if they are to succeed. And beyond their individual interests and viewpoints, each of these actors is also likely to have a very different idea of what "good design" means.

Three different perspectives on "good design"

Perspective one: High design versus everyday function

Not everyone looks at design the same way. Numerous studies in the field of environmental psychology stretching back to the 1960s have continued to illuminate why architects and non-architects experience buildings differently. For example, Devlin and Nasar (1989, 333), asked architects and non-architects to look at photographs of two styles of residential architecture: "high" and "popular." Both groups favored novelty and coherence (or clarity) but non-architects favored simplicity and "popular" attributes while architects favored complexity and "high" attributes.

Another study by Gifford et al. (2002) asked architects and non-architects to consider buildings in terms of their "objective characteristics" and their "subjective meanings." The architects in the study focused on the objective characteristics – physical details such as arches, balconies, columns, and so on – while non-architects cared most about the "subjective meanings" – their ideas about the properties of a building and whether it was, for example, clear, complex, friendly, meaningful, rugged, or original. Hershberger (1988) found that experts respond more to representational, physical meanings of architecture – objective physical cues – while lay groups respond more to responsive, ethno-demographic meanings, or subjective ideas about buildings and places. Several studies (Groat 1982; Devlin 1990) found that architects are more likely to see buildings through stylistic and formal category systems while laypeople rely on functional categories, and Gifford et al. (2002) speculated further that architects might be more influenced by materials while non-architects are influenced by form.

Wilson and Canter (1990) concluded that architects have different ideas about aesthetics because their professional training socializes them in ways that create or widen the aesthetic gap between themselves and the public. Similarly, Hubbard found that an architect's education "inculcates a distinction knowledge structure" that laypeople lack (Hubbard 1996, 81). Gifford et al. (2002, 146) proposed that architects should be specifically trained to understand how non-architects think about buildings, concluding, "the greatest architects will be those with the creativity to design buildings that are delightful to design professionals and the public." But if design professionals and laypeople have such different ideas about what good design is, then what do the actual users of buildings think?

In a study by two scholars in the field of design, Forsyth and Crewe (2009) focused on the user's experience of design by examining differences in opinion between architectural critics and building users that centered on the question of style. They studied three planned communities that were designed in different styles, evaluating them on four dimensions: "objective aesthetics," "style," "place," and "satisfaction." Forsyth and Crewe compared the results of a "post occupancy evaluation" – a survey of many users – and the architectural criticism that each project had received. While a single expert writes an architectural critique, usually after a building has just been completed and often before anyone has moved in, a post-occupancy evaluation is based on a much larger set of data – surveys taken by many users after having lived or worked in a place for some period of time. Forsyth and Crewe found that the actual residents felt that all three of their communities were aesthetically good, but because their designs were based on different stylistic principles, all three projects had also received criticism from architects and planners promoting different styles. In other words, architects judge buildings based on objective or formal styles but since they have different individual stylistic preferences, they each judge buildings differently. But when surveyed, the users cared less about style and more about satisfaction and popularity. Style has to do with "taste" and in turn, "class" but satisfaction has to do with how good a place is in which to live or work and how well it fulfills a user's needs. Forsyth and Crewe concluded that debates about style are not the same as debates about aesthetic quality because a project that users find to be aesthetically good can still be critically condemned by an expert who has different stylistic preferences.

When thinking about good design, architects care more about style while the non-professionals, who represent the majority of buyers and users of real estate, care more about objective aesthetics and satisfaction. In other words, the user is concerned primarily with whether or not a place looks and feels good to them and if it will be a good place for them in which to live or work.

Perspective two: Exchange value versus use value

While these actors have different views of what good design means, they also have different opinions on land use. In their book *Urban Fortunes*, Logan and Molotch (1987, 1–2, 33) argued that scarce urban land has both "exchange value" and "use value." Exchange value refers to goods that have a monetary value that can be easily calculated. Use value refers to goods that can only be valued for their intrinsic properties. Developers and community members see the design of a real estate project from these two, very different perspectives – developers for the economic value a new project will create and community members for how it will change their experience of their neighborhood and community (see Table 23.1).

In the case of urban land, exchange value is the market value of a piece of property, based on the assumption that it will be developed to its highest and best economic use. Exchange value is influenced by planning and zoning codes and regulations that determine how densely a piece of property may be developed – the maximum buildable envelope – and for what uses. From the developer's perspective, land cost must be an acceptable share of the project cost – in total and per unit – if the project is to make economic sense. Exchange value is also affected by other market forces that tell sellers and buyers what it will cost to develop, design, and construct a project, how much demand there is likely to be for the product, and at what price the developer will be able to sell or rent the product. By considering this combination of factors, a seller or buyer can easily determine the rough monetary value of a piece of property.

Unlike exchange value, use value can be very difficult to quantify. The neighbors value a piece of urban land based upon how they experience and use it every day and different members of the community perceive it and use it in many different ways. Some may use a vacant site as a dog park while others may be concerned that a new building will block views and sunlight from their units or increase congestion on the streets.

It is at the intersection of exchange value and use value where most conflicts over urban real estate development occur. Developers must always keep an eye on costs and returns but cities are typically required to ignore economics when considering variances and conditional use permits and most community members see project economics as the developer's problem – a problem that they are not obligated to understand or appreciate.

Because the community members do not understand or simply do not care about the developer's exchange value perspective, public discussions about development projects focus mostly on use value. The common language in those discussions is that of design and the key words include density, massing, height, articulation, setbacks, style, materials, colors, uses,

Table 23.1 Exchange value vs use value

Exchange value – developer	Use value – community
Land cost ($ per square foot or $ per unit)	Current use of site (loss of informal dog park?)
Zoning (allowable uses, density, height, maximum building envelope)	Views across the site (blocked?)
	Daylight on site and nearby (shading?)
Site size and shape (efficiency of design)	Existing historic or valued structures (demo?)
Environmental contamination (clean-up cost)	Natural habitats (displacement?)
Approvals required (entitlement risk)	Environmental contamination (disturbance?)
Political/bureaucratic culture (entitlement risk)	Current traffic patterns (increased congestion?)
Timing within economic cycle (market risk)	Existing neighbors (future neighbors?)
Potential future phases (market opportunity)	Current tax base (real estate tax increase?)
Proximity to transit, retail, office, amenities	Subsidy to developer (taxpayer cost?)
What can it be tomorrow?	*What will I lose tomorrow?*

Source: Author

historic preservation, environmental impact, habitat preservation, congestion, parking, greening, and public space. As with the very idea of use value, these words and their meanings are qualitative rather than quantitative and because they are based more on subjective tastes and opinions rather than simple arithmetic they are also open to broad interpretation.

The developer focuses on the potential economic value of a property in the future while members of the community almost always value it for its current intrinsic qualities and how they experience it today. Community members often have good ideas that will help a developer improve the quality of a project and smart developers know that these community members are potential future buyers and that their ideas reflect the desires of the marketplace. But some suggestions from community members are Trojan Horses for positions that people know they cannot voice in public because they are blatantly self-interested, legally indefensible, or just politically incorrect. For example, the neighbors may oppose density and massing, arguing that it will be "out of character with the neighborhood," when their real concern is that the new building will block their views and access to sunlight although they do not own the rights to either. Similarly, the neighbors may claim that "what the neighborhood really needs is more for-sale housing and condos, not apartments or affordable housing," when their real concern is that apartments and affordable housing will attract the wrong kind of neighbors and reduce property values. And often the neighbors simply like the property the way it is – as an undeveloped lot where they can walk their dogs or enjoy a bit of open space. But while community members focus primarily on the exterior design of a project, the building's interior design is also important, since it must attract a large group of potential buyers. And those buyers care less about what the outside of a building looks like and more about what it will be like to live, work, or play on the inside.

Perspective three: What do the buyers want?

Discussions of style, satisfaction, objective aesthetics, exchange value, and use value all leave out one key ingredient in real estate development that the developer cannot neglect: price. A developer earns a profit by producing a product that can be sold for a price that is greater than the cost of developing it. But the developer is not operating in a vacuum – there is competition – so one major objective is to design a product that is competitively priced yet distinctive enough to cut through the clutter of the marketplace and stand out.

To succeed, the developer must not only balance these potentially conflicting views of good design but must also resolve them within the context of a more important concern: Who will buy the finished product? Despite opinions from architects, planners, critics, politicians, neighbors, and many others, a good design from the developer's perspective is one that will attract a broad and deep market: It must sell or the developer is at risk of losing large sums of money.

Many of the physical cues that give a design character – features such as style, details, material quality, massing, setbacks, and façade articulation – and that make a building look better, both to design professionals and laypeople, also cost more money. The developer must balance the costs of these various elements and features of the design with the price for which he or she hopes to sell the finished product.

"Design quality definitely has an impact on the marketing and sales of condominiums," says Chicago market research expert Gail Lissner (interview). Design also helps differentiate a product from its competition and this becomes even more important when markets weaken. "A cookie-cutter building is simply not as valuable as a well designed building." At the same time, the design must have lasting power – projects that are too trendy or cutting edge do not translate well to the broader market and projects that evoke a "love it or hate it" reaction are even more difficult to sell, particularly when the market softens. "While architects and developers may think and talk about 'target markets,' the fact is that the market is diverse and the developer's first priority is to create a product that will attract the largest possible pool of potential buyers."

Good design does matter to buyers, in part because it can enhance future resale value, but the bottom line remains price. The developer simply cannot afford to do something that is architecturally remarkable if it is going to cost 20 percent more to build than a comparable product in the same market. "Real estate is all about price and there is no amount of good architectural design that is going to cure pricing problems." So when they begin a new project, a developer starts by looking at their own location, what has worked, and what might work a little better. "But nobody gets too cutting edge," says Lissner. "It is very hard to deviate from the norm and no one does."

Integrating all three perspectives: The developer

A typical criticism of developers is that they do not care about good design. But that perspective dismisses the impact that other actors have on the design process. Once those interests are accounted for, it is easier to understand the developer's perspective: A good building design is one that attracts capital, secures entitlements, offers a marketable product, makes economic sense, and reduces the developer's risk. The developer must balance all of these factors to achieve a successful outcome.

Access to capital

A good design is one that attracts capital. The developer's investment typically represents only a fraction of the total cost of a development, although their cash is first in and 100 percent at risk. Once the idea has proven to be viable, the developer will need to obtain more capital from investors. And once the project has received city approvals and is ready to begin construction, the developer will need to obtain debt from a lender. Rather than taking a chance on something original or unique, however, the lender will want proof that there is already a market for similar products and will also require presales, signed leases, or a positive market study, at the very least. The lender's underwriters and loan committee will carefully compare the proposed project to other similar projects, scrutinizing everything, including the overall design, the mix and layout of spaces or units, interior finishes, exterior building materials, features and amenities, parking arrangements, estimated costs, and projected prices. If the project compares favorably the loan committee will approve the loan. A project that cannot demonstrate that it will be attractive to buyers or users at a price that will exceed its costs cannot gain access to capital – investor equity and debt – and is dead.

Securing of entitlements

A good design is one that attracts the support of local community members and groups or at least it does not generate insurmountable opposition. Community support, in turn, makes it easier for planning staff and elected politicians to promote the project if it must come before the planning commission, the historical commission, the zoning board of appeals, or the full city council on appeal, and improves the chances that the project will receive necessary entitlements and approvals from the city. A project that cannot secure entitlements is dead.

Product marketability

The developer is typically not building a highly unique project; rather they are producing a stock product that is only slightly different from the products being offered by their many competitors. Projects will vary from one to the next in terms of location, design, market timing, and price. But each project will be designed to offer an attractive and distinctive product that stands out in the marketplace. For the buyer this means a product that provides a good place to live, work, or play in a good location and for a good price.

Project economics

Developers engage in what economists call joint production, by buying sets of resources – land or property, construction materials and services, and professional services – and combining them into a single new product that can be sold for higher price that will produce a profit. But whether or not they intend to sell or own for the long run, developers must look beyond these capital costs towards life-cycle costs and the total cost of ownership by making design decisions that will minimize operating and maintenance costs in the long run. Interest, taxes, and utility costs add up quickly, too, so prudent developers do not build more product than they can reasonably hope to sell within a certain time period. Rather, they must balance maximizing buildable area with rates of "absorption" or else their profits will be eroded by the carrying costs of holding unsold or vacant units for an extended period. Time is money.

Development risk

In the early stages of a project, developers try to reduce risks and uncertainty associated with the property as much as possible by completing thorough due diligence, including environmental studies, geotechnical studies, and studies of existing building conditions. They also start meeting with community members and city officials and begin to assess the chances of receiving approvals as a way to mitigate entitlement risk. Developers also manage their construction risk by controlling fees and construction costs throughout the entire duration of the project. And developers use design to mitigate market risk by creating flexibility and allowing for multiple exit strategies so that they can adapt to changing conditions. The building may be designed so that it can be converted from one product type to another – from apartments to condominiums, for example – should market conditions change. Or the design may allow for phasing to reduce financing costs and avoid building product faster than the market can absorb it.

Successful developers manage risk by incorporating lessons learned from their previous projects, making marginal improvements to things that have worked well and avoiding past mistakes. They use derivative design to manage risk by making new buildings that are not very different from their own previous projects and the projects of their competitors, which makes it more likely that the market will accept the finished product.

Good design for real estate development: One example

One example of designing to manage risk by addressing the interests of all actors can be found at 600 Lakeshore Drive North, a condominium project in Chicago that was developed by David "Buzz" Ruttenberg of the Belgravia Group and completed in 2009 (Brown 2015, pp. 55–60; see Figures 23.1, 23.2, 23.3). The previous owner of the property had planned to build

Figure 23.1 Designed to manage risk: The two-tower design for 600 Lake Shore Drive was built in phases, which allowed the developer and the design team to improve the quality of the product on the second tower and made financing easier and cheaper

Source: Courtesy of Pappageorge Haymes Partners

a single, long, residential building of over 1,000,000 square feet facing the lake but he had difficulty obtaining financing and encountered serious resistance from the community because the building would block views and sunshine. So Ruttenberg came up with an idea for a two-tower design that would be easier and cheaper to finance while better responding to the community's concerns. For example, the design was based on local precedent, echoing an important nearby architectural landmark, the two 1960s Mies van der Rohe apartment towers two blocks to the north. But more importantly, Ruttenberg employed a number of other design strategies to reduce his market risk and increase his chances of success.

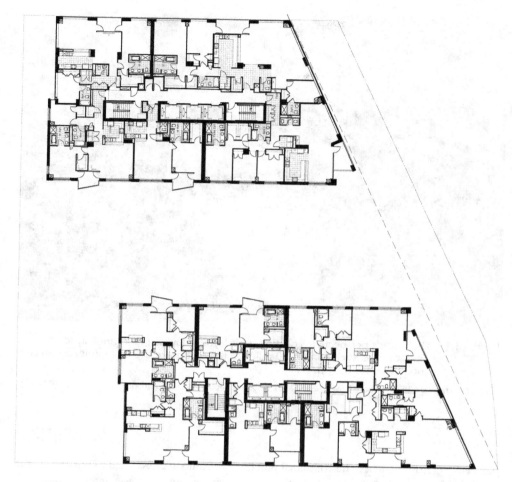

Figure 23.2 No bad units: All units have lake views to the east (to the right in this plan) so there are no undesirable units, ensuring a faster sell-out and reducing market risk

Source: Courtesy of Pappageorge Haymes Partners

First, the orientation and smaller size of the floor plates for the two towers ensured that each of the six units per floor (per tower) had lake views, so there were no undesirable units, which mitigated market risk by ensuring a faster sell-out. The design also allowed for phased construction of the 700,000 square foot project (40 percent in the first tower and 60 percent in the second) and the ability to make improvements to the second tower based on buyer feedback from the first tower as it was completed and occupied. Phasing also made financing easier, improved cash flow, and reduced debt. "We sold rapidly in 2005, 2006, and 2007 and delivered completed units in 2007, 2008, and 2009," says Ruttenberg (interview, 2016). "We sold the first 300 units in the first four and a half years and it took three more years to sell the last 100 units." But because Ruttenberg was able to sell completed units and use the sales

Figure 23.3   The buyer's perspective: Location, unit layout, daylight, and unobstructed views of Lake Michigan for 80 miles to the east make the condominium units a good value for the price

Source: Courtesy of 600 Lake Shore Drive LLC, a Sandz/Belgravia Group Ltd Development

proceeds to repay his debt as he continued building new units, he never had to borrow for the whole project at one time, which meant a smaller loan amount and lower interest costs. "We paid our loan off in mid-January of 2009 and even though we had some unsold inventory at the time, we had no debt and we made a profit in the end." Ruttenberg also hedged against a potential market downturn by choosing not to build to his maximum allowable density or height. This decision reduced costs and shortened the schedule, allowing him to get fewer units to market faster, reducing his market risk related to slow absorption and having too much product left on the shelf in a downturn.

But Ruttenberg also focused on providing value to his buyers by driving his architects, Pappageorge Haymes Partners, to produce efficiently designed units that could compete on total price with the larger, less efficient units that other developers were developing based on dollar-per-square-foot pricing. "So if you provide the same unit as your competitor," says Ruttenberg, "but you blow a little less air into the floor plans and you make them slimmer and a little more efficient, then you are hedging against the day when the market slows down. And if that means as a buyer you are paying $1.7 million to live in a three-bedroom, three-bath unit on the lake but others are paying $2.5 million to $5 million, then that feels pretty good to you." From the buyer's perspective, location, unit layout, daylight, and unobstructed views of Lake Michigan for 80 miles to the east make Ruttenberg's condominium units a good value for the price.

Finally, like his predecessor, Ruttenberg still faced concerns from neighbors and the surrounding community that the new building would cast shadows on nearby Ohio Street Beach. Community members were not comforted by the architect's analyses, which showed that the two towers would only partially shade the beach and only on the shortest days of the year, in December. But the 60-foot, street-width space between the towers allowed the neighbors to see in between, so Ruttenberg's plan was "like two fingers rather than one hand." The two-tower design was a big improvement over the previous developer's single-building design and Ruttenberg's decision not to build to the maximum allowable height helped too, but good design alone was still not enough. In the end, says architect David Haymes, "the community members' opposition was publicly focused on the shadow issue but it was really driven by the fear of change, and while the two-tower concept aided the argument in a material way, many were not convinced, and it took proactive, objective, political leadership to push it past the goal line to obtain the needed entitlements" (personal communication).

Conclusion

Design for real estate development is not a solo act but rather a group performance. The developer facilitates a process involving many different actors with varying degrees of power and influence over the design of a project. The developer must take into account the preferences of politicians, neighbors, and other interest groups who care about how a building's exterior design will impact the community while ensuring that the interior spaces will work well for users and attract buyers and tenants. The project must also be economically viable and capable of obtaining entitlements and attracting capital from investors and lenders. The finished building will be a three-dimensional manifestation of the influence of all of these actors and their preferences.

The best developers manage their risk and increase their chances of success by actively soliciting feedback from all of the actors who may potentially have an influence on a project. They then integrate that feedback into the design in an ongoing effort to constantly improve the quality and value of the product. The developer must constantly direct this process towards a good design – one that will satisfy and address all of the actors and interests as much as is practical – because in the end, a good design in real estate development is one that gets built and meets with success in the marketplace.

Author's note

This chapter is adapted from the research and ideas described in my book *How Real Estate Developers Think: Design, Profits, and Community* (University of Pennsylvania Press, 2015). My thanks to Michael Byrd, Curt Gunsbury, and Murray Kornberg, for their encouragement and for reading and making helpful comments to this chapter.

References

Brown, P. H. (2015) *How Real Estate Developers Think: Design, Profits, and Community*. Philadelphia: University of Pennsylvania Press.

Devlin, K. (1990) An Examination of Architectural Interpretation: Architects vs Non-Architects. *Journal of Architectural Planning and Research* 7(3): 235–244.

Devlin, K. and Nasar, J. L. (1989) The Beauty and the Beast: Some Preliminary Comparisons of "High" Versus "Popular" Residential Architecture and Public versus Architect Judgments of Same. *Journal of Environmental Psychology* 9: 333–344.

Forsyth, A. and Crewe, K. (2009) New Visions for Suburbia: Reassessing Aesthetics and Place-Making in Modernism, Imageability, and New Urbanism. *Journal of Urban Design* 14(4): 415–438.

Gifford, R., Hine, D. W., Miller-Clemm, W., and Shaw, K. T. (2002) Why Architects and Laypersons Judge Buildings Differently: Cognitive Properties and Physical Bases. *Journal of Architectural Planning and Research* 19(2): 131–148.

Groat, L. (1982) Meaning in Post-Modern Architecture: An Examination Using the Multiple Sorting Task. *Journal of Environmental Psychology* 2: 3–2.

Hershberger, R. G. (1988) A Study of Meaning and Architecture. In: Nasar J. L. (ed.) *Environmental Aesthetics: Theory, Research, and Applications*. Cambridge: Cambridge University Press, pp. 175–194.

Hubbard, P. (1996) Conflicting Interpretations of Architecture: An Empirical Investigation. *Journal of Environmental Psychology* 16(2): 75–92.

Logan, J. and Molotch, H. (1987) *Urban Fortunes: The Political Economy of Place*. Berkeley: University of California Press.

Wilson, M. A. and Canter, D. V. (1990) The Development of Central Concepts during Professional Education: An Example of a Multivariate Model of the Concept of Architectural Style. *Applied Psychology: An International Review* 39(4): 431–455.

Interviews

Haymes, D. Architect. Personal email communication, May 7, 2016.

Lissner, G. Interview by author, July 1, 2010, Chicago.

Ruttenberg, D. Interview by author, July 1, 2010, Chicago.

Ruttenberg, D. Telephone interview by author, May 5, 2016.

24

Good design for real estate development
Moving beyond the generic city

Alex Duval

Abstract

This chapter examines the role architects, landscape architects and urban designers play in real estate development. The global real estate market and project delivery methods within those markets vary greatly. However, the increasing internationalization and institutionalization of real estate capital have put pressure on design to standardize in order to mitigate risk. Simultaneously, cultural and ecological requirements are causing project design to respond with increased sensitivity to local conditions. For a project to succeed as an investment, respect local community values in order to obtain approval, and promote sustainability to benefit future generations, its designers must integrate these often-competing elements.

There is a broad spectrum of approaches to achieving successful outcomes. In some cases, a development firm will choose to focus exclusively on creating value by perfecting the delivery of a single building type, or it may work across a variety of types but limit itself to a specific region. At the other end of the spectrum, firms choose to specialize in creating value by resolving complex special situations across global markets that are not repeatable. Each approach requires a design team that is well suited to the value creation model of the real estate developer. This chapter puts forward a set of principles for design that can be used across this spectrum as a framework for creating unique developments that differentiate themselves from other projects. Through design differentiation, developers can gain competitive advantages, generate value premiums, and ultimately create a more heterogeneous built environment that meets the unique needs of a wider range of diverse users.

Introduction

My colleagues, friends, and family often say to me that so many of the new development projects in cities, whether New York or Shanghai, Atlanta or Minneapolis, feel *the same*. Cities have "collected" the signature, often stylistically repeated cultural and civic buildings designed by a handful of global "starchitects." Many cities now have a "Gehry" or a "Renzo." But museums, like the art within them, have become commodities in the global exchange culture. The case is even worse for commercial development of office buildings, retail malls, houses, apartments, hotels, and industrial buildings that make up the vast majority of the urban fabric.

Because cities have drawn from the same limited well, the differentiator that could exist between them has been neutralized. Cities are making themselves equivalent. Office towers look the same, glassy boxes that could be anywhere with the same familiar mass and bulk. Well-meaning retail environments have brands we know and love. But once you are at the mall, you could be *anywhere*.

Families with children need a yard and a good school district. They can pick from a few options in the model sales center in a suburban subdivision. But unless they are wealthy, it's hard to find a home that expresses the unique character of a place or individual taste and style. Recent graduates or empty nesters want the convenience, amenities, and walkability of an urban high-rise rental apartment or condominium, and there are plenty to choose from, but they all seem *the same*.

Travelers are increasingly faced with a similar experience. They choose to travel to cities for leisure or business and often select a hotel because of rewards and points, not because of the inherent quality of the building as something they *experience*. So many opt to go to the "historic district" of a city to experience what is unique about it. New developments are having a hard time tapping into what is culturally, climatically, and technologically unique about a place; its *genius loci*.

Rem Koolhaas asked the question in his essay the "Generic City," "Is the contemporary city like the contemporary airport - 'All the same'" (Koolhaas and Mau, 1995, p. 1248)? How can the development community, which is responsible for defining the vision, program and financing of buildings, respond? Why are cities becoming *more generic* when we have greater capacity than at any other time in history due to technological advancement, wealth, and stability to take risks and generate mass customized approaches to the creation of commercial real estate?

Causes of the generic city

Cities are ecologies. Like a swamp or a forest filled with specialized species of animals that have evolved into their perfect forms, buildings within cities have also been evolving into their highly calibrated typologies over time. Historically, climate, locally available materials, and local know-how produced a distinguishable "indigenous" style. In the last 20 years, the world has globalized and with it has come the globalization and standardization of real estate investment markets. This phenomenon can be clearly seen by the continued growth of the global REIT market. The global market capitalization of REITs now stands at approximately $1.7 trillion, up from $734 billion in 2010 (Ernst & Young, 2016).

The real estate market is the largest asset class in the world. The total value of world real estate reached $217 trillion in 2015. This represents 2.7 times world GDP and 60 percent of global mainstream assets (Savills, 2016). In an effort to price risk, real estate investors seek to standardize assets; buildings, in order to compare apples to apples. Gerald Hines observes, "The movement [has been] to very large organizations and capital structures. And who would have thought that pension funds would be doing joint ventures with developers?" This has had the net effect of causing architecture to respond to the requirements of lenders and investors who may be able to appreciate the qualities of one-of-a-kind works of architecture, but find themselves bound to a system that has evolved into a "check-the-box" regime of risk mitigation. It's not to say that this system does not exist for good reason.

Many people view the majority of commercial real estate development as boring. Boring aesthetically, boring programmatically, and boring from the point of view that they are just happy to get a place that meets their bare minimum requirement for a price they can afford. But what are people looking for in a place in which to live or work, or to go for cultural enrichment? The exercise of determining what the market wants is dependent on the view of a limited handful of market research firms that lenders and investors trust. While the data is useful it is as one saying puts it, "footprints in the snow." It tells you where you've been but not necessarily where to go.

Most developers and designers have come to adopt a strategy of specializing in a building typology and replicating it in a geographical area that they dominate. In the USA, most firms serve a local or regional client base, with 51 percent serving a single metro area, 34 percent serving a multistate or multi regional area, 6 percent are national and 9 percent international (AIA, 2006). Firms use data and experience from past projects and make minor innovations within the margin of risk that the process will allow. Data is useful. My view is that our increased ability to segment data opens the possibility for more mass customization. The sameness of our cities is due to a lack of detail or nuance in our analysis of real estate market demand and program. But it is only one arrow in a quiver of potential strategies for bringing more diversity and authenticity back to cities.

This chapter is meant for people who are interested in an aspirational vision for the built environment. We have to understand and question the underlying forces and logic of a system that is presently causing a self-replicating specialization and the resulting homogenization of our cities. The chapter will offer a set of principles that can be explored in moving beyond the generic city.

Design approach

When a developer is engaging in a design process, what do they need to know? There are different approaches to development. Some developers follow a formula and impose it on a site. For example, a developer may be very successful creating a specific brand of hotel. They look for an opportunity where the market conditions pre-exist that will support what they know will sustain the project and then execute a formula. The opposite approach is to start with a site and allow a custom, locally based project program to evolve organically from it.

The legendary developers throughout history have always had a very strong idea about the design of their projects. The developers in Table 24.1 were committed to good design in terms of creating projects that have been considered by industry experts and the general public as beautiful, vibrant, authentic and in some cases, ultimately iconic. What are the principles of these developers? What are the different strategies of each? What can we learn from their approaches? What are younger firms doing and thinking about design? What experiments are they conducting and what are the outcomes? Most importantly, what approaches are relevant for conditions that exist today?

Table 24.1 Successful design oriented developers

John C. Portman: An architect and developer.
Gerald D. Hines: Hired many different architects of high caliber.
William Zeckendorff: Hired I.M. Pei and forged an ongoing relationship.
J.C. Nichols: A place based community developer.

Source: Author

329

The developers who rose to the top of the game care about design and have been rewarded in terms of financial return, prestige, and longevity. This chapter will provide examples of projects and design principles for development that promote the creation of unique places and generate quality and differentiation.

Specialization versus diversification: A dynamic tension

What is the development process? Development usually starts from one of several points: either someone has a site, a tenant, or investment capital allocation. Each starting point has a different impact on choosing a design approach. Often, a local design firm will have political influence or networks of relationships that can facilitate the development of a specific site. But using a design firm because they have political influence alone may not generate the best design project from an aesthetic point of view.

The same is true when a tenant who has worked with a design firm before insists that they be used on a project or they will not proceed. This can put the developer in a difficult position. The last example, is perhaps the most challenging. Financing due diligence almost always includes the relevant experience of the architecture firm in having designed the same type of building as a prerequisite for approval. Investors and lenders want to know that the architect has done "this type of building" before.

The logic of qualification, experience and pre-existing relationships dominate the real estate development world. It is a self-reinforcing system of "unto him who has, shall be given." The current system is logical, but at the expense of logic and risk mitigation is innovation and diversification. It is analogous to the logic that drove the expansion, domination, and homogenization of the food industry.

There are hundreds of potato varieties grown globally (College of Agricultural and Life Sciences, 2008), but due to the logic, efficiency, and economy of scale that the fast food industry generated in the latter half of the twentieth century, the US potato crop had essentially been reduced to a single species, the Idaho russet that now comprises 57 percent of the annual yield (Miller, 2009). Fortunately, the monoculture approach to food production has been declining as the food industry seeks to diversify.

The food industry has been exploring ways to generate more diversity. The consumer is demanding it. Likewise, today's consumers of space are looking for unique, locally based and inspired typologies. The current trend of mass customization in the third industrial revolution is moving away from the generic to the unique. Real estate must respond by finding ways to enable a multiplicity of voices in an industry that is currently consolidating, according to the logic of global finance.

The designer's role

What is the role of the designer? The designer's primary responsibility is to translate the developer's program into a plan and materials that visualize an idea or aspiration. Good designers with experience in the type of project a developer is undertaking can offer a lot of insight into the program creation.

What is the relationship between the developer and the designer? How does a developer select a designer? And what value does design bring to a development? The success of a project is absolutely dependent on having the correct designer and good chemistry. Good chemistry can mean a dynamic tension wherein the developer and designer challenge each other to consider things beyond the limits of their own individual imaginations. The developer must also select a designer who has demonstrated an ability to problem solve. Each project will have a unique set of challenges and require a unique solution.

A good design has the potential to command a premium in rent or total value the same way a better view, higher floor, or better location can command a premium. In real estate development, a primary factor affecting rate is location. Increasingly, the brand of a designer plays an important role in certain real estate asset classes, most notably luxury residential and hotel projects that tend to be more directly end-user or consumer facing. Developer William Zeckendorf made this concept clear, "I recognized that we could afford to lose ground income if, by giving tenants a prestige building, we could ask for rents fifty percent greater than those in the immediate vicinity."

Innovation

Development starts with market demand and design must respond correctly to that demand. Understanding how demand triggers design is simple. People need space. Everyone needs a place in which to live, space to work in, and facilities for shopping, culture, wellbeing, and other activities. Designers often want to create novelty and innovation at a rate that is not logically aligned with the rate of transformation taking place in society at large or a specific segment of the real estate market. It is important for developers and designers to understand where the opportunity for innovation exists due to increased population growth, real estate supply and demand, credit availability, technological progress, and the evolution of cultural values and tastes and price the risk of innovation correctly.

Creating novelty for novelty's sake can work sometimes, but it is not a rational or prudent approach to creating diversification or differentiation in a project. Innovations need to be grounded in analysis, empirical process and creativity. A number of principles illustrate thematic areas or approaches that guide a rigorous approach to arriving at an innovation in the development design process.

Principles

The following principles have the potential to increase positive outcomes for projects in the development and design process. They take into consideration the realities and constraints of the current market demand context while also creating space for innovation.

I've defined these principles for use in our own practice (see Table 24.2). They act as a touchstone that we return to as we evaluate a project opportunity, our own performance, or the performance of the team. I believe the overarching principle for good design was best articulated by John Portman as he observed that, "Buildings should serve people, not the other way around" (personal communication). I would extend this idea to the entire built environment including public space, and infrastructure, which increasingly are being integrated into the design and development of the total environment.

The principles are small and simple concepts that can be used to instigate approaches to projects that have the potential to generate something unique. They are rooted in a premise that decisions about projects are never imposed arbitrarily, but grow naturally and organically out of the conditions, people and aspirations of a specific time and place. They are something to meditate on when engaging in a project. They are meant to serve as guidance and not as a formula.

Table 24.2 Design principles for differentiated real estate development

1. Unique – Recognize that every project is unique.
2. Patience – Don't impose design solutions too soon.
3. Clarity – Make sure that how you talk about a project makes sense to the public.
4. Solvers – Select designers who are problem solvers, not past repeaters.
5. Context – Gather lots of context information before starting design.
6. History – Take time to learn the history of the site, the politics, and how the community feels.
7. Program – Create a solid space program before you come up with a building form.
8. Simplify – Simplify, simplify, simplify.
9. Understand – Know in detail what your end-user is looking for.
10. Let go – Be prepared to eliminate a design feature you love if the tenant or community doesn't.
11. Fight – Fight for the design even if it isn't immediately embraced.
12. Equality – Our cities must serve everyone. Sex, gender, race, age, sexual orientation, origin, caste or class, income or property, language, religion, convictions, opinions, health, or disability must not result in unequal treatment.
13. Ecology – Buildings and communities must regenerate and sustain the health and vitality of all life. Its bond with nature complements and enriches this relationship with its environment.
14. Beauty – The production of beauty in our communities inspires us to work, and work raises us up.
15. Teamwork – Creation involves teamwork. In addition, research has become consolidated as a new feature of the architectural creative process.
16. Empirical – Good design asks questions. It is the record of one's direct observations and provides a repeatable system for innovation.
17. Culture – Regional design as a style is an expression of its own geographical and cultural context as well as its design traditions.
18. Computation – As has occurred in most fields of human evolution through the ages, new technologies are a resource for the progress of design.
19. Materials – All materials have the same architectural value, regardless of their price.
20. Media – Design must consider how new forms of media, communication, and machines are affecting our interaction with one another, our communities and the natural world.
21. Scalable – Architectural innovations may start small at the scale of an individual room or building and expand to address the needs and problems of larger segments of society.
22. Iterative – A project solution, whether architectural or financial should be created quickly, tested and refined in a rapidly evolving and repeating process that constantly responds and adjusts to changing conditions.
23. Experiments – Some experiments succeed and some fail, but ideas from them can be saved and recycled to solve other problems.
24. Global – Real estate is always a local business, but concepts, capital, technologies, and other forces that significantly impact a project are increasingly global.
25. Heterogeneous – Cities are moving away from the mono-functional planning systems of the twentieth century and demanding more complex mixes of uses and programs.
26. Health – The impact of a project on human health is an important consideration.

Source: Author

Project examples

As previously mentioned, as a single asset class, real estate is the largest. But within the overall class, there are five main subcategories that are widely recognized and used by the development community. Primarily, the types of activities or uses that occur within the building define the categories: residential, office, lodging, industrial, and retail. One can see this classification system mirrored in NAREIT's system of REIT sectors (NAREIT, 2016). It is important to draw attention to the pre-existing system of order that governs and controls the development world. Ideas about design innovation, mixing of uses, creating interesting programmatic combinations, and ultimately "breaking the mold" remain in the realm of naïve speculation without developing an understanding of the system of investment and financial logic that ultimately constrains or sets free new ideas.

The following examples of projects demonstrate the principles for good design. They are organized by the five categories for simplicity but also to highlight that these projects are realized examples of innovations that were executed within their categories. These example projects are not intended to represent the entire category, but rather serve to illustrate the principles and demonstrate positive, innovative outcomes.

Residential

Residential is the largest segment of all real estate asset classes. The cumulative value of all homes in the USA at the end of 2013 was approximately $25.7 trillion (Hopkins, 2013). Because the residential sector of real estate is so large, there is significant potential to scale innovations and have wide reaching impacts on the built environment.

There is wide variation within the residential real estate sector. There are locational differences, whether urban, suburban or rural. There are typological differences, whether single-family homes, multifamily, low-rise, or high-rise. And there are differences in how people occupy real estate, whether as owners or renters and now, as participants in the sharing economy that is emerging.

The overwhelming trend that is reshaping the global landscape of residential development is the migration from rural to urban. It is predicted that by 2050 about 64 percent of the developing world and 86 percent of the developed world will be urbanized (*The Economist*, 2012). The landscape of urbanization is varied. Cities are striving for diversity in housing typology, recognizing that housing stock will inevitably reflect the heterogeneous population that occupies it. Cities understand how high-density high-rise apartments and condominiums use land and infrastructure more efficiently and advocate for this development pattern. But users continue to be motivated by other factors such as style, cultural predispositions, dreams, amenities, relationship to nature, communal living, and a host of other symbolic considerations that drive demand for different types of housing.

Developers and the financial partners that invest with them carefully study what buyers and renters are looking for. Home styles have evolved slowly over time and have been designed to respond to timeless needs of human inhabitants, but in every generation there are advancements in technology and lifestyle that drive transformations in the design of domestic space. One of those evolutions was the introduction of what we now refer to stylistically as "mid-century modern." But what this type of housing indicates is a move away from traditional conceptions of what housing should be or look like, towards something new that is meeting a demand for a differentiated form of domestic space.

Post-war America was best defined by the rapid suburban expansion and production of single-family homes of traditional styles in new subdivisions. These homes were built using a mass production methodology that focused on efficiency, affordability, and a design that responded to historical conceptions of what the ideal home should "look like." But in 1956, in Palm Springs California, a new type of development and design emerged.

The Alexander Construction Company, working with architects Dan Palmer and William Krisel in Palm Springs, produced an approach to delivering a new housing type that anticipated and met the demands of a new way of living. Together, they designed and developed over 2,500 modern homes, known as the Alexander Tract, which is the largest modernist housing development in the United States (Newman, 2009). They took their cues from custom modernist homes that had been built for wealthy clients a generation earlier, and made them accessible to the burgeoning middle class.

Through design, they pioneered a method of construction that was affordable and simultaneously had the effect of appearing custom made. "The Alexander homes have very similar floor plans but the unique rooflines, variety in front finishes, the placement of the properties on the lots, make the neighborhoods look more like a collection of custom built homes" (midcenturypalmsprings.com, 2016). Today homes that were sold for $19,000 in 1956 – adjusted to present dollars, about $260,000 – are selling for over $1,000,000 (midcenturypalmsprings.com, 2016).

The Alexander Construction Company had been a more traditional home builder. It was their architects, Palmer and Krisel, who persuaded them to conduct an experiment and test the market to determine if there was enough demand for a radically different home. George Alexander and Robert, his son, agreed to build ten homes designed by Krisel in a new subdivision. The original ten homes sold quickly and as a result, the Alexanders developed, on average, 250 homes annually for the subsequent ten years in this small idyllic pocket of the USA. It is difficult to determine how far-reaching the Alexander model could have extended. George and Robert, with their spouses and friends, were killed in a plane crash in 1967.

Office

One of the most successful office developers is Hines. Gerald D. Hines founded the company in 1957 and has gone on to create some of the world's most iconic office buildings. He revolutionized the way developers partnered with institutional capital and also revolutionized the way developers work with architects. "Hines raised the bar by showing that quality and financial success can be mutually attainable. He was one of the first developers to hire world-famous architects, believing tenants would flock to top-quality buildings, even in a down market" (Sarnoff, 2007).

Hine's current development, T3, in the North Loop of Minneapolis, is innovative in multiple ways and embodies several of the principles. The project is situated in a post-industrial warehouse district. The local urban context and existing building fabric inspired the development team in the creation of its program and design approach. This area of Minneapolis has evolved from a collection of dilapidated warehouses made of brick and timber into a burgeoning creative office and residential district.

The North Loop is one of three top neighborhoods in the city for millennials. Tenants within the target market have a strong desire for authenticity and quality of materials that evoke the environments of Soho in New York City and BOMA (a warehouse district in San Francisco), places where tech start-ups and creative firms moved into old warehouse space that had a casual, yet "real" aesthetic. The success of these firms, such as Facebook, and others, has generated a demand for comparable space.

Since there is a limited supply of historic brick and timber warehouses with locational advantages and amenities such as transportation access and pedestrian-friendly context, Hines began the exercise of developing and designing the new construction of a seven-story office building of about 230,000 square feet made of post and beam timber construction with a "raw" space aesthetic sought by millennials.

The approach of Hines to the T3 project embodies many of the principles, most importantly that the design of the building grows organically and relates to the urban context in which it is located. It reimagines a new warehouse for the third industrial revolution; that of the knowledge-based, shared economy. The design takes cues from the timber structural system and large, open, polished concrete floor plates of the previous industrial revolution. As an office building, it stands in sharp contrast to the glass and steel office towers with gypsum wallboard, acoustical tiles, drop ceilings, and synthetic carpeting of the latter twentieth century.

Industrial

Industrial buildings are not typically thought of as having a lot of potential for innovation or creativity. The vast majority are utilitarian structures that are designed and engineered to provide the most economical enclosure possible while meeting the requirements of the user.

Architects Herzog and De Meuron have approached the design of the typical warehouse and created an artful work of architecture in the creation of the Ricola warehouse in Laufen, Switzerland.

Retail

The Design District in Miami, Florida, is a unique retail environment being developed by Craig Robins and his firm Dacra, with L Real Estate. Robins worked with Duany Plater-Zyberk to create a master plan for the district that codified guidelines promoting a development that responds to its local culture and climate, while taking risks on new avant-garde architects and designers to reimagine the shopping experience in the twenty-first century.

The development is located in the Buena Vista neighborhood of Miami. The Design District encompasses approximately 18 city blocks and when complete will accommodate about 200 retail shops, restaurants, and galleries. The area has evolved from a blighted warehouse district into one of Miami's premier shopping and art destinations with global luxury brands such as Tom Ford, Hermes, and Louis Vuitton locating flagship stores within the development.

The developer took an unconventional approach to the design of the project by assembling a consortium of up-and-coming architects and designers to each design individual buildings. The result and effect has been the creation of an authentic and organic place with great aesthetic, visual, and material diversity. This would contrast with previous approaches wherein a singular developer would define a program for a shopping mall and engage a single architecture firm to assume responsibility for the design of the entire area and then leave the individual shop interiors to be designed by the tenant's designer.

Lodging

Hotel development varies in complexity. Limited service economy hotels in suburban locations are very different from full service luxury hotels in city centers. John Portman is widely considered to be the most successful and innovative developer of hotels globally in the modern era. An architect by training, he is also credited with reimagining the role of the architect as a developer, and in effect, becoming his own client. He has also made numerous other pioneering innovations in the world of design and development.

Portman's rise to prominence began with his development and design of the Hyatt Regency in Atlanta. What is not well known about this project is that it began, conceptually, as a design for affordable senior housing. Portman was engaged by a public agency as an architect, to design a modest building for seniors. He came up with a design for a building with a central atrium of about seven stories, naturally ventilated, with apartments ringed around it. The building was built but Portman took the atrium concept experiment he developed for the housing project, iterated it, recycled it, and enlarged it to create the project that made him famous: the Hyatt Regency.

As Portman states, "Architects in the past have tended to concentrate their attention on the building as a static object. I believe dynamics are more important: the dynamics of people, their interaction with spaces and environmental condition." This approach to viewing a project as a dynamic experiment that relates and adjusts to changing conditions, is one of the principles that it is important to meditate on. All too often, architects and engineers become frozen in a process that is inherently dynamic. Successful developers tend to be highly adaptive.

Summary

The expansion of globalization and the institutionalization of capital on one hand, and the increased localization of political, cultural, and ecological forces on the other is creating a new paradigm for real estate development. Developers are faced with defining a strategy that differentiates what they do from others in order to gain competitive advantages and increase successful outcomes. Design is a discipline that is intrinsically based in problem solving. It is an aesthetic pursuit, but at a deeper level it is a process that synthesizes and resolves often seemingly contradictory conditions. The principles set out in this chapter are a framework for more consciously engaging in a process that is occurring regardless of whether the participants are actively managing it or not. Meditating on these principles and applying them, testing them and adapting them to the specific conditions of a project can serve the developer and design team by providing a flexible framework around which the development and design process can be organized.

References

AIA. (2006). *The Business of Architecture*. Washington, DC: AIA. Retrieved December 5, 2016, from: www.aia.org/aiaucmp/groups/ek_public/documents/pdf/aiab083575.pdf

College of Agriculture and Life Sciences. (2008). Potato Fast Facts. Retrieved December 7, 2016, from: www.cals.uidaho.edu/edcomm/magazine/summer2008/facts.htm

The Economist. (2012). Urban life: Open-air computers. *The Economist*, October 27. Retrieved April 4, 2017, from: www.economist.com/news/special-report/21564998-cities-are-turning-vast-data-factories-open-air-computers

Ernst and Young. (2016). *Global Perspectives 2016 REIT Report*. Ernst & Young. Retrieved December 5, 2016, from: www.ey.com/Publication/vwLUAssets/global-perspectives-2016-reit-report-ey/$File/ey-global-perspectives-2016-reit-report.pdf

Hopkins, C. (2013). Combined value of US homes to top $25 trillion in 2013. Zillow Porchlight, December 13. Retrieved December 7, 2016, from www.zillow.com/blog/value-us-homes-to-top-25-trillion-141142

Koolhaas, R. and Mau, B. (1995) *S, M, L, XL*. New York: Monacelli Press.

midcenturypalmsprings.com. (2016). About Palm Springs Alexanders. midcenturypalmsprings.com. Retrieved December 7, 2016, from: www.midcenturypalmsprings.com/alexander-homes-palm-springs.php

Miller, J. (2009). McDonald's the holy grail for potato farmers. Associated Press, September 23. Retrieved December 7, 2016, from: www.nbcnews.com/id/32983108/ns/business-us_business/t/mcdonalds-holy-grail-potato-farmers/

NAREIT. (2016). REIT Sectors. Washington, DC: NAREIT. Retrieved December 7, 2016, from: www.reit.com/investing/reit-basics/reit-sectors

Newman, M. (2009). Masters of modernism: The butterfly effect. Palm Springs Life, January 21. Retrieved December 7, 2016 from: www.palmspringslife.com/masters-of-modernism-the-butterfly-effect/

Sarnoff, N. (2007). Hines a towering influence on Houston landmarks. *Houston Chronicle*, June 10. Retrieved April 13, 2016, from: www.chron.com/business/article/Hines-a-towering-influence-on-Houston-landmarks-1611940.php

Savills. (2016). World real estate accounts for 60% of all mainstream assets. London: Savills. Retrieved December 5, 2016, from: www.savills.co.uk/_news/article/72418/198559-0/1/2016/world-real-estate-accounts-for-60—of-all-mainstream-assets

25

The design of real estate

A framework for value creation

Bing Wang

Abstract

Design and real estate are inherently intertwined, but this linkage is not easy to dissect or analyze without encountering the obvious danger of oversimplification. Design is often closely associated with an emphasis on form and aesthetics, while real estate has a focus on implementation and process. The interaction between these two fields inevitably creates new perspectives and methods, and sometimes, possible hybrids that bring valuable additions to the existing repertoire of design and development vocabularies, prompting new ways of thinking as we construct a workable, even innovative, urbanism for the future. Despite the differences in focus, real estate and design share certain intrinsic commonalities. Both contribute powerfully to building physical environments that reflect vision and imagination. Both offer a unique array of knowledge, operational skills, and intelligence that must be synthesized to produce a suitable outcome or product, whether a set of drawings or a fully finished building (Wang, 2015: 16).

As the built environment is increasingly becoming a medium that shoulders and reflects issues in wider social and cultural spheres, it is time to deepen the understanding of value creation in real estate by exploring its relationship with design from multifaceted angles and to expand value creation through its engagements with a larger scale of urban environments and a longer time frame beyond the mere time span of completion of a development or project investment. This chapter analyzes value-adding strategies through selected case studies and dissects conceptual overlaps and differentiations between design and real estate. By examining divergent perspectives in evaluating design and value creation, this chapter elucidates the underlying principles of value creation and offers the design of real estate as an effective framework for capturing and operationalizing the complex relationship between design and value creation in real estate.

Introduction

Understanding "the design of real estate" requires a scrutiny of the nature and process of design and real estate. As the two disciplines of design and real estate both possess the plurality of meanings and multilayered interpretations, this process of scrutiny poses a challenge.

Nevertheless, their close linkage with each other and potential significance to the production of the built environment, including the prospect of value creation, provide a promise and imperative for us to encounter this challenge.

> Design and real estate are inherently intertwined, but this linkage is not easy to dissect or analyze without encountering the obvious danger of oversimplification.
>
> Design is often closely associated with an emphasis on form and aesthetics, while real estate has a focus on implementation and process. The interaction between these two fields inevitably creates new perspectives and methods, and sometimes, also, possible hybrids that bring valuable additions to the existing repertoire of design and development vocabularies, prompting new ways of thinking as we construct a workable, even innovative, urbanism for the future.
>
> *(Wang, 2015: 16)*

For the following chapter, design refers only to the design of the physical environment that in general includes architectural, urban, and landscape design. For real estate, although conventionally many activities in practice involving the development, investment, maintenance, leasing, financing, and trading of physical buildings are singularly or collectively called real estate, the use of the term *real estate* here in this context focuses on the development aspects of the practice, which includes creating physical objects and space that form shelters and places for the working, playing, and living of human beings, from non-existence to being.

There are two possible meaningful ways to dissect the complex and multilayered relationship between design and real estate. One is to utilize a process point of view, articulating design as one of the specific actions in real estate, thereby situating real estate as an overarching process, leading and coordinating various players along the steps of building a physical environment. The other approach is to highlight design as a methodology and effective instrument for strategizing the business of real estate.

This chapter focuses on the former, and thus unfolds and interprets the complex relationship between design and real estate through three conceptual frameworks, the sequence of which guides the structure of this chapter and constitutes its three sections. The first section articulates the interrelation between design and real estate with brief case studies illustrating concretized scenarios of value creation by design. The second section dissects conceptual overlaps and differentiations between design and real estate. The third establishes a definition of the design of real estate and elucidates its underlying principles, highlighting why this concept possesses critical importance for both disciplines in the built environment.

Design in the process of real estate

From a processual perspective, design is a specific action and critical component of real estate. The role that design plays shifts at different temporal junctures of the real estate process. The generalized division of four stages of design in a project – conceptual design, schematic design (SD), construction drawings (CD), and design and development (DD) – illustrates the various levels of details that design thinking needs to address, with drawings as a communicative means, appropriate to the specific stage of the project. For example, at the inception stage of a real estate project, conceptual design helps to define the physical scale and scope of the project and evaluates programmatic options. At the feasibility stage, a schematic design (the "SD") forms the basis for estimates of construction cost and is critical for completion of an environmental site assessment (ESA), at least in the US context.

Starting with the task of helping create a vision, design specializes in concretizing this vision through optimization in the composition of the formal languages of architecture, urban design and planning, or landscape design that incorporate possible systematic thinking, including but not limited to aesthetical, cultural, structural, mechanical, and environmental components. Using visual representation, the task of design harnesses creativity to construct a narrative through the detailed organization of functionalities and spatial sequences and searches for a tangible form that provides a physical framework and suitable platform to define the everyday lives of future users. This process ultimately responds to and reflects the ideas and concepts of investors, developers, the public, and the designers themselves.

Throughout this process, collecting relevant information, interpreting the intentions of clients (be they developer, owner, or investor), dissecting users' behavioral patterns and their needs, and analyzing contexts (e.g., physical, social, cultural) all constitute initial due diligence that brings forth the designer's potential design propositions. At this initial stage, any related information, analyses, interpretation, imagination, and evaluation could trigger a potential design *parti* and inspire a unique design approach.

Based on due diligence, design creativity aims to frame questions and establish approaches that are unique to the issues at hand and addresses possibilities by transforming conceptual parameters into the concrete and visual languages of architecture, urban design, or landscape design through diagrams, programmatic allocations, architectonic logics, and compositions of various building elements to create form and space so as to resolve the defined problems based on propositions and conditions. In this process, creativity penetrates various steps, filtering as a result of possibilities, finding and adjusting foci, establishing formal approaches, and visualizing the form. These multiple steps are never arranged in a linear fashion; rather, the sequence of steps constantly changes through experimentation, visualization, and prototyping as well as through feedback from clients and the public. It is an iterative process filled with adjustment, redefinition, and revision.

As the design concept is further formalized, design and its visual representation help establish a concrete vision of a building, a complex, a district, or a city, thereby forming a tangible projection of a real estate product. This vision then functions going forward as a visual reference and a concretized foundation for the project and provides a reference for future steps of estimation and projection of financial and economic potentials and operational organization. The developer can then proceed with a detailed establishment of the project, perform feasibility underwriting, estimate a schedule, and evaluate organizational and executional strategies.

Aspects of value creation

The role of design in real estate goes beyond the visual representation of concretizing a vision; it also helps convene and coordinate the imagination of the public and launch future possibilities, thereby being persuasive toward constructing collective wills, and presents projected solutions in forming consensus and provides blueprints for buildings and urban scenarios.

As one of many steps in the implementation of a real estate vision, at a macro level, design aims to foresee and lead the construction and production of space and the built environment that ultimately shapes a society's behaviors and aspires to its possible future. That process of constructing and transforming the blueprint into reality is real estate. Throughout this process, design carries many value creation possibilities at different temporal or thematic junctures of a project. Real estate professionals increasingly feel the need to understand how design can add value and how evaluating design can be done effectively. The following examples illustrate how design thinking and design maneuvers might be addressed and applied when dealing with some different physical aspects of real estate.

Site

As real estate is location-bound, intelligent design can help overcome the intrinsic constraints of a site. Instead of having a site with reduced usability due to its physical irregularity, a thoughtful design can expand the potentials of the site, thereby helping to optimize existing conditions. Design and design thinking offer many possible applications in dealing with a site, including how to best maximize the view of a site or reduce the wind tunnel impact by strategically orientating the building(s) located on it; how to make decisions on volume, shape, façade, and physical typology of a building based on the urban context of a given site, and how to organize vehicular and pedestrian circulation on the site to minimize impact on users. Managing the spatial and visual complexity of a site provides a premise for value creation via design.

Program composition/usability

For real estate to be valuable, it first and foremost needs to be functional. Although definitions of "usability" vary with perspective, the logic underlying a program's composition usually guarantees the basic usability of real estate. Program denotes the compositional content, use, and function of buildings in physical design and real estate. Creativity embedded in design can help challenge conventional thinking in program composition while still maintaining usability in real estate. For example, when Gerald Hines decided to include an ice-skating rink in the 600,000-square-foot indoor shopping center of the Houston Galleria in the 1970s, an upscale mixed-use urban development, his ideas were opposed by many on his team, including his architect. However, this unconventional component, as the first ice rink centered inside a mall, functioned as a successful magnet for families within the retail-focused mixed-use development. The rink not only became a significant draw for the entire development, but also helped dramatically increase the visibility of the stores on the mall's lower level. The rents charged to retailers of the lower level became equivalent to those of fashion retailers at the prime location on the first floor. The annualized gross retail sales at the Houston Galleria reached $200 million in 1981, nine times that of average retail sales of approximately 30,000 malls across the US during the 1980s, in terms of measuring the profitability of mall operation (Feinberg and Meoli, 1991: 426).

One of the more recent cases of adding value through creative programmatic composition is the case of the 1111 Lincoln Road development in Miami Beach. By placing retail stores at the street level, a penthouse apartment and a boutique as well as an event space/viewing platform on the fifth floor and roof deck of a parking structure, the developer Robert Wennett and his architects turned an otherwise banal parking garage with fixed functionality into a vibrant "design-driven, experience-based venue" of mixed-use development that has gained national fame since its completion and brought financial success with a retail rental income of $110 per square foot (Wennett, 2013), almost triple what was expected (i.e., $30 per square foot), all occurring during the economic downturn of 2009. The rent increase generates an additional income stream of $1.6 million on an annual basis.

Infrastructure and public space

Determining how infrastructure and public space can help enhance the value of private development through design creativity is both a long-held and novel proposition. Since its inception, Central Park in New York City has offered sound testimony of what infrastructure and a thoughtful public space can do to real estate values. In 2015, the twenty most expensive residential real estate transactions in New York City took place on the blocks surrounding Central Park (Marino, 2015). And the highest retail rents in the world – $3,000 per square foot as of the end of 2016 – were along upper Fifth Avenue, south of Central Park (C&W, 2016: 6).

Another significant infrastructure project that has proved to be a catalyst for revitalization of the surrounding neighborhoods in New York City is the High Line project, a 1.45-mile-long landscaped linear park winding through the city like a ribbon, linking small-scale parklands and civic amenities, built on a disused and dilapidated elevated rail line dating from the 1930s. The High Line project has become an engine of economic development, revitalizing property values and civic spirit, even within the trough of a financial crisis. It is a manifestation of how urban transformation can take place by utilizing derelict public infrastructure and how infrastructure and public space created through design formulates value in economic and social realms. Initiated in 2009, by 2014, the High Line had already attracted approximately four million tourist visits per year, with a peak of 60,000 visitors per day during the summer season. It generated total revenue of $32 million in 2013 alone, of which public subsidies accounted for less than 1 percent, with the rest generated through membership fees, donations, and leasing income from event activities. With nearly 30 projects currently under construction and newly built along the High Line, the city estimates it will generate approximately $900 million in tax revenues over 20 years (Tate and Eaton, 2015: 45). In this specific case, the visionary marshaling of public space design to reposition a derelict rail line winding through neighborhoods is as innovative as the landscape design itself on top of the structure.

Façade

Architectural façades are layers that transition the spatial organization of buildings into their exteriors, helping establish identities and the interface of buildings with a larger audience. The choice of a facade goes beyond mere aesthetic considerations; it is also technical and functional. A good example of applying creativity in a façade design that adds value to a project in multifaceted dimensions is the John Lewis department store development located in Leicester, UK, designed by Foreign Office Architects. The façade of the department store is constructed of four panels of glass with a decorative pattern reminiscent of the city's rich textile heritage and the store's history of fabric production (Wang, 2014: 21–22). The four layers of glass and its final glazing create a visual perception of a moiré pattern that can be viewed from the exterior and enriches the urban environment while enabling people from the inside to view out, thereby maintaining privacy while allowing the penetration of natural light and the exterior views inside. The combined façade material and its fabric pattern emerge as the single design element that uses a curtain strategy to add visual, contextual, and iconic value to an otherwise typically mute department store façade. It is difficult to quantify to what extent this façade has created value visually, historically, and from the public perceptional recognition, but its thoughtful application of design creativity to the façade alone indisputably meets multiple aspects of the owners' needs, as well as those of users, the public, and the city.

Construction delivery

The application of design permeates various aspects of the process of real estate, including the choice of construction delivery methods, which is often considered a mundane production aspect of a project. A good case is Carmel Place in New York City, a multi-family housing development that comprises loft-like rental apartments, ranging in size from 260 to 360 square feet. This project optimizes the process of housing delivery as one of the first multi-unit

Manhattan buildings using modular construction that included the fabrication, transportation, and stacking of 65 individual self-supporting steel framed modules, 55 of which serve as residential micro-units with the remaining ten as the building's core. The overall design and production are highly coordinated in that the dimensions of furniture, rental units, and the entire building are based on an 11-foot grid module, including but not limited to, the size of beds, height of windows, interior dimensions of corridors, and stair widths, that in turn were generated from the proportions of nineteenth-century residential brownstones in NYC. Even the entire building's massing stems from the module, with four 11-foot wide towers being staggered in an east–west orientation. This utilization of modular design and construction delivery enables the entire construction (excluding the foundation and ground floor) to be constructed offsite and then individually "installed" on site within a two-week time frame.

Thus, the overall design, aesthetics, and construction redefine methodologies of housing delivery and reflect a process of optimization through pre-fab construction that reduces on-site construction time and costs, while avoiding noise and disruption to urban neighborhoods while maintaining quality and delivery efficiency and consistency. This case illustrates that construction efficiency and architectural quality are not necessarily at opposite ends of a continuum. In fact, creativity in design is reinforced when the limits and constraints of construction delivery and economic and financial means are considered.

The above cases are a few examples of how creativity is applied in real estate design and development. They are testaments to adding value through design, which can be multivalent in terms of program, form, spatial organization, physical scale, construction delivery and strategy that meet the many physical aspects of real estate.

In addition, the application of design knowledge also permeates the various stages of real estate valuation. Among the three valuation methods for real estate, the sales comparison method, as well as the cost approach, often requires a truthful understanding of the physicality of building elements. It is critical to determine in the process of valuation the extent to which a tangible feature of the comparables can either add to or subtract value from the subject property and the extent to which a building component (whether it a column design, the size of a room, or the spatial sequences of a building) is regarded as functional or trendy, thereby conveying a sensible design and being more costly in the calculation of the replacement cost of the property. A trained eye and learned sensitivity to design play a critical role in valuation of the physical characteristics of a property and lead to better judgment in real estate valuation. Perceiving spatial organization and interior tastes helps in understanding the market's acceptance of certain design characteristics, and thus, helps gauge value differences among comparables. Finally, gaining insights into the surrounding urban quality of a location and adopting a macro perspective on urban planning can facilitate the detection of potential long-term possibilities for value appreciation of a site.

In real estate development, design is embodied through final products that are the tangible representation of accumulated wealth via income generation and/or price appreciation that together are greater than the cost of efforts applied during the process of assembling land, obtaining permits, and designing and constructing building(s) with inputs of labor and material. Reducing costs, enhancing efficiency while maintaining quality, creating positive public perception, attracting the market, and thus increasing the income stream can all help create monetary value for owners, investors, and developers. In doing so, design, as the previous cases indicate, creates unique experiences for users. This uniqueness can occur through spatial composition, such as by combining a garage with a viewing platform functioning as a public gathering space in the above case of 1111 Lincoln Road in Miami Beach. Design creativity can also relate to an innovative and efficient construction method that saves time, thereby reducing the budget without compromising design quality. It can also occur at the urban scale, creating a public amenity by taking advantage of an existing infrastructure to sequentially increase property values of the surroundings. Adding value can also mean more than just establishing a brand for a building; it can place an entire city on the global map, such as with the Guggenheim Museum in Bilbao and Burj Kalifa in Dubai. In sum, the possibilities for value creation by design are as endless as the imagination of a creative mind with a solid grip on reality, a sensitivity to demand, and knowledge of execution.

Value creation in design and real estate

In real estate terms, value is reflected through market pricing and rental yields from buildings being built, purchased, or disposed of, and rent collected. As financial return is taken as an objective measurement of real estate performance, value is calculated and forecast through mathematical formulas and expressed with spreadsheets as a medium. Real estate is invested and traded based on such concepts of value revealed through fixed industry terminologies such as net present value (NPV), equity multiples, and internal rate of return (IRR).

There are four categories of research approach for measuring real estate values in relationship to design. The first approach utilizes the methodology, borrowed from social sciences, of surveys, interviews, and descriptive categorizations via presentations of visual images. By asking the target audience to review two-dimensional images or by visually comparing photos of a built environment and ranking the order of these images from "most liked" to "least liked" or with any other specified categories, a preferred design image is selected (Meinig, 1976; Steinitz, 1995). This approach links a descriptive design preference to an implicit value of real estate. It is often used to elicit public opinions on design and help achieve public consensus. Its descriptiveness is based on subjective experiences and opinions of design rather than on analytical evaluation of value created by design.

The second approach to exploring the relationship between design and real estate measures the effectiveness of the value-adding design through quantitative economic modeling. Often the focus is to gauge the extent to which selected amenities of a building or an urban environment contribute to the monetary value of real estate. In this situation, the selected amenities are often treated as variables of hedonic modeling, which involves a large collection of data and mathematical economic calculation. The measured relationship between value and the tested amenities (variables) is numerically presented, providing an objective analysis with regard to the linkage between the two (Vandell and Lane 1989; Asabere et al. 1989; Kain and Quigley 1970; Kaufman and Norman 2006; Guttery, 2002; Matthews and Turnbull, 2007). However, this method has been limited to amenity-focused variables, and the design characteristics tested in this approach are often either urban in nature, such as urban parks, schools, neighborhood convenience stores nearby, or common amenity-like architectural features such as balconies, numbers of bedrooms or bay windows. In these cases, amenities often do not represent the creative design thinking in dealing with visual languages and spatial compositions that are utilized by professional designers with an aim toward spatial uniqueness and the aesthetic characteristics of design. Instead, resultant design features are compartmentalized and established as independent variables of a mathematical formula, rather than as connected physical features based on coherent design thinking.

The third approach of delving deeply into value-added design is through Michael Porter's classic value chain model. This approach helps to dissect the multiple steps and activities involved in the real estate development process and identify the productive contributions of each step in maximizing the final and total value of a set of assets (Porter, 2008). This analytical approach shifts the emphasis of understanding real estate product from what things look like to how they work and the interconnection of the various activities.

The fourth and often used approach in delineating and understanding the tangible value created by design involves analytical descriptions of case studies (Pawlicka, 2014; Tiesdell and Adams, 2011). As design is specific to context, good design customization, based on situated conditions and users' needs, highlights the value created by design in qualitative terms, which can be captured through case studies, and is therefore less generic. This approach emphasizes design tact by expedient pragmatism in situating design thinking in its process and a suitable context. Nevertheless, this method often achieves empirical richness at the cost of a case-by-case evidence base.

Each of the four aforementioned existing approaches in the literature addresses the value creation of real estate by design, where the interrelationship between design and real estate is determined through visual preference and consensus, statistical modeling, generic value chain analysis, or analytical case studies. As real estate and design are increasingly produced within a complex set of contexts and are contingent upon various social, cultural, regulatory, and financial conditions, their impacts on human lives through the embodiments of the physical environment are increasingly becoming multilayered. It is time to widen the meaning of real estate by expanding it from a mere production of the physicality of the built environment to a potent platform shouldering the economic and spatial logics of city making and social responsibilities.

As design and real estate can each be viewed as products and processes of production, they share overlaps and differences. While the exploration below bears generalization, it is an attempt to provide possible conceptualizations of the evident linkages between the two fields.

Experiential construct

The creation of a unique experience through design has increasingly become the goal of real estate. It is no longer about the same typology of big-box shopping malls and repetitive cubic workspaces. Individualized, flexible, and effective spaces that foster creativity and differentiation are increasingly standard measurements of the quality of environment that we strive for through real estate development. Just as Boatwright and Cagan vividly described:

> while we may look at a picture for seconds or minutes, experience a movie that lasts an hour or two, or read a novel over a few days, we may inhabit buildings for months, years, or lifetimes. Our individual experiences of architecture are complex, and are often constructed over extended periods of time, so our emotional responses are correspondingly diverse, multilayered, and evolving.
>
> *(Boatwright and Cagan, 2010: 114–117)*

The multifarious and long time span of possible emotional associations between human beings and buildings highlight designers' and developers' overlapping need to attend to experiential constructs when dealing with the environment through a physical framework of spaces and the organization of everyday functions. During this process, design is often closely associated with an emphasis on form and aesthetic intention, while real estate has a focus on vision, implementation, and sensitivity to market dynamics.

As real estate is capital intensive and is in essence an operation that connects physical product with capital investment, the real estate process is driven by economic logic and business rationale. The quest for efficiency and desire for risk reduction often leads to the continuing application of repeated building types that have tested results from the market. Meanwhile, in comparison, richness in experience and unpredictability is often one of the many aims for a successful designer to achieve, especially in urban and landscape design and in public building design. In this sense, the logic of real estate is contradictory to the pursuit of a unique experience; and it explains the singular and monotonous office parks and suburban shopping dispersed across the landscape.

Today, however, thanks to technological advancements, "what was once exotic is now commonplace" (Postrel, 2003: 47). Thus, it is no longer the products or services being celebrated by the market, but rather the experience that comes with the product or service that is consumed. Under this premise, real estate increasingly must push the boundaries of its "product" – building forms and spatial quality – by utilizing creative design as a strategy. Design helps construct a narrative of the experience. As Ellen Lupton put it, "part of what design does is sensual storytelling, emotionally storytelling" (Green, 2016: ST10).

Complexity and constraints

Both design and real estate are procedural, embodying organizational complexities that involve multiple players and engage multidisciplinary knowledge domains. "In design, one needs to deal with the structural, environmental, and acoustical aspects among others, of a building, landscape, or city, while in real estate development one must cope with design, financing, construction, environmental concerns, and much more" (Wang, 2015: 16). The sequences of the involved tasks of both fields are never linear, and neither the relationship among the multiple tasks of design nor that of real estate necessarily follows a clear temporal progression. A decision in one particular knowledge domain in the design and real estate process often impacts other related tasks and even likely becomes a constraint. For example, a change in the air duct path of a new construction may lead to the relocation of a beam, and, consequently, may alter the spatial effects of the design intentions for a room.

Risk and return

Since unpredictability is embedded in the implementation and realization of most ideas, visions, and products and can be further reinforced by market frictions, the design and real estate processes intrinsically bear risk. Due to the different nature of their operations, the design and real estate processes as well as the consequential products encounter starkly different types and levels of risk. These lead to contrasting perceptions and attitudes among designers and developers toward risk.

Modern design (especially the Modern Movement of architecture) advocates creativity and challenges the status quo. While design aims to envision the future, real estate endeavors to turn the future into reality and, thus, it is about buildability. This fundamental difference defines the respective risk and return profiles of design and real estate. Good designers strive to be distinctive and provocative by taking risks. In stark contrast, risk aversion is an imbedded tenet of real estate. As real estate is a capital-intensive business and investors' and developers' capital is at stake upfront, risk is carefully measured and managed and financial return becomes a quantitative gauge of the effectiveness of risk management. Therefore, building upon an established and tested building type that has been approved by the market is often regarded as one measurement for significant risk reduction.

Since risk and reward are correlated based on economic principles of efficient market theory, the gap between financial rewards of design as a business venture versus those of real estate has recently increased exponentially. The financial rewards of design are based on labor and materials, with either a fixed rate for the estimated hours of time input by the designer or a meager percentage of the project's construction cost. Rewards for real estate development are established based on the investment of contributed capital by the developer as equity or "sweat equity." As evinced in the historical data of Piketty's book *Capital in the Twenty-First Century*, investment through capital has proven to be significantly more effective in wealth accumulation than contribution of labor and time, which is proved to be a foundational mechanism of the capitalist society (Piketty, 2014).

With the dwindling total public investments across the globe, the real estate of private investment is increasingly becoming a medium of city building. By taking more risks with capital through market mechanisms, private real estate developers and investors possess more decision-making power over the physical and spatial constructs of the built environment. Meanwhile, in the context of the dominating power of capital, designers are withdrawing from their previous decision-making position, or, at the very least, are unwillingly to compromise it and, thus, become sidelined.

The design of real estate: A framework in value creation

The *design of real estate* is a relatively novel concept that originated as a title for a real estate course focusing on the interrelationship between design and real estate development and investment at the Harvard University Graduate School of Design. How to incorporate design thinking in the real estate process and how to better understand real estate through the lens of design have increasingly gained in popularity within the professions of the built environment and with the general public. The interaction between these two fields has the potential to create new perspectives and innovative methods. Importantly, hybrids may bring valuable additions to the existing repertoire of design and development vocabularies, prompting new ways of thinking as we construct a workable, even innovative, urbanism for the future (Wang, 2015).

The proposition of the design of real estate is more of an attitude, method, and intellectual framework that intends to guide the way we conceptualize and operationalize real estate. In many ways, this can be viewed as a response to the increasing critique of real estate operations that pursue the singular goal of profit maximization and short-term political and economic advantages without adequate consideration of cultural, social, and environmental impacts. This proposition is also a prompt call for action to provocatively denounce monotonous development prototypes across the landscape and to proactively mediate the disciplinary separation between real estate and design in education. As built environments are increasingly becoming a medium that shoulders and reflects issues in wider social and cultural spheres, it is time to deepen the understanding of value creation in real estate by exploring its relationship with design from multifaceted angles. In this way, it becomes possible to expand value creation in real estate through its engagements with a larger scale of urban environments and a longer time frame beyond the mere time span of completion of a development or project investment.

With their possible leadership in the initiation and production of the built environment, entrepreneurial spirits in the constant search for opportunities, and potential cross-disciplinary knowledge domain in nature, real estate developers and investors have the potential to organize and construct "a more complex form of creativity" (Kamin, 2015: 30), one that diffuses the design-driven culture in the real estate process and in its business decision mechanism.

Characteristically, real estate is rational and logic. Real estate's engagement with capital requires the decision process to be goal-oriented. Such a process requires efficiency and judgment of the market that is sober, data-oriented, and analytical. The pro forma itself establishes economic logic in the foundation of real estate finance and development. However, with the increasing scarcity of land resources, awareness of ecological needs, and unprecedented mobility, real estate requires a more suitable way of practice and a profound change in implementation. With this context, the proposition of the design of real estate provides a possible and promising future.

Emotional narrative

The underlying principle of the design of real estate is to expand the scope of real estate into three aspects. First, it is to understand that real estate is developed using a physical framework embodying an emotional narrative through specificities of design language, spatial construct, and material sensibility. The developed products need to tell a story. Establishing an emotional connection and empathy between humans and the built environment is increasingly critical for real estate development. Good real estate projects create a balance between the rational and the emotional – rational in terms of how the development process maximizes efficiency while meeting the needs of functionality for users and economics for owners, and emotional in terms of speaking to humans in a vivid and thoughtful manner and helping construct an experiential narrative. An effective real estate project needs to please the senses and provide a framework for imagination as well as mind exploration that requires design sensitivity and creativity in the development process.

Critical thinking and cultural production

The second principle of the design of real estate is to approach real estate projects with critical thinking; that is, to examine real estate development from multifaceted dimensions, and with calm reasoning and possible humanistic scrutiny beyond simply the financial measurement of quantitative returns. This approach requires viewing real estate discipline through a lens of cultural production that has hitherto been lacking in the real estate arena. Real estate is not only a means of providing concrete and quotidian necessities of everyday life and a medium from wealth accumulation through investment in and trading of an asset, but also a physical framework in the context of people's lives, behaviors, and thus, thoughts. This implies imbued cultural production and therefore invites scrutiny and analysis by critical thinking to view the trends, process, and production in the context of history and culture, with the capacity of abstract insight and theorization.

Awareness of urban scale and context

In the process of real estate development and investment, the focus on financial returns has rendered real estate products to be viewed as autonomous objects, often separated from their surrounding urban environments. The concept of the design of real estate calls for the active mindset of framing real estate in an urban context and understanding how the urban environment and physical characteristics can help increase the value of real estate that applies to a longer term and generates direct or indirect benefits that go beyond the mere stakeholders of individual projects alone. The example of the High Line project in NYC is a good example of this perspective. That being said, as the characteristics of urban design and quality of urban life are difficult to measure in quantitative manners and then translate to a line on a spreadsheet, there exists a dilemma for developers and investors in identifying investment viability. To this point, it would be practical and tactical to introduce public and private joint ventures or mechanisms that can facilitate the consideration of the public realm of cities by the private sector.

Measures to combine design thinking and economic logic can be incorporated into various stages of the development process. The criticality of urban contexts entails one key method in creating design value through the "value-grading" method in allocating the land use of a site. When one encounters a site, a useful value-grading process can ensure maximization of the value for various parts of a site through analysis of the immediate urban surroundings. This is crucial in determining which orientation and component of the site have the highest economic value for a particular use that can offer amenities, uniqueness, and overall better functionality. Then the part of the site that can be secondarily suitable for a use that possesses potential market premium, although less than the highest value, can be determined. This method of matching the locational urban characteristics with a specific programmatic use should continue until the least valuable part of the site is matched with a selected use. Based on this logic, urban design framework is key, along with inherent economic logic, in determining the allocation of functions on the site. This method will help create active frontages of retail use along major thoroughfares to enhance foot traffic, for example. It could locate a residential area next to a scenic view in order to increase rent premiums or sales price, thereby creating a pleasant environment for the community. It might also help place corporate signage on top of an office building along a highway as a visible advertisement. In reality, there are numerous possible alternatives for designing a site with allocated programmatic uses. However, it is critical to first value-grade the site based on the understanding of urban contexts and potential market premiums defined by the site's physical characteristics.

Value framework

Based on the above principles established with the concept of the design of real estate, the following three frameworks are further positioned to provide guidance of specific actions in addressing the linkage between design and real estate. First, it is crucial to highlight the role of design in the process of real estate development and investment via the focuses of value creation, innovation, and design thinking in general (Table 25.1). Second, in addressing the design of real estate, it is critical to highlight the linkage between real estate and city marking (Table 25.2). Third, with the design of real estate framework, it is important to establish a position in linking real estate with social responsibilities and cultural production that can be addressed through the inclusion of mutual impacts (Table 25.3).

Summary

In short, the design of real estate is a timely proposition. This overall chapter initiates a discussion of this proposition by addressing three aspects embedded within its multi-layered conceptualization. The first aspect involves analyzing how design thinking can be incorporated into various temporal stages and physical dimensions of real estate development and investment process. The selected case studies included in the chapter highlight the idea that value-adding possibilities in real estate can be multivalent and have numerous possibilities, including but not limited to qualitative and quantitative interpretations in the long- and short- terms. The possibilities for value creation by design are as endless as the imagination of a creative mind, with a solid grip on reality, a sensitivity to market demand, and knowledge of execution. In addition, this chapter addresses current research approaches utilized in the literature to dissect the relationship between design and real estate and help elucidate the overlaps and differentiations of the two to provide a perspective for future possible development on the convergence and divergence between design and real estate. Lastly, based on the discussion of the applications and research possibilities of linking real estate and design, the concept of the design of real estate is proposed. In simple terms, this concept addresses the design dimension of real estate. However, in essence, this new proposition intends to establish a framework that expands the urban, cultural, and social scopes of real estate and project an approach for how we should think of, practice, and teach real estate through multiple lenses that help construct real estate as a more stable and long-lasting practice and, more importantly, a profound knowledge domain and critical academic discipline of building communities and cities.

Table 25.1 Key components of real estate and a framework for guiding value creation through design and design thinking

Key components of real estate	A framework for guiding value creation through design and design thinking
Product anatomy	Dissect design products with a real estate perspective to comprehend the anatomy of complex buildings and mixed-use development, including programmatic composition, density, ecosystems, and infrastructure.
Product differentiation	Position design with an understanding of development processes and financing structures to improve efficiencies, economic returns, time to market, as well as trademarks and differentiation of design services.
Valuation strategy	Establish familiarity with valuation techniques of real estate to emphasize design alternatives and to highlight the impact of design on capital values and financing possibilities.
Performance impact	Integrate design with real estate products through the application of detailed strategies of building form and typology to improve the relevance, functionality, productivity, sustainability, and aesthetics of real estate, as well as to strengthen the branding of real estate products through design.
Implementation process	Understand and harness design thinking, design theory, and the creative process through a multidisciplinary approach to address competing interests and special constituencies of real estate.

Table 25.2 Key components of city making and a framework for guiding value creation through the linkage between real estate and city making

Key components of city making	A framework for guiding value creation through the linkage between real estate and city making
Urban vitality	Realize real estate development and investment as a driver of urbanism and urban vitality. Dissect urban characteristics and place making through the lens of real estate operations.
Urban infrastructure	Understand real estate in relation to urban systems (including transportation and natural resource planning).
Urban history and formation	Consider the implications of urban history and formation for the real estate process and its consequent impacts.
Urban players	Understand the implications of urban governance structure and urban politics among real estate players.

Table 25.3 Key components of social responsibilities and a framework for guiding value creation through the linkage between real estate and social responsibilities

Key components of social responsibilities	A framework for guiding value creation through the linkage between real estate and social responsibilities
Social change	Address real estate in relation to urban growth and social change.
Social equality	Contribute to affordability and social equality through specific actions of housing and community development.
Environmental ecology	Construct a conceptual framework and practice specifics about environmental real estate to address the needs of ecological urbanism.

References

Asabere, Paul K., George Hachey, and Steven Grubaugh. 1989. "Architecture, Historic Zoning, and the Value of Homes," *Journal of Real Estate Finance & Economics*, Vol. 2, Issue 3: 181–195.

Boatwright, Peter, and Jonathan Cagan. 2010. *Built to Love*. San Francisco: Berrett-Koehler Publishers.

C&W. 2016. *Main Streets Across the World 2016/2017*. Cushman & Wakefield Research.

Feinberg, Richard A., and Jennifer Meoli. 1991. "A Brief History of the Mall," *Advances in Consumer Research*, Vol. 18, Issue 1: 426–427.

Green, Penelope. 2016. "Cooper Hewitt's Triennial Asks: What Makes Something Beautiful?" *The New York Times*, February 14, ST10.

Guttery, Randall S. 2002. "The Effects of Subdivision Design on Housing Values: The Case of Alleyways," *Journal of Real Estate Research*, Vol. 23, Issue 3: 265–275.

Kain, John F., and John Quigley. 1970. "Measuring the Value of Housing Quality," *Journal of the American Statistical Association*, Vol. 65, Issue 330: 532–548.

Kamin, Blaire. 2015. "Place and Profit: Assessing the Plans of Global Leadership in Real Estate and Design," in *Global Real Estate in Real Estate and Design*, edited by Bing Wang. Cambridge, MA: Harvard University Graduate School of Design, pp. 29–33.

Kaufman, Dennis A., and Cloutier Norman R. 2006. "The Impact of Small Brownfields and Greenspaces on Residential Property Values," *Journal of Real Estate Finance and Economics*, Vol. 33, Issue 1: 19–30.

Marino, Vivian. 2015. "In 2015, Shattering Records in New York City Real Estate," *The New York Times*, December 27, RE1.

Matthews, John, and Geoffrey Turnbull. 2007. "Neighborhood Street Layout and Property Value: The Interaction of Accessibility and Land Use Mix," *Journal of Real Estate Finance & Economics*, Vol. 35, Issue 2: 111–141.

Meinig, D. W. 1976. "The Beholding Eye, Ten Versions of the Same Scene," *Landscape Architecture*, Vol. 66, Issue 1: 47–54.

Pawlicka, Kinga. 2014. "The Significance of Architectural Attractiveness in Creating Property Value: A Case Study of Poznań," *Poznan University of Economics Review*, Vol. 14, Issue 2: 118–137.

Piketty, Thomas. 2014. *Capital in the Twenty-First Century*. Translated by Arthur Goldhammer. Cambridge, MA: Belknap Press.

Porter, Michael. 2008. *Competitive Advantages: Creating and Sustaining Superior Performance*. New York: The Free Press.

Postrel, Virginia. 2003. *The Substance of Style*. New York: Harper Collins Publishers: 47.

Steinitz, Carl. 1995. "Design is a Verb; Design is a Noun," *Landscape Journal*, Vol. 14, Issue 2: 188–200.

Tate, Alan, and Marcella Eaton. 2015. *Great City Parks*. London and New York: Routledge.

Tiesdell, Steve, and David Adams. 2011. *Urban Design in the Real Estate Development Process*. Chichester: John Wiley & Sons, Ltd.

Vandell, Kerry D., and Jonathan Lane. 1989. "The Economics of Architecture and Urban Design: Some Preliminary Findings," *Journal of the American Real Estate & Urban Economics Association*, Vol. 17, Issue 2: 235–260.

Wang, Bing. 2014. "Retail and Globalization: John Lewis, Leicester Case Study," in *Prestige Retail: Design and Development of Luxury Goods Market*, edited by Bing Wang and Richard Peiser. New York:

International Council of Shopping Centers (ICSC), pp. 21–22.

Wang, Bing. 2015. "Innovative Juncture: Real Estate and Design," in *Global Real Estate in Real Estate and Design*, edited by Bing Wang. Cambridge, MA: Harvard University Graduate School of Design, pp.16–22.

Wennett, Robert. 2013. "Presentation on the Development of 1111 Lincoln Road Project," lecture delivered to the Master of Design, Real Estate and the Built Environment program, Harvard University Graduate School of Design. October, 2.

Part VII

Land use policy and governance of real estate development

26

Developing vibrant centers
The public–private development perspective

Emil Malizia

Abstract

Public–private partnerships (PPPs) are the preferred vehicle in the USA to foster urban revitalization. Properly structured, PPPs can increase vibrancy in urban employment centers by offering public subsidies commensurate with public benefits. The logic and methods to structure PPPs are explained and illustrated, focusing on ways to lower costs and reduce risk. Cost reduction methods have been widely discussed whereas ways to reduce near-term market risk and long-term financial risk have been largely ignored. Strategies to promote vibrancy in weak or strongly vibrant centers are quite different. Weak centers need strategic public and private investments to change perceptions of risk. Increasing the attractiveness of some places requires ignoring other places given resource constraints. Stronger centers need ways to increase density, affordability, and diversity without undermining the vibrancy already attained. Providing workforce housing, energizing the pedestrian, increasing critical mass without damaging uniqueness, preserving iconic places and open space, and resolving conflicts between different downtown activities are the approaches suggested. Supporting export/traded sectors should take priority over adding urban amenities. Increasing the vibrancy of urban centers, especially downtowns, by attending to export services is becoming more urgent as these sectors, often located in urban centers, come to constitute the most viable part of the metro economy's economic base.

Introduction

After decades of decentralized, low-density, auto-oriented suburban development in the USA, urban centers in metro areas are making strong comebacks by attracting or growing professional, technical, financial, medical, and educational services, which are emerging as leading export sectors of the US economy (Jackson 1985; Storper 2013). Cities offer intense face-to-face interaction, dense organizational networks, and overall critical mass (agglomeration economies) that increase the productivity of this economic base, especially when city centers are vibrant (Dreier, Mollenkopf, and Swanstrom 2001; Florida 2008; Glaeser 2011; Kelly 2016; Moretti, 2012; Storper, 2013).

This chapter focuses on public–private partnerships (PPPs), designed to enhance the vibrancy of metro-area employment centers in the USA, including primary and secondary downtowns and more peripheral urban centers. Vibrant centers are dense, diverse, connected, and walkable places often well served by public transit (Campoli 2012; Crankshaw 2009; Haughey 2008; Jacobs 1985; Levy and Gilchrist 2013; Paumier 2004; Talen, 2012). More vibrant centers, especially ones in larger metro areas with good rail transit have outperformed comparable less vibrant centers having higher rents, lower cap rates, higher occupancy, and greater transaction volumes (Kelly et al. 2013; Malizia 2014; NAIOP 2014; Malizia and Song 2015; Kelly and Malizia 2016). High-growth firms more densely populate places with greater vibrancy in the Washington, DC consolidated area (Malizia and Motoyama 2016).

Vibrancy can be increased by boosting private investment and associated public investment through PPPs. Ideas in the next two sections are guidelines for local jurisdictions with relatively weak vibrant centers. These guidelines emphasize the economic and financial dimensions of PPPs at the project and strategic levels. PPPs can stimulate private investment in vibrant centers by reducing costs or more importantly by exploring ways to lower the perceived near-term market risk and long-term financial risk associated with redevelopment. In cities with relatively strong vibrant centers, the task is to overcome six common challenges to sustain progress. To simplify, the chapter will focus on downtown revitalization (for further examination of this general topic, see Frieden and Sagalyn 1989; Sawicki, 1989; Kotin and Peiser 1997; Adair et al., 1999; Gyourko and Rybczynski 2000; Meyer and Lyons 2000; Sagalyn 2007).

Project-level approaches

At the outset of the development process, developers typically use "back-of-the-envelope" techniques of financial analysis to compare value to cost in order to gauge the attractiveness of alternative projects. Project value is found by dividing estimated net operating income (NOI) by the appropriate market-determined capitalization (cap) rate. NOI is gross revenue (current market rent multiplied by the number of units or square footage of the project) reduced by vacancy allowance and operating expenses. The cap rate comes from transactions involving comparable properties. The cap rate can be considered the overall return investors expect per dollar invested. In other words, a 6 percent cap rate reflects the expectation that investors will receive six dollars for every hundred dollars invested in a property.

Project cost is estimated by adding up site acquisition and development costs, construction costs including the structure and systems (hard costs), and associated cost of services (soft costs). Soft costs include architectural and engineering fees, and fees resulting from legal work, accounting, surveying, land planning, appraisal, market research, and marketing. Costs associated with the construction loan are included; construction-period interest is an important line item. All costs incurred to take the project through the development review process (entitlement process) are added, such as fees, cost of impact studies, community relations efforts, etc. Contractor fees, overhead, and contingency can be soft costs but can also be built into the hard cost estimates. Soft costs also include the developer fee (profit) and project contingency. The size of the latter reflects the perceived risk and complexity of the redevelopment. In complex urban redevelopment projects, soft costs tend to be 25 percent or more of hard costs (Graaskamp 1981; Ciochetti and Malizia 2000).

For downtown redevelopment projects, the lead entity representing the public sector is usually a non-profit. This publicly funded organization receives input on redevelopment options from the public and fashions plans or visions for the land uses to be included in the redevelopment of small areas and specific sites. Charrettes are often used to structure public input. The plan/vision is expected to be consistent with legal/institutional requirements and the site's physical constraints, leaving the economic and financial dimensions to be addressed.

The public sector usually solicits interest in specific redevelopment sites from real estate development firms with Requests for Qualifications (RFQs) and/or Requests for Proposals (RFPs). RFQs are quite common because real estate developers in the USA have more experience developing new product on "greenfield" sites than redeveloping urban infill sites. Infill sites may be vacant or have buildings that the public wants to preserve through adaptive reuse. Once the qualified developer is selected, a properly structured PPP is needed to ensure that the selected developer will be able to carry out the redevelopment project. Otherwise, another developer may be able to replace this developer by making a nominally more attractive financial offer after all of the feasibility work and technical studies are completed (an upset bid).

Feasibility research is conducted or commissioned to analyze potential uses including combinations (mixed use) that would implement the plan/vision. As in private development, alternative uses are evaluated to find the most productive use of the site – its highest and best use. Highest and best use calls for the identification of the most economically profitable use given financial, legal and physical constraints (Appraisal Institute 2013; Vandell and Carter 2000; Malizia 2009; Godschalk and Malizia 2013).

Appraisers have difficulty valuing downtown redevelopment projects accurately due to thin markets (few transactions). The absence of good comparable properties increases the influence of opinions and other subjective factors. Existing property owners who derive income from current land uses have a vested interest in having an optimistic outlook on the long term because future growth usually increases property appreciation. Real estate developers may envision viable redevelopment of downtown sites but only if the properties can be secured in the near term at reasonable prices. Investors have a wider range of assessments about risk and uncertainty compared to developers or property owners. Some are attracted to urban redevelopment projects whereas others have little interest in them. Because transactions are infrequent and projects are complex, markedly different investment outlooks may be defensible. However, such differences of opinion form a major barrier to attracting the resources needed for downtown revitalization because they ultimately lower the value of potential projects. In essence, more variation in expected returns translates into higher risk premiums (for elaboration, see Luscht 1997, Chs. 14–15).

Public subsidies

Non-profit entities promoting redevelopment on behalf of the public often invite private developers to select from the menu of subsidies available, anything from façade improvement programs to concessionary financing. To fashion cost-effective PPPs however, the public needs to know the economic and financial details of the deal in order to estimate how much subsidy may be needed. Thus, cost and value estimates should first be made without the inclusion of subsidies.

Discounted cash flow analysis (DCF) is the method used to compare cost to value and to evaluate different forms of public participation in the project. Most projects in less vibrant downtowns will initially show total development cost to be greater than market value. The basic question is: Does the redevelopment project provide sufficient public benefits or positive externalities to justify the public subsidy required to make it feasible (Sagalyn 1997; Malizia, 1999). The DCF methodology for estimating public subsidies is described and illustrated in the Appendix.

Downtown redevelopment projects can be subsidized to reduce cost or risk. The most frequently used cost-reducing subsidies are land write-downs, tax abatement, and concessionary financing (Hodge 2004). Lowering or eliminating site acquisition cost is one of the most popular subsidies, whereby publicly owned land is contributed at no cost. Although eminent domain can be used to purchase parcels that are subsequently contributed to private developers, public development entities avoid using this government power, which has become more controversial in many parts of the USA since the Kelo case.

Although less common, the public can subsidize the long list of soft costs including professional fees and studies needed to plan and review the project. For example, public entities could conduct surveys, appraisals, or land planning. Or the public entity could encourage private consultants to provide services pro bono. The public may also waive fees or pay for impact studies. Timely building inspections and issuance of the certificate of occupancy (CO) can reduce the construction period somewhat and lower construction interest.

With respect to tax abatements, local jurisdictions in some states in the USA have the authority to reduce, rebate, or eliminate ad valorem property taxes. Although the positive impact on the redevelopment project can be significant, the reduction of tax revenues and the potential increase in public borrowing have opportunity costs that should be explicitly addressed. For many projects, tax abatements during the early years of operations are gradually eliminated over time as NOI increases.

Finally, many public entities participate in redevelopment through concessionary financing often provided by Community Development Financial Institutions (CDFIs). In comparison to private financing available in the market, the CDFI's financing is offered at lower cost (interest rate), longer term, longer amortization period (or interest only), and in some instances with flexible repayment schedules. CDFI financing is subordinated to private debt. With flexible debt financing, most public objectives can be achieved without having local government assume the complications and risks associated with ownership. In addition to lowering project cost, concessionary financing lowers the permanent lender's financial risk because a subordinated second mortgage loan is treated as near-equity, which reduces leverage. The private lender then may offer better permanent loan terms – lower interest, longer term, and/or longer amortization period. The amount of loan may also be increased modestly thereby replacing more expensive equity with less expensive debt. This outcome reflects the fact that debt is cheaper to secure than equity because creditors assume less risk and earn less than equity investors (Gordon 1997; Meeker 1996; Malizia 1997).

Tax increment financing (TIF) is a popular way to add public subsidies to urban redevelopment projects. Formal TIFs often float debt to finance infrastructure and then use the tax revenue stream to repay the debt. Informal TIFs (synthetic TIFs) can be established by mutual agreement as part of a PPP and used as a source of public subsidy. These options vary by state, only being available for publicly owned facilities in some places.

The importance of reducing project-level risk is not widely recognized among the public participants of PPPs. The basic point is that real estate developers, equity investors, and construction and permanent lenders all price their services in relation to risk; the higher the risk, the higher the expected return. Real estate developers and equity investors form expectations by comparing risk and return for different development opportunities. They often view urban redevelopment projects in less vibrant centers as carrying more risk than projects in locations where the market is more established and robust. Therefore, their return requirements are higher. Public entities can work with developers to help educate their equity sources about the many ways that project risk is expected to be lowered through the PPP. These include public investments that improve infrastructure and services, small-area plans that make area build-out more certain, commitments from anchor institutions, like corporate headquarters, major hospitals, or universities, to guarantee the rent payments of prospective tenants or to co-sign leases, and education of local investors about local redevelopment opportunities and the details of redevelopment ideas. The problem is that most US equity investors, like most developers, primarily have experience with new development in suburban locations which appear to be lower risk than urban redevelopment, especially in less vibrant downtowns. Since risk assessments are ultimately subjective, any reduction in risk that lowers equity return requirements would increase project value with no additional public subsidy of the project. Project value increases when terminal-year NOI is divided by a lower going-out cap rate that anticipates less risk (see Appendix).

When the redevelopment project finally achieves stabilized occupancy, it is an operating enterprise that should have positive cash flow with public subsidies included. Effective property management is essential to maintain and improve this enterprise and protect or increase value over time.

Strategic-level approaches

Strategies that promote urban vibrancy are designed to influence the expectations and behaviors of debt and equity capital (Malizia and Accordino 2001; Malizia 2003; Farris 2001; Leinberger 2007). In general, lenders and other sources of debt capital underwrite urban redevelopment projects in less vibrant centers more conservatively than more robust projects and commit less debt capital to them. In addition to lowering financial risk through more conservative underwriting, commercial lenders often try to lower their financial exposure by including their community development corporation (CDC) subsidiary since bank CDCs often have the contacts needed to bring multiple financing sources together for redevelopment projects. Unfortunately, commercial lenders do not have access to secondary markets when they lend to downtown redevelopment projects. Therefore, the capital devoted to these unique projects remains committed and is not replenished by another financial source (Gordon 1997).

Unlike investors that can seek higher returns by accepting greater risk, prudent lenders have to take a different approach to the deployment of debt capital. Lenders determine a comfortable, presumably modest range of risk in identifying "credit worthy" projects. They charge similar interest rates to these projects when they make loans. If the project exceeds its expected rate of return, the owners benefit, but lenders earn the same rate of interest. Therefore, lenders worry much more about downside risk than upside earnings potential. They ration credit to the projects they prefer. If lenders pursued projects with the highest expected returns and charged higher interest rates afforded by these high return expectations, they would end up holding a portfolio of loans that would be high risk. The conventional wisdom is that greater potential earnings from high-risk loans would not compensate for the additional risk assumed (Wood and Wood 1985).

With respect to equity capital, investors consider both amount and timing of equity infusions as well as holding period and exit strategy in determining return requirements. They should earn more than tax-credit investors who invest late in the development process, after construction and substantial leasing have been completed. When investors are asked to infuse equity during early stages of development or provide the balance sheet that is used to guarantee the construction loan, their equity return requirements naturally increase.

Given these expectations and behaviors, public sector entities involved in PPPs should formulate strategies to lower risk and uncertainty in order to promote downtown vibrancy, especially in centers with relatively low vibrancy. Four strategies are discussed in this section: develop and steadfastly implement small-area revitalization plans, concentrate infrastructure improvements, use triage when necessary, and promote competition. On the other hand, urban centers with relatively high vibrancy face very different challenges. Specific strategies for more vibrant downtowns are proposed later.

Small-area plans

As a risk reduction technique, small-area planning is critical because it significantly reduces the number of alternative futures for places within the downtown. Good graphics and three-dimensional representations of what downtown areas could become are important ingredients in changing attitudes about places. However, risks will not be decreased with visioning or general land-use proposals alone. The best plans identify which public improvements will be made, where they will be located, and when they will become operational. In other words, planning and zoning need to be linked to capital improvements programming to carry small-area plans forward. Good planning includes implementation with identified sources of financing.

Considerable political will is needed to implement such specific small-area plans, but jurisdictions that reduce risk and uncertainty in this way should be rewarded with increased private investment and more successful PPPs. At the same time, local governments should pursue strategic land banking to help provide future redevelopment sites since land values in locations earmarked for revitalization will probably increase.

Concentrated infrastructure investments

Local political representation by ward/specific district increases the likelihood of dispersed resource allocation. Widely spread public investments rarely change perceptions of any particular area enough to reduce risk there. It is much more effective but politically difficult to concentrate resources in strategic locations where investments can encourage larger projects that are more impactful. Projects located near strong institutional anchors – universities, medical centers, large government facilities, as well as sports facilities, civic centers, performing arts centers, museums, religious buildings, and public/civic open space – are likely to be more successful. Anchor institutions hire workers and subcontractors, purchase local products and services, demand good public services, provide for physical safety and security, etc., which strengthens the surrounding area. Providing infrastructure within public rights of way is the traditional way the public sector enables private development. Infrastructure spending becomes a subsidy if extended onto private property.

In less vibrant centers, urban redevelopment resembles an invasion that must first establish a beachhead in one location before it can spread. Without adequate infrastructure, project scale, and concentrated development, this beachhead cannot be firmly established. Success in one location increases the odds that redevelopment will spread to proximate sites. As projects accumulate, deals tend to become more self-sustaining and market information improves. As the result of sustained redevelopment, risk levels assessed for future projects go down. Over time, property values in targeted vibrant centers should increase.

Triage strategy

Certain areas may be too deteriorated to remediate given limited resources and the need to concentrate public investments spatially to increase critical mass. Private investors will come to understand that the public sector intends to ignore these areas for some period of time. As a result, property values may decline.

The triage strategy will be a hard pill for local politicians to swallow, but a benefit exists that may make this strategy more palatable. Deteriorating areas are quite attractive if they are physically safe because many individuals and businesses seek space in buildings charging modest rents. Budding entrepreneurs pursuing commercial or social ventures, small business owners, artists, and lifestyle businesses want cheap space. Areas not targeted for focused redevelopment can offer space for these prospective tenants/owners. Business incubators and business accelerators can also be located in these properties.

Competition

The public sector can promote competition among real estate developers and investors when vibrant centers, in fact, provide viable investment opportunities resulting from strategic public investments and subsequent revitalization. In less vibrant centers however, most redevelopment opportunities represent "buyers' markets" where private developers are scarce and interested ones are able to exact major concessions from the public in return for making investments. With increasing vibrancy, downtowns can gradually become sufficiently attractive to create "sellers' markets" where private developers are lining up to buy or option sites to participate in downtown revitalization. The competition among private developers should drive up asset values, make redevelopment projects in these locations more feasible, attract even more private investment, and generate more property tax revenues from appreciating property values.

Challenges and strategies for more vibrant centers

More vibrant downtowns face very different challenges. Unlike the general strategies appropriate for less vibrant centers, more vibrant ones evolve specific needs. Although every downtown is unique, six challenges appear to be most pressing. These challenges and possible responses are examined in this section.

Vibrancy can be enhanced by providing high-density, urban style, multifamily rental housing in various forms. The goal is to provide housing for all market segments wanting to live downtown. Market-rate housing will serve the upper-income segment, and subsidized housing will serve low-income residents. However, many middle-income households will find affordable housing increasingly scarce. One approach is to conserve C-grade multifamily housing near downtown that is especially valuable for larger households. Another is to construct mid-rise or high-rise workforce housing in downtown locations attractive to young singles and couples. Unit size is reduced to make rents affordable (micro units). Using modular construction, designing apartments that are easier and cheaper to build, and reducing regulatory burdens, especially parking requirements, are other ways to keep housing more widely accessible as land values appreciate.

Except in large, transit-rich metro areas, most cities in the USA rely on automobiles. Yet, vibrant downtowns are places where pedestrians, cyclists, and public transit should dominate. The key is to put the pedestrian first and promote ways to make walking useful, safe, comfortable, and interesting (Speck 2012). One way to discourage auto use is to increase the cost of driving downtown, through tolls and expensive parking, and to provide peripheral parking decks. Auto speeds can be reduced with tactics that include removing driving lanes, adding on-street parking, bike lanes, planters, etc., narrowing lanes, changing one-way to two-way streets, and providing dedicated lanes for transit (buses, taxis, trolleys).

Vibrancy is increased with more dense, compact development, but strong downtowns have unique places and historic structures that make the place attractive, memorable, and valued. Outstanding building and landscape design is needed to add new construction while preserving existing structures, unique places, and public open space. Well-designed, high-rise construction can be carefully located on appropriate infill sites. Low-rise buildings adjacent to sidewalks with continuous facades and ample windows at street level encourage pedestrian activity (Ewing and Clemente 2013). Density can be added to the interior of these sites. Large buildings with glass skin reduce awareness of their mass and can call attention to iconic or historic buildings they may be designed to reflect. Striking architecture can draw attention to the public realm and away from building bulk and size.

Vibrant downtowns are increasingly challenged to accommodate growing numbers of workers, residents, and visitors. Without careful planning, conflicts will arise. Although many tactics exist for reducing land use and activity conflicts, the simplest approach is to use districting. Many vibrant downtowns have dedicated areas for finance, entertainment, shopping, dining, and other uses. Quiet areas where downtown residents can sleep should be separated from loud areas where people congregate to play. Vibrant downtowns should serve different income levels, occupations, ethnicities, races, and sexual preferences, with minimal conflict. Artists, musicians, entertainers, and others that contribute to the local culture need performance venues and spaces for events (Markusen and Gawda 2010; Isserman and Markusen 2013).

Finally, land becomes increasingly scarce and expensive as vibrancy increases since downtowns have edges and boundaries that contain development. Economic activities that export services or goods outside the region are far more important than sectors that predominantly serve the local market. If these traded sectors are located downtown, it is critical to meet their expansion needs. Zoning and other land use regulations can be used to protect commercial land and space from conversion. Selective land acquisition can also become part of PPPs to assist traded-sector growth. Property taxation based on use-value is a proven tactic to deflect conversion pressure. Although primarily used to protect prime agricultural land on the urban fringe, use-based valuation can be applied to any class of properties located centrally.

In summary, we can enhance vibrancy in stronger downtowns by: providing different housing types to sustain diversity, especially workforce housing; supporting multiple modes of transportation to promote walking; adding density and compactness to increase critical mass without damaging uniqueness; preserving iconic buildings, public places and open spaces; resolving conflicts between different downtown activities with districting; and providing adequate land for traded-sector expansion. Taken together, addressing these six challenges successfully should lead to positive real estate outcomes and sustained economic development in vibrant centers.

Conclusions

Effective public interventions through PPPs at the project and strategic levels can promote vibrancy in weaker urban centers. At the project level, financial feasibility is bolstered by reducing project cost and mitigating risk. At the strategic level, small-area planning, public infrastructure investments and consistently applied priorities should make selected locations more attractive to private investors. In more vibrant centers, PPPs implementing specific redevelopment strategies and tactics can further strengthen vibrancy. As a result, density, diversity, connectivity, and walkability should be enhanced, further increasing agglomeration economies.

Greater vibrancy generates private and public benefits – higher real estate returns, positive economic development outcomes and positive fiscal impacts. Vibrant centers also afford greater social tolerance and smaller carbon footprints per capita than suburban development alternatives.

Urban employment centers are adding vibrancy by affording places to live and play. But their traditional role as workplaces must be emphasized because the economic base of metro areas in the USA increasingly relies on competitive firms providing export services, often located downtown, as well as innovative firms traditionally centered in denser core areas. Metro areas with more dynamic, competitive and innovative economies will foster vibrancy in their urban centers, and vibrant urban centers, in turn, will strengthen the regional economy.

Appendix

Discounted Cash Flow Analysis (DCF)

DCF is used to determine financial feasibility by comparing cost to value and can also be used to estimate the subsidy needed to make redevelopment projects feasible.

Basic assumptions and estimated project costs are in Table 26.1. The aim is to develop the 0.8 acre site by constructing a 115,000 square foot building with retail on the ground floor and office above. The land development costs, hard costs and soft costs are shown. Total development cost without financing is $19.6 million and $1 million more with financing costs included. The land development costs include $112,000 to demolish existing improvements on the site. The 75% loan-to-cost ratio provides a $15.5 million loan. Equity of $5.2 million is needed to cover the remaining 25% of development costs.

The DCF analysis in Table 26.2 adds assumptions needed to estimate cash flows: office and retail rent, vacancy allowance, operating expenses including real estate taxes, selling expense, and the going-out cap rate which is used to estimate the sales price of the project at the end of the holding period.

This simplified analysis assumes that equity is invested in year 0 and the project achieves stabilized occupancy in year 1. NOI for the 10-year holding period is driven by the rents, expenses, and growth rates. Debt service is found by amortizing the $15.5 million loan for 240 months at an annual interest rate of 6.5%. Debt service is almost $1.4 million annually. Subtracting debt service from NOI gives before-tax cash flows for the 10-year period. The project is expected to be sold at the end of year 10; estimated NOI in year 11 divided by the going-out cap rate is done to find sales value.

The bottom part of Table 26.2 – Investment Performance – is set up to find the present value of annual cash flows and the residual which is the net proceeds from the sale after covering selling expenses and paying the mortgage balance of $10.2 million. A present value analysis with 9.0% as the discount rate shows that the share of returns from cash flow over the holding period is less than the share from the residual. Almost 59% of total returns come from the residual.

The project performance in Table 26.2 indicates that return on cost is 7.7% and cash-on-cash return at stabilized occupancy (year 1) is 4.0%. Before-tax IRR turns out to be 9%. IRR is the rate that equates equity invested in year 0 to the present value of all subsequent cash flows. Therefore, net present value (NPV) equals zero when BTIRR is 9%.

The DCF can be repeated using the 12.0% BTIRR that equity investors expect as their hurdle rate (discount rate). At this hurdle rate, net present value (NPV) is in the red by $1.013 million. Thus, equity investors would not invest in this project because it fails to meet their return requirements. However, the project would become feasible if the public sector generated $1.013 million in subsidies. With these subsidies, investors would achieve the desired BTIRR of 12.0%.

Table 26.1 Capital budget

Assumptions

	Totals
Gross Buildable Area (SF)	115,000
Net Leasable Area (SF)	103,500
Project Contingency	10.0%
Operating Expenses (per SF)	$4.50	$517,500
Taxes (% of total development cost)	1.5%	$310,070
Cost of Equity (%)	10.0%
Cost of Debt (%)	6.5%
Loan Term (years)	10
Loan Amortization (years)	20
Loan to Cost (%)	75%	$15,503,478
Debt Service Coverage Ratio	1.15
Equity Contributed	$5,167,826

Land Development Costs	Square Feet	$ / SF	*Budget*
Acquisition of Site	35,000	$65.00	$2,275,000
Demolition	45,000	$2.50	$112,500
On-site Improvements	21,000	$15.00	$315,000
Total Land Development			$2,702,500

Hard Costs	Square Feet	$ / SF	*Budget*
Shell & Interior Construction			
Office	90,000	$120.00	$10,800,000
Retail	25,000	$155.00	$3,875,000
Total Hard Costs			$14,675,000
Total Acquisition & Hard Costs (before financing)			*$17,377,500*

Financing Costs		
Construction Loan		
Origination/Commitment Fee (1%)	$137,510	
Interest	$446,909	
Permanent Loan		
Origination/Commitment Fee (1%)	$155,035	
Tenant Improvements and Maintenance Reserve	$287,500	
Total Financing Costs	$1,026,954	
Total Development Costs (after financing)	*$20,671,304*	

Soft Costs	
Architecture & Engineering	
Survey	$12,000
Phase 1 Environmental Report	$5,000
Building Design	$115,000
Geo-technical Engineering	$14,000
Total Architecture & Engineering	$146,000
Legal	
Partnership Organization	$15,000
Loan Documents	$45,000
Entitlement	$55,000
Total Legal	$115,000
Marketing	
Marketing Studies	$14,000
Advertising and Public Relations	$7,500
Commissions	$69,000
Total Marketing	$90,500
Taxes, Title, Appraisal	
Construction Period Taxes	$48,000
Title/Closing/Escrow	$32,000
Property Appraisal	$12,000
Total Financial	$92,000
Government Relations	
Application Fees	$25,000
Total Government Relations	$25,000
General & Administration	
Insurance	$12,500
Project Contingency	$1,785,850
Total General & Administration	$1,798,350
Total Soft Costs	$2,266,850
Total Development Costs (before financing)	*$19,644,350*

Table 26.2 Discounted cash flow

Assumptions

Office rent	$26/sf
Retail rent	$22/sf
Weighted rent	$25.13/sf
Vacancy	7.0%
Rent Growth	2.0%
Expense Growth	2.0%
Operating Expenses per SF	$4.50
Going-Out Cap Rate	10.5%
Selling Expense	6.0%

Year	1	2	3	4	5	6	7	8	9	10	11
Gross Rent Receipts (GRR)	$2,600,955	$2,652,974	$2,706,034	$2,760,154	$2,815,357	$2,871,664	$2,929,098	$2,987,680	$3,047,433	$3,108,382	$3,170,550
– Vacancies (VAC)	$182,067	$185,708	$189,422	$193,211	$197,075	$201,017	$205,037	$209,138	$213,320	$217,587	$221,938
Total Income	$2,418,888	$2,467,266	$2,516,611	$2,566,943	$2,618,282	$2,670,648	$2,724,061	$2,778,542	$2,834,113	$2,890,795	$2,948,611
– Operating Expenses (OE)	$517,500	$527,850	$538,407	$549,175	$560,159	$571,362	$582,789	$594,445	$606,334	$618,460	$630,830
– Real Estate Taxes	$310,070	$316,271	$322,596	$329,048	$335,629	$342,342	$349,189	$356,172	$363,296	$370,562	$377,973
Net Operating Income (NOI)	$1,591,319	$1,623,145	$1,655,608	$1,688,720	$1,722,494	$1,756,944	$1,792,083	$1,827,925	$1,864,483	$1,901,773	$1,939,808
– Debt Service (DS)	$1,387,077	$1,387,077	$1,387,077	$1,387,077	$1,387,077	$1,387,077	$1,387,077	$1,387,077	$1,387,077	$1,387,077	$1,387,077
Before Tax Cash Flow (BTCF)	$204,241	$236,068	$268,531	$301,643	$335,417	$369,867	$405,006	$440,848	$477,406	$514,696	$552,731
Debt Service Coverage Ratio	1.15	1.17	1.19	1.22	1.24	1.27	1.29	1.32	1.34	1.37	1.40

Total Development Cost (TDC)	$20,671,304
Return on Cost (NOI/TDC)	7.70%
Cash on Cash Return	4.0%
Before Tax IRR	9.0%

Cash Flows from Disposition
Property Cash Flows

Gross Selling Proceeds	$18,474,366
Selling Expenses	$1,108,462
Net Selling Proceeds	$17,365,904

Table 26.2 continued

Financing Cash Flows

Net Selling Proceeds	$17,365,904
Unpaid Mortgage	$10,179,817
Before Tax Equity Reversion (BTER)	$7,186,087

Investment Performance

Year	0	1	2	3	4	5	6	7	8	9	10
Equity Contributed	$(5,167,826)										
BTCF		$204,241	$236,068	$268,531	$301,643	$335,417	$369,867	$405,006	$440,848	$477,406	$514,696
BTER											$7,186,087
Total Cash Flows	$(5,167,826)	$204,241	$236,068	$268,531	$301,643	$335,417	$369,867	$405,006	$440,848	$477,406	$7,700,783

Required BTIRR	12.0%
Before Tax IRR	9.0%
NPV (discounted at required BTIRR)	$(1,023,872)

Partitioning NPV at BTIRR

Year	Cash Flow	PV CF	PV Disposition
0	$(5,167,826)		
1	$204,241	$187,407	
2	$236,068	$198,756	
3	$268,531	$207,453	
4	$301,643	$213,826	
5	$335,417	$218,170	
6	$369,867	$220,748	
7	$405,006	$221,797	
8	$440,848	$221,526	
9	$477,406	$220,123	
10	$7,700,783	$217,756	$3,040,265
Total		$2,127,561	$3,040,265	$5,167,826
% of Total NPV inflows		41.2%	58.8%

Outflow (at BTIRR)	$(5,167,826)
NPV (at BTIRR)	$–

Table 26.3 Summary of public options to improve feasibility

	Return on Cost	Cash on Cash	Before Tax IRR	Net Present Value
Baseline Case: No Subsidies	7.7%	4.0%	9.0%	$(1,023,872)
Option	**Return on Cost**	**Cash on Cash**	**Before Tax IRR**	**Net Present Value**
Donate Site Land = $0.00/SF	9.0%	9.2%	16.2%	$1,394,775
Subsidize Land Acquisition Costs Land Cost = $37/SF Write down = $28/SF	8.2%	6.1%	12.1%	$18,007
Use Abatements to Subsidize Real Estate Taxes Real Estate Taxes = 0.98% of Total Project Costs	8.2%	6.0%	12.0%	$6,838
Use Concessionary Financing/Guarantee Loan to Cost = 75%, Interest Rate = 5.05%	7.7%	6.9%	12.0%	$12,926
Mitigate Risk with Public Strategies Going-Out Cap Rate lowered by 160 bps = 8.9%	7.7%	4.0%	12.0%	$(18,685)

Table 26.3 shows the base case in the first row: return on cost of 7.7%, cash on cash return of 4.0%, BTIRR of 9.0% and negative net present values calculated using the equity return requirement of 12.0%. Three typical ways to lower project costs are presented. If the public sector donated the redevelopment site, it would generate $1.4 million more subsidy than needed and BTIRR of 16.3%, well above the required 12.0%. Reducing site cost to $37 per square foot should enable the developer to do the deal. If the public sector abated taxes from 1.5% to 0.098% of total development cost, which is treated as assessed value, the project becomes feasible. Similarly, concessionary financing at 5.0% interest provides slightly more subsidy than is needed.

In addition to public subsidies that reduced project costs, Table 26.3 also shows how changes in risk perceptions can significantly impact returns. Since risk perceptions can change without incurring direct subsidies, public participants should try to influence private perceptions in positive ways. If the going-out cap rate were reduced by 160 basis points to 8.9% to reflect growing investor optimism about the redevelopment area, the project becomes feasible without direct subsidies that lower costs. At this cap rate, the residual value (sales price) of the project increases significantly from $7.7 million to $10.3 million. Improvements in the market provide the needed returns.

Although the numbers in these spreadsheets are all based on assumptions and represent forecasts, DCF is a powerful tool to use in negotiating PPPs. First, the public entity can calculate the cost–value relationship without subsidies. Then, the impact of various tactics for lowering project cost or project risk can be assessed. This analysis can be used to estimate an appropriate subsidy needed to make the project financially feasible when negotiating the final development agreement.

References

Adair, A., McGreal, S., Deddis, B., and Hirst, S. 1999 "Evaluation of investor behaviour in urban regeneration," *Urban Studies*, vol. 36, no. 12, 2031–2045.

Appraisal Institute 2013 *The appraisal of real estate*, 14th edn, Appraisal Institute, Chicago.

Campoli, J. 2012 *Made for walking*, Lincoln Institute of Land Policy, Cambridge, MA.

Ciochetti, B. A. and Malizia, E. 2000 "The application of financial analysis and market research to the real estate development process," in DeLisle, J. and Worzala, E. (eds) *Essays in honor of James A. Graaskamp*, Kluwer Academic Publishers, Boston, pp. 135–165.

Crankshaw, N. 2009 *Creating vibrant public spaces*, Island Press, Washington, DC.

Dreier, P., Mollenkopf, J., and Swanstrom, T. 2001 *Place matters: Metropolitics for the twenty-first century*, University of Kansas Press, Lawrence.

Ewing, R. and Clemente, O. 2013 *Measuring urban design: Metrics for livable places*, Island Press, Washington, DC.

Farris, J. T. 2001 "The barriers to using infill development in achieving smart growth," *Housing Policy Debate*, vol. 12, no. 1, 1–30.

Florida, R. 2008 *Who's your city: How the creative economy is making where to live the most important decision of your life*, Basic Books, New York.

Frieden, B. J. and Sagalyn, L. B. 1989 *Downtown Inc: How america rebuilds cities*, MIT Press, Cambridge, MA.

Glaeser, E. 2011 *The triumph of the city*, The Penguin Press, New York.

Godschalk, D. and Malizia, E. 2013 *Sustainable development projects*, APA Press, Chicago.

Gordon, D. L. 1997 "Financing urban waterfront redevelopment," *Journal of the American Planning Association*, vol. 63, no. 2, 244–265.

Graaskamp, J. A. 1981 *Fundamentals of real estate development*, Development Component Series, ULI, Washington, DC.

Gyourko, J. and Rybczynski, W. 2000 "Financing new urbanism projects: Obstacles and solutions," *Housing Policy Debate*, vol. 11, no. 3, 733–750.

Haughey, R. 2008 *Getting density right: Tools for creating vibrant compact development*, The Urban Land Institute and National Multi Housing Council, Washington, DC.

Hodge, G. 2004 "The risky business of public–private partnerships," *Australian Journal of Public Administration*, vol. 63, no. 4, 37–49.

Isserman, N. and Markusen, A. 2013 "Shaping the future through narrative: The third sector, arts and culture," *International Regional Science Review*, vol. 36, no. 1, 115–136.

Jackson, K. 1985 *Crabgrass frontier: The suburbanization of the United States*, Oxford University Press, Oxford.

Jacobs, J. 1985 *Cities and the wealth of nations*, Vintage Books, New York.

Kotin, A. and Peiser, R. 1997 "Public–private joint ventures for high volume retailers: Who benefits?" *Urban Studies*, vol. 34, no. 12, 1871–1997.

Kelly, H. 2016 *24 hour cities*, Routledge, London.

Kelly, H. and Malizia, E. 2016 "The influence of 24-hour cities and vibrant centers on the value of office properties and apartments in large U.S. markets," *Real Estate Finance*, vol. 32, 129–139.

Kelly, H., Adair, A., McGreal, S., and Roulac, S. 2013 "Twenty-four hour cities and commercial office building performance," *Journal of Real Estate Portfolio Management*, vol. 19, no. 2, 103–120.

Leinberger, C. 2007 *Footloose and fancy free: a field survey of walkable urban places in the top 30 US metropolitan areas*. Brookings Institution, Washington, DC.

Levy, P. and Gilchrist, L. 2013 *Downtown rebirth: Documenting the live–work dynamic in the 21st century US cities*, Center City District Report, Philadelphia.

Luscht, K. M. 1997 *Real estate valuation*, Richard D. Irwin, Chicago.

Malizia, E. 1997 *Economic development finance*, 2nd edn, American Economic Development Council, Chicago.

Malizia, E. 1999 "The Garvey retail center case: Redeveloping an inner-city site," *Journal of Real Estate Practice and Education*, vol. 2, no. 1, 63–120.

Malizia, E. 2003 "Structuring urban redevelopment projects: Moving participants up the learning curve," *Journal of Real Estate Research*, vol. 25, no. 1, 463–478.

Malizia, E. 2009 "Site use in a redeveloping area," *Journal of Real Estate Practice and Education*, vol. 12, no. 1, 81–104.

Malizia, E. 2014 "Office property performance in live–work–play places," *Journal of Real Estate Portfolio Management*, vol. 20, no. 1, 79–84.

Malizia, E. and Accordino, J. 2001 *Financing urban redevelopment projects*, Community Affairs Office, Federal Reserve Bank of Richmond Research Report, Richmond, VA.

Malizia, E. and Motoyama, Y. 2016 "The economic development–vibrant center connection: Tracking high-growth firms in the DC region," *The Professional Geographer*, vol. 68, no. 3, 349–355.

Malizia, E. and Song, Y. 2015 "Does downtown office property perform better in live–work–play places?" *Journal of Urbanism*, vol. 9, no. 4, 372–387.

Markusen, A. and Gawda, A. 2010 *Creative placemaking*, White paper for the Mayors' Institute on City Design, Washington, DC.

Meeker, L. 1996 *Doing the undoable deal*, Federal Reserve Bank of Kansas City Monograph, Kansas City, MO.

Meyer, P. and Lyons, T. 2000 "Lessons from private sector brownfield redevelopers: Planning public support for urban regeneration," *Journal of the American Planning Association*, vol. 66, no. 1, 46–57.

Moretti, E. 2012 *The new geography of jobs*, Mariner Books, Boston.

NAIOP 2014 *Preferred office locations: Comparing location preferences and performance of office space in CBDs, suburban vibrant centers and suburban areas*, NAIOP Research Foundation, Washington, DC.

Paumier, C. 2004 *Creating a vibrant city center: Urban design and regeneration principles*, Urban Land Institute, Washington, DC.

Sagalyn, L. B. 1997 "Negotiating for public benefits: The bargaining calculus of public–private development," *Urban Studies*, vol. 34, no. 12, 1955–1970.

Sagalyn, L. B. 2007 "Public/private development: Lessons from history, research, and practice," *Journal of the American Planning Association*, vol. 73, no. 1, 7–22.

Sawicki, D. S. 1989 "The festival marketplace as public policy," *Journal of the American Planning Association*, vol. 55, no. 3, 347–361.

Speck, J. 2012 *Walkable city*, North Point Press, New York.

Storper, M. 2013 *Keys to the city*, University Press, Princeton, NJ.

Talen, E. 2012 *City rules: How rules affect urban form*, Island Press, Washington, DC.

Vandell, K. D. and Carter, C. C. 2000 "Graaskamp's concept of highest and best use" in DeLisle, J. and Worzala, E. (eds) *Essays in honor of James A. Graaskamp*, Kluwer Academic Publishers, Boston, pp. 307–319.

Wood, J. and Wood, N. 1985, *Financial markets*, Harcourt Brace Jovanovich, New York.

27

The challenges for effective governance of real estate development in a future urban world

Kathy Pain

Abstract

The twenty-first century is a time of dynamic global change and cities are the places where this change is most fundamental. Cities are increasingly interconnected in a networked world in which they are now the major locations for international commerce and economic production. Understanding the part that real estate development plays in supporting their functioning is vitally important. Real estate development has an increasingly powerful role in facilitating the circulation and accumulation of financial capital in a global economy that is transforming structurally with most of the growth coming from the service sector. In a world of volatile international capital flows, real estate-led development of commercial offices occupied by multinational firms in the business districts of major cities has come to be regarded by entrepreneurial governments worldwide as critical in creating both the physical infrastructure and the "global" atmosphere essential for inward investment, employment and economic growth. This century is the first in which humanity is becoming a predominantly urban species and, increasingly, it is cities that promise opportunities for work, access to resources and services, and for prosperity and well-being; in turn, urban work increases global productivity and innovation, and boosts GDP. As the world continues to urbanize, real estate development challenges will be increasingly concentrated in and around large business cities. This chapter focuses on the major global and local drivers that real estate and cities are caught up in, in a changing world, and the consequent pressing challenges for effective real estate development governance.

Introduction

Real estate development has an increasingly powerful role in facilitating the circulation and accumulation of financial capital in a globalizing economy, especially in an economy that is transforming structurally with most of the growth coming from the service sector. In a world of increasingly mobile international capital flows, real estate-led development of glittering commercial office towers occupied by multinational firms in the financial districts of major business cities has come to be regarded by entrepreneurial governments worldwide as critical in creating the physical infrastructure and the "global" atmosphere essential for inward investment. Indeed, in a neo-liberalizing political-economy, governments in fierce competition for foreign direct investment (FDI) have become active agents in the co-production of iconic "starchitect"-designed financial landscapes and the commodification of urban space (Knox and Pain, 2010). Deregulation, tax breaks, and land use zoning are often employed to encourage FDI through real estate development. Meanwhile, oversight of how fiscal, taxation, and monetary policies, and spatial and economic planning together influence the development process in a macroeconomic context, is generally absent.

The twenty-first century global space of cross-border real estate activity contrasts with the territorially bounded space of nation-state government institutional structures. Furthermore, the challenge for effective processes of real estate development "governance," as distinct from those formal structures (Heywood, 2000, p. 19), is compounded by a lack of coordination of institutional interventions.

A major challenge for sustainable development is an appreciation of the role of real estate as a conduit for flows of international finance. First, as a physical entity, real estate provides the city-based offices where business actors as international financial and linked business actors generate major capital flows in the global economy as property occupiers. Second, as a service provider, real estate brings together the economic agents that globalize urban space; it facilitates the international flows of finance capital required to create that space and to trade within that space. Third, as an asset class, real estate has in itself become a financial product that generates growing international flows of capital through increasingly fragmented property investment and ownership (Lizieri, 2009).

By considering the interrelated roles of real estate in the case of commercial office property, this chapter will illustrate the key challenges for the governance of development in a globalizing and urbanizing world and why these matter. As argued by Ball (1998), in commercial property research institutional analysis, the interactions of all of the social agents involved in the real estate development process are relevant and must be taken into account. First, the chapter focuses on global macroeconomic changes that are leading to the financialization of real estate and its integration in international financial services. Second, the role and concentration of commercial office real estate development in global cities is considered as an outcome of social practices that need to be taken into account by real estate research. Finally, new spaces for the governance of real estate development are examined against the backdrop of an unfolding new geometry of the global economy in an era of unprecedented world urbanization.

The globalization of real estate investment

Consideration of structural macroeconomic changes is necessary to understand the drivers of contemporary real estate development and the challenges for its governance. However, up until now, the interrelationship between economic "globalization" and real estate has mainly been of interest to real estate and management consultancies as opposed to academics in the real estate discipline.

A globalizing economy

In the social sciences beyond real estate however, the concept of globalization has been a subject of much discussion for more than half a century (Murray, 2006). The meaning, existence, and implications of globalization have been keenly debated in the international literature by academics with diverse "globalist," "traditionalist," and "transformationalist" perspectives (Cochrane and Pain, 2000, p. 23), reflecting "continuity" and "disjunction" positions on global change (Pain and Van Hamme, 2014, p. 4). Of particular relevance for real estate development, the consequences of globalization for systems of economic governance remain contested. A "strong globalist" perspective points to a paradigm change in local–global economic relations, arguing that due to late twentieth-century intensification of international trade and investment, "distinct national economies have dissolved into a global economy determined by world market forces" (Thompson, 2000, p. 89). On the contrary, a more skeptical "traditionalist" perspective argues that while the development of multinational enterprises (MNEs) and increased international trade have led to growing global economic integration, this has not undermined the power of states to manage national economies (Thompson, 2000, p. 89). A third, transformationalist, perspective of global change, on the one hand, acknowledges the interdependency of capital flows and global economic integration promoted by increasing openness to international trade and FDI but, on the other hand, suggests that this does not necessarily mean the loss of all state control of economic management within countries (Thompson, 2000, pp. 85–126). It is this last view of global economic change that proves especially pertinent when considering real estate development governance. Ultimately, this chapter will make the case that local real estate markets and economies are increasingly intertwined globally, yet the local sphere of real estate development and its governance remains critically important. In consequence, the words of Barkham are salient:

> real estate research is only likely to produce accurate forecasts when it is fully cognisant of the influence of geopolitics on asset market performance. As globalization proceeds, real estate outcomes at the city or even neighbourhood level are ever more influenced by politics and economics on the other side of the world … Real estate research that does not imaginatively and creatively deal with these themes runs the risk of being irrelevant.
>
> *(Barkham, 2012, p. 17)*

Global economic change and real estate financialization

Economic globalization has been accompanied by the internationalization and financialization of real estate development, especially in the commercial office sector that is concentrated in major business cities. Attention to macroeconomic market conditions and development drivers is therefore essential when seeking to understand local real estate markets and investment behaviors. Geographically fixed office towers have become progressively more liquid and tradable internationally as innovative financial vehicles such as real estate investment trusts (REITs) have become popular in recent decades, posing the analytic a priori question whether listed real estate equities are actually physical real estate or intangible stocks (Ling et al., 2015). Examining the impact of globalization on the real estate industry in the USA, Bardhan and Kroll observe that real estate,

> is among the least "tradable" of products, in the sense of being physically unmovable, even though it can be bought and sold both domestically and internationally. This combination of local knowledge and predominantly local tradability was the primary reason why discussions of globalization in the 1990s and earlier, overlooked the real estate industry as a possible participant in the ongoing phenomenon of increasing global economic integration.
>
> *(Bardhan and Kroll, 2007, p. 1)*

The creation of specialist financial services, increased data availability, the digitization of capital transfers and the securitization of real estate through REITs, have together opened up access to property as an asset to investors worldwide. These developments have swept up real estate in new types of cross-sector and cross-border interdependencies in a more integrated global financial system, especially in the commercial office sector. Important in comprehending their significance for real estate governance are the changes facilitated by advances in information and communications technology (ICT) that have been in progress in the organization and operations of city-based international financial and business services, including real estate services, since the latter decades of the twentieth century. Real estate development makes possible investments in the office space that advanced producer services, including real estate services, occupy and own, directly and indirectly (Lizieri, 2009), and this space constitutes the infrastructure of the networked world economy in which cities are the nodes for capital flows.

The virtualization of trade in these "advanced producer services" that include wholesale finance, insurance, accountancy, management consultancy, and advertising, has been theorized as generating a global inter-city "space of flows" (Castells, 1996) in which real estate has become increasingly integrated. When commercial offices are transformed into tradeable securities and futures, cross-border, liquidity, and fungibility are lent to the urban built environment. However, the financialization of real estate thereby presents a conundrum for the governance of development by nation states whose powers are limited by their boundaries within a geo-political "space of places" (Castells, 1996). In his famous turn of millennium Reith lecture, sociologist Giddens described the loss of national sovereignty associated with globalization as heralding an out of control, "runaway world" (Giddens, 1999). The urban organization of the world financial system in economic globalization, calls into question the authority and power of states to manage and control development that leads to cross-border capital flows facilitated by real estate's three interrelated conduit roles.

Financial services and real estate development integration

While the financialization of real estate development would seem to present transformative challenges for governance, it can be argued that this process has in reality been in emergence for at least half a century. Radical 1960s and 1970s UK publications documented the growth of insurance companies and pension funds as "powerful groupings of financial interests" in property development (CIS, 1973, p. 1). Marriott's 1967 *The Property Boom*, described the formation of a new UK property development "system" as insurance companies and development companies between 1945 and 1967 became "interlocked":

> With the end of the credit squeeze straightforward borrowing became easier again and the insurance companies decided that they would climb on to the developers' bandwagon and, in return for lending the vital finance, share in the profits ... From 1959 onwards they made more and more equity links with the developers in return for a commitment to provide finance.
>
> *(Marriott, 1989, p. 40)*

A decade later, in their 1975 book *The Property Machine*, Ambrose and Colenutt drew attention to the increasing interdependence between property development and finance capital in the UK. The authors traced the growth of large property companies measured by their share valuation on the stock exchange in the changing development industry post-World War II, and new associations between financial institutions, property companies, and property investment companies: "it is almost impossible to draw the line between property companies, construction companies, banks, insurance companies and investment trusts" (Ambrose and Colenutt, 1975, p. 38). The deepening interrelationship between finance capital and property development described reflected the growth of complex networks of economic actors in the development process: "construction firms, land owners and financial institutions trading in stocks and shares, government bonds, and currencies, and … providing services such as insurance, pensions, hire purchase and mortgages" (Ambrose and Colenutt, 1975, p. 41). Ambrose and Colenutt also pointed to ties between property development and finance capital that go beyond the completion of a development project, challenging government intervention in the development process: "Commercial buildings are extremely valuable financial assets and like any others are traded for investment and profit. Such is the value of buildings that they are sometimes sold in pieces to different investors" (Ambrose and Colenutt, 1975, p. 41). They argued that, "Any intervention by the government in the property market is closely followed by the financial institutions and any move which threatens their interests is quickly and vigorously blocked" (Ambrose and Colenutt, 1975, p. 59).

These accounts of early financial services and real estate development integration suggest that in advance of later ICT developments, the UK "opening up" to foreign competition in the 1986 "Big Bang," and increased take up of innovative real estate financial products, finance–real estate independencies were already eroding the power of the state to manage the development system within its national borders. Post twentieth century, real estate development is more interlinked than ever before in global economic changes (growth and recession, and interlinked financial boom and crisis) and macroeconomic–real estate interlinkages are most evident in cities. Although inter-city flows of finance are challenging to measure and map (Pain and Van Hamme, 2014) and the real growth of FDI since 1980 has been debated (Thompson, 2000, pp. 107–109), large business cities are the locations where openness to MNE and transnational corporation (TNC) capital flows and FDI are most notable (Pain and Van Hamme, 2014). The urban origins and destinations of major international financial flows in the networked world economy coincide with the geographic locations of international commercial office real estate concentration. Hence, an appreciation of the global drivers that increasingly shape, and are shaped by, real estate development in cities, is critically important in informing the governance of that development and of the capital flows that it facilitates.

Real estate development as social practice

Viewing real estate as having interrelated conduit roles within a world network of financial cities, offers a framework for understanding how global capital flows and the local development of urban space intersect through the working practices and interactions of economic agents – financial and real estate actors. Property consultancies and banking, brokerage, wealth management, insurance, and construction companies are increasingly globally networked yet, together with other advanced producer services, their key actors and functions are clustered in the financial districts of major cities; furthermore they are concentrated globally in a select group of those cities with deep social networks.

Real estate development and global cities

In 1991, Sassen labelled such cities, "global cities," noting for the first time the strategic role of certain cities in the world in dual geographic global dispersion and concentration of advanced producer services. Sassen thereby contributed to the understanding of key structures of the global economy beyond the earlier theorization of "world cities" by Hall (1966) and Friedmann (1986) simply as the world locations for important economic, financial, business, and government functions. In the 1990s, Taylor and co-researchers further advanced understanding of these key structures by developing a methodology to measure quantitatively the interlinkages between cities that are generated by advanced producer services in a "world city network" (Beaverstock et al., 2000; Taylor, 2001; Taylor, 2004). The methodology involves computing data on the importance of functions undertaken in each of the city offices of global firms in each of the advanced producer services sectors, to calculate the connectivity and position of cities in the world network associated with the organizational and functional activities of these economic agents. Real estate is not only the physical fabric of this network of city-based commercial offices but, on account of its interrelated conduit roles, real estate is also a powerful facilitator of economic globalization within that network. Real estate allows that physical fabric to be traded globally as an asset through pension funds, private equity, and investment trusts, and it facilitates interactions between the economic agents in the network who generate financial flows beyond property investment.

Theorization of the economic role of global cities by Sassen and of the global network structure of those cities by Taylor and colleagues, allows the powerful roles of real estate in the development of the world city network in a macroeconomic context, to be recognized. Importantly, this world city network is shaped not only by tangible fabric and intangible flows in commercial offices but by the presence and social practices of real estate actors in the network. Urban real estate development has a powerful influence on the functioning and vitality of the world city network and the world service economy. Towering high-spec "Class A" office blocks are cultural symbols of that capitalist economy and they are the places where advanced producer services command and control functions in global production networks and supply and value chains are located. As a geographically situated infrastructure and asset, real estate development is at once a major component of Castells' space of places and of the space of capital flows (Castells, 1996). It opens countries up to FDI and it shapes the contours of the networked economy that is located in cities. Densely clustered in global cities, real estate development and practices make possible interactions between multiple actors in financial and linked advanced producer services within deep social networks.

The co-location in cities of advanced producer services, including real estate services, that are traded globally, occurs because their business operations are critically interlinked; they not only provide services to other businesses and governments, but they are dependent on the services of each other and thus on close proximity (Pain, 2007). In contrast to firms in manufacturing, their capital is specialized knowledge that is vested in highly skilled people, who are also present and concentrated in global cities. The urban and sociology literature of the 1980s coined the term "FIRE" to describe the interdependency between "finance, insurance, real estate" in post-industrial, global cities of London and New York (Tabb, 1982; Fainstein, 1994). It is through the interlinking of real estate and finance in the advanced service economy, which is especially concentrated in global cities, that commercial office development has been turned into a divisible asset, both as equity and debt, that can be purchased indirectly via REITs (Lizieri, 2009). REITs provide the funding for and spread the risk for large-scale global city developments. Furthermore, REITs are "heavily involved in many activities related to the physical asset, from land assembly to design and construction to building management" (Bardhan and Kroll, 2007, p. 5). The conduit roles of real estate activity in the development process are enabled by the immersion of its key actors in those cities in the world city network where concentration provides access to multi-actor economic and social networks. The global organization of these locally situated networks diffuses information, tacit knowledge, and innovations in the public and private real estate–financial services supply chain. Recent real estate analysis of REITs and private direct investments performance in the USA by Ling et al. (2015) appositely focuses on illustrating how the performance of fund managers shapes investment flows and fixes capital in economically successful cities. Along with other actors necessary to this process of flow-shaping and territory-fixing, fund managers are shown by Lizieri and Pain (2015) to be concentrated in "elite" global cities.

Global cities real estate concentration

Lizieri and Pain's (2015) analysis of data on the top 1000 commercial office deals in the world recorded by Real Capital Analytics (RCA) for each year from 2007–2014 (in all, 8,000 deals with a capital value of US$1.25 trillion and an average value of $158 million per building) shows investment to be dramatically concentrated in just a few cities worldwide. Even though, as a consequence of the global financial crisis, there was a fall in sales values by 65% from 2007 to 2009, investment that did take place concentrated on global cities strongly connected to capital markets in the world city network (Lizieri, 2009; Lizieri and Pain, 2015). Notably, inward investment flows converged on London in the years 2007–2010 (Pain et al., 2012) suggesting a "flight to safety" that reflects investor confidence in ongoing London international business concentration and liquidity in global capital markets (Lizieri and Pain, 2014).

The dominance of a small number of brokerage firms worldwide is reflected in a significant geographical concentration of the headquarters (HQs) of firms involved in 53% of sales activity and 96% of (known) sales brokerage in just ten cities. Importantly, these HQs hold the command and control functions in firms' worldwide networks and beneficial ownership of profit streams revert to them (Lizieri and Pain, 2015). Nearly 30% of all deals occurred in London, New York, and Tokyo, and 11.5% of transactions, worth $107 billion, occurred in London alone. Fifty percent of transactions by value took place in just nine global cities; some $437 billion of investment and 76% of the total value of sales ($648 billion) took place in 30 cities that represented just 11% of all cities included in the database (Lizieri and Pain, 2015). The data provide evidence of a global city-centric concentration of real estate financial intermediaries. In the years 2007–2014, 276 brokers were acting for sellers but, by value, activity was dominated by a small group of global firms. The top ten brokers by value of activity were global chains of real estate agents and surveyors; they were involved in 45% of all sales and 79% of sales where a seller broker was recorded (Lizieri and Pain, 2015). Just two firms acted for nearly a quarter of all transactions by value and for 43% of the sales where a seller broker was recorded. Furthermore, the analysis may under-record subsequent concentration levels as a number of independent national-focused agents have joined global chains in consolidation since the global financial crisis. Only the US broker HFF had a predominantly domestic focus indicating that Ling et al.'s observation of a fixation of US fund managers on investments in US "gateway cities" does not reflect the wider global geometry of real estate investment in the commercial office sector (Ling et al., 2015). Investments in local office markets take place within a global capital markets system that is shaped by the concentration of global occupiers of urban space within that system.

Ling et al. have investigated which US Metropolitan Statistical Areas (MSAs) are the focus for real estate investment portfolios, National Council of Real Estate Investment Fiduciaries (NCREIF), and REITs, by property type. Data on quarterly means of NCREIF NPI returns, and the difference in means, for gateway and non-gateway cities were analyzed for the periods 1996–2001, 2002–2007, 2008–2013, and 1996–2013 for four property types that included commercial offices. Over the 1996–2013 period, gateway markets were found to outperform non-gateway markets by 26 (106) basis points quarterly (annually) for all property types. However, the most significant difference in performance between gateway and non-gateway markets was in the commercial office sector, which outperformed non-gateway offices by 44 (177) basis points quarterly (annually). During the period 2002–2007 when there was rapid expansion in commercial real estate markets, the return difference was significantly larger, with gateway office markets outperforming non-gateway markets by 96 (387) basis points quarterly (annually) (Ling et al., 2015, pp. 14–15). It can be speculated that this finding reflects the globalization of real estate commercial office capital markets, their managers and their occupiers.

Despite increasingly fragmented investment, the contemporary commercial office real estate development process is made possible by proximities of economic actors who are co-located in global city clusters and are embedded in social networks in those clusters. However, it is important to be mindful that those local social networks are transnationally constituted since real estate finance and investment actors are economic agents within international firms with global office networks. An inter-city, macroeconomic perspective is therefore essential in understanding and comparing local office markets.

Real estate economics and urban sociology: Mind the research gap

As already discussed, up until now, research into the interrelationships between globalization and real estate has generally not been prominent in the academic real estate discipline. As Lizieri (2009) complains, real estate economics literature has come to focus on increasingly elaborate quantitative micro-economic and econometric analysis and modelling; these do not take wider globalization processes and global cities research into consideration. The actual reasoning behind decisions of different actors that are determinants of market change, are therefore necessarily speculative in much real estate research. There is a failure to take into account and explain, holistically, the global drivers and influences that shape the behaviors of real estate finance actors and their responses to the practices of other global economic agents. Commercial office real estate as a physical stock, an asset class and a service, is concentrated in the same cities as other specialized international financial and business services. As global city commercial office suppliers, occupiers, and investors, real estate and financial services are linked into the complex of transnational economic agents in the world city network and into the mix of global and local drivers that shape FDI. Evidence from qualitative interview research on London–Frankfurt relations for the UK European Union (EU) Economic and Monetary Union (EMU) fourth economic test on the likely effects of entry to EMU for the competitiveness of London, demonstrated just this point (Pain, 2008). The research found that, as theorized by Bourdieu in his 1980 *Logic of Practice*, social capital is commutable to economic capital in the advanced producer services networks that link global cities, giving those cities complementary economic relations (Pain, 2008). Understanding the economics of real estate development and the challenges for governance, demands an appreciation of its "soft" social co-construction by interactions between multiple actors in global cities, to complement sophisticated micro-economic modelling.

In the same 2014 issue of *Regional Studies* in which Lizieri and Pain illustrate the aforementioned London flight to safety response of investors during the recent global financial crisis, Henneberry and Mouzakis speculate that investor "familiarity" with the London market explains apparently economically irrational investment decision-making that ignores real estate economics evidence on the value of regional yields in UK cities (Henneberry and Mouzakis, 2014). Drawing on quantitative world city network analysis, Lizieri and Pain (2014) point to a correlation between the geography of financial services and office real estate investment concentration in response to the global financial crisis. Evidence from qualitative research eliciting views of senior actors in transnational advanced producer services firms located in London since the year 2000, reveals the critical importance for their global operations of access to specialized international labor, and nurturing relationships, trust, and knowledge exchange (Taylor et al., 2003). These are fundamental requirements to be a "global player" and are made possible by local proximities in places where scale and depth of infrastructure are the basis of agglomeration economies in worldwide business networks (Pain, 2007). Social networks through which heuristic information is gleaned to inform real estate investment that continues to focus on London, discussed by Henneberry and Mouzakis (2014), extend beyond real estate actors, encompassing actors in other advanced producer services.

In contrast to advantages of sectoral specialization in business clusters for some other types of economic activity, multi-sector clustering is essential in vibrant financial districts. In the case of London, real estate investment market intelligence comes from the connectivity of its social networks within the transnational informational cluster where global coordinating functions shape revenue streams in the world city network. Lizieri describes the "interlocking of occupation, ownership and finance" in global cities; firms that occupy office space "are the same firms that acquire offices as an investment asset (directly and indirectly) and who provide funding for the creation of new office space" (Lizieri, 2009, p. xi). Governance of real estate development requires an understanding of investment decision-making in local contexts as not simply economically but socially constructed, by complexes of multiple urban actors.

New spaces for sustainable real estate governance

As real estate and finance firms, and investment flows, are globalizing in response to macroeconomic changes, global economic structural change is shaping unparalleled world human settlement mega-trends that have critical relevance for real estate development and its governance. First, world urbanization has escalated during the past century (Harrison and Pain, 2012) and this is set to continue because, increasingly, it is cities that promise opportunities for work, access to resources and services, prosperity and well-being. Second, physically extensive "mega-cities," and emerging "mega-city regions" consisting of a number of spatially separate but functionally interconnected cities and towns of different sizes, are becoming the worldwide poles for mushrooming populations of 10 to 20 million people. The twenty-first century is experiencing and will continue to experience historically unprecedented global clustered settlement change.

Urbanization implications for real estate development

As the global population grows towards ten billion by mid-century, with increasing human movement from rural to urban areas, it is estimated that up to three-quarters of the world's inhabitants will become urban dwellers, mostly living in developing countries, representing a doubling of urban population in the past 100 years (UN DESA, 2014, 2015). United Nations projections indicate that the proportion of city dwellers in the world will increase from 54% in 2014 to 60% by 2030 and 66% by 2050 (UN DESA, 2014, 2015). Between 2014 and 2050, 2.5 billion people are projected to be added to the world's urban population, with the major growth occurring in emerging markets; India alone is projected to add 404 million urban dwellers and China 292 million by 2050. Together, China and India are contributing more than one-third of world urban growth and generating a growing middle-class population (UN DESA, 2014, 2015). A consequence of this urbanization mega-trend is that 700–900 million square meters of residential and commercial floor space has been predicted to be required in India and 1.6–1.9 billion square meters in China, every year for the next 20 years (McKinsey & Co., 2011, 2013). Eighty percent of global GDP (gross domestic product) is already generated by the world's urban population and as rural to urban migration proceeds, growing urban employment is expected to increase productivity and to boost GDP, contributing further to unprecedented poverty reduction and social progress since the beginning of the century (UN DESA, 2016a, p. 4). However, the delivery of the physical infrastructure, services, financial mechanisms, sustainable construction, etc., to support this major urbanization trajectory presents a huge challenge for real estate development and its governance in the context of changing patterns of urbanization, associated consumption and pollution, natural resource depletion, and environmental and climate change.

A new geometry of the global economy: The example of China

Pre-dating the paradigm change presented by twenty-first century urbanization, Friedmann's "world city hypothesis" (1986) depicted the geography of the world financial system located in cities as the outcome of a Northern Atlantic framed "core–periphery" model of industrial capitalism (Pain, 2017). However, the rise of the tertiary advanced service economy is changing this geography substantially. In contemporary economic and real estate globalization, the development of cities is less linked to their domestic hinterlands than to their integration in a worldwide network of informational and financial flows that were not conceived of in early twentieth century urban location and systems theories (Pain, 2011a). As a consequence of structural changes and integration in the world economy with the growing dominance of city-based advanced producer services, and linked neo-liberalizing approaches to urban development by governments competing for FDI, there has been an unprecedented repositioning of cities in the world city network.

Running alongside major Chinese urbanization and "open door" policy reforms, the rank positions of Shanghai and Beijing in terms of their world city network connectivity (Pain et al., 2012), increased from positions in the top 30 cities in the world in the year 2000, to rank 7 and rank 12 respectively for all advanced producer services, and to rank 5 and rank 8 respectively for financial services, by the year 2010 (Pain and Van Hamme, 2014). State central and local government agencies determining the openness of China to economic globalization influence the pace and the location of real estate development that supports this urban advanced producer services growth trajectory. The rising concentration of advanced producer services both requires and leads to a critical mass of office space in a selection of globalizing Chinese cities. The case of China exemplifies how government bodies seeking to promote economic growth through urban development do retain agency in the space of inter-city investment flows despite being territorially grounded. Regulatory and tax regimes give the state leverage in encouraging inward investment by city-based advanced services that add value in wider primary and secondary sector global production networks and value chains (Pain and Van Hamme, 2014). Since 1979, China has introduced a series of reforms and measures, and a collaborative governance model that has rolled back state regulation in order to attract FDI and upscale national exports. Initially, the priority once given to territorially balanced development has been replaced by an efficiency strategy that has allowed the concentration of FDI by advanced producer services in the Chinese cities of Shanghai and Beijing (Shi et al., forthcoming). In 2006, prior to the most recent world economic recession, China's wholesale financial services sector was estimated to be growing at nearly three times the pace of that of developed economies (Corporation of London, 2006). Political-economy change and increasing global integration that are repositioning China in the world through urbanization, demand continued agile governance to assure the delivery of sustainable real estate development and growth. Sustainability priorities are environmental, natural resource, food and human health security, and the effects of development on the socioeconomic gradient of China's population as a whole. Moreover, these governance considerations apply to countries in developed as well as emerging economies.

Governance of new urban landscapes for real estate development

The second major trend associated with global economic change which has critical relevance for the governance of real estate development is that the global urban demographic just described is focusing in on large cities in the world that are spreading out physically and, in many cases, merging them with proximate cities, functionally. In 2014, one in eight people lived in 28 "mega-cities" in the world, each with more than 10 million inhabitants (UN DESA, 2014, p. 1). It has been predicted that the number of cities in the world with a population of 10 million or more will increase to 41 cities in 2030, and the number of cities with a population of up to 5 million inhabitants will increase from 417 in 2014 to 558 cities in 2030 (UN DESA, 2014, p. 1). Furthermore, the macroeconomic changes that are occurring as nation states worldwide embrace the post-industrial global economy by opening up to direct foreign competition are impacting at the local level around big cities that are gateways for inward FDI.

Since the latter decades of the twentieth century, there has been recognition that the economic growth of metropolitan areas can extend across large city-regions (Jacobs, 1969), crossing government administrative jurisdictional boundaries. In 2001, Scott predicted the coming of a new reality: the economically developed "global city region" constituted by global flows active at a city-region scale. Such urban formations will have a "deepening role" in the economy because they are places where globalization conspicuously "crystallizes out" on the ground; in other words, the global role of cities noted by Sassen (1991) is conferred on global city regions. Scott foresaw that these urban entities would have increasing agency to shape the geography of economic globalization and would be the principal world locations through which the world financial system operates. Building on the global city region concept, empirical research by Hall and Pain (2006) introduced the term, the "global mega-city region" (Pain, 2008), to describe such post-industrial regions with an emergent "polycentric" or multi-center spatial and functional form. Now identified across the world, such regions are characterized by interdependent globalization and regionalization processes (Pain, 2011b, c). In these regions, global, national, and local markets and firms intersect as networks and investment strategies of advanced producer services firms unfold across urban landscapes where separate towns and cities are becoming functionally connected and interdependent. Urban and economic growth at this new scale is introducing new divisions of labor, socioeconomic gradients, and land use, mobility, and work–life patterns with environmental, social, and human health costs. Cross-cutting inter-city travel, urban congestion, and carbon emissions are presenting major challenges for coordinated land use and economic planning for sustainable real estate development. Global mega-city regions span multiple metropolitan and regional government administrative areas and policy silos; meanwhile, policies for regulation and taxation, and land use and transportation planning in mega-city regions are commonly fragmented and policy objectives are often contradictory. Strategic planning and policy oversight is needed to frame local planning initiatives.

In China, mega-city region development and socioeconomic change are being assisted by centrally planned entrepreneurial government strategies and direct investment (DI). Investment programs in education, skills, IT, high-tech and high-speed transit development, are intended to upgrade economic activity with advanced services and logistics supporting advanced manufacturing and advanced agriculture, through sustainable real estate development, for example in the vast Beijing Bohai Economic Rim (cctv-america, 2015). A 2016 United Nations report on the state of the world economic situation, post-2007/8 global financial crisis, highlights the enormous challenges faced by governments worldwide to stimulate investment and global growth more than seven years later (UN DESA, 2016b, p. vi). While the tertiary service sector, which adds value to primary and secondary production and trade in mega-city regions of emerging economies, has remained relatively constant post-crisis, macroeconomic "major headwinds" present challenges worldwide. Notably, the report highlights a growing "disconnect between finance and real sector activities" as significant in holding back revival of the world economy. Summers is cited as blaming low real interest rates "which encourage excessive risk-taking by the financial sector" for this disconnect (Summers, cited in UN DESA, 2016b, p. 23). In consequence, the report identifies a critical need for policy coordination at national, regional, and international levels and integration between monetary, exchange-rate, and fiscal policies in the context of "ever greater complexity in the financial market." Silo policy approaches to objective setting for economic, social, and environmental policies need to be resolved at the domestic level (UN DESA, 2016b, pp. 40–41). But despite the economic slowdown reported by UN-DESA in China, East and South Asia are expected to remain the world's fastest-growing regional grouping (UN DESA, 2016b, p. 23) with this growth continuing to focus on global mega-city regions (UN DESA, 2014), reinforcing its need for policy coordination.

Conclusion

Globalizing forces constrain, but do not annihilate, the agency of states in the governance of real estate development; states remain sites for national and sub-national decision-making that shapes the development process. Domestic and international fiscal, taxation, and monetary policies that provide the context for investment in real estate, both as an immobile physical and a fungible asset, are generated by national governments. Equally, the location and form of development remain to a considerable extent socially and culturally produced and determined by public and private agents and actors with a presence in local settings. However, government interventions are generally not framed by a holistic appreciation of the multi-faceted macro and micro influences on local development outcomes. Meanwhile, attempts to model the development process from an institutional perspective (Ball, 1998) have lacked an appreciation of the complex global–local social construction and fluidity of real estate investment drivers and flows.

The pressing agenda for the governance of sustainable real estate development is the coordination of policies, regulatory mechanisms and decision-making to address interrelated urbanization and economic globalization dynamics through collaborative state institutional structures and public–private partnership arrangements. The commodification of urban space in which entrepreneurial states are complicit, requires an overview of the sustainability of real estate development in new landscapes of urbanization. Real estate research has an important role to play in informing the governance agenda by unravelling for decision-makers the multi-dimensional ways in which real estate acts as a conduit for changing agglomeration geometries on a macro-scale and investment strategies at a micro-scale. The warning of Ambrose and Colenutt when discussing the research needed to inform key decision-makers on the outcomes of the planning and property development process in 1975, remains pertinent: "What we need is research which questions the present structure and which focuses on the operation of the system, not on its output" (Ambrose and Colenutt, 1975, p. 187).

Acknowledgments

Writing this chapter has brought together themes from across the many research projects I have contributed to and led during the past 15 years. People who I have worked with closely and whose ideas have inspired my thinking along the way, include: Peter Taylor, Founder and Director of the Globalization and World Cities Research Network and Emeritus Professor of Geography at the University of Loughborough and Northumbria University; the late Sir Peter Hall, Emeritus Professor of Urban Regeneration and Planning at University College London; and Colin Lizieri, Professor of Real Estate Finance at the University of Cambridge. Thanks must also go to my late son, Andrew Pain, Portfolio Analyst with Prudential Property Investment Managers (PRUPIM), London, now M&G Real Estate, whose personal reflections on how deep social networks in the London real estate market shape rational investment behavior, must be credited here.

References

Ambrose, B. and Colenutt, P. (1975) *The Property Machine*. Harmondsworth: Penguin.

Ball, M. (1998) Institutions in British Property Research: A Review, *Urban Studies*, 35(9), 1501–1517.

Bardhan, A. and Kroll, C. A. (2007) Globalization and the Real Estate Industry: Issues, Implications, Opportunities. Haas School of Business, UC Berkeley Industry Studies Working Paper WP-2007-04. Prepared for the Sloan Industry Studies Annual Conference Cambridge, MA. Available at: http://isapapers.pitt.edu/122/1/2007-04_Bardhan.pdf (accessed February 22, 2016).

Barkham, R. (2012) *Real Estate and Globalization*. Chichester: Wiley-Blackwell.

Beaverstock, J. V., Smith, R. G., and Taylor, P. J. (2000) World City-Network: A New Metageography? *Annals of the Association of American Geographers*, 90(1), 123–134.

Bourdieu, P. (1990) *The Logic of Practice*. Cambridge: Polity Press.

Castells, M. (1996) *The Information Age: Economy, Society and Culture, Vol. I, The Rise of the Network Society*. Oxford: Blackwell.

cctv-america (2015) The Heat: The Beijing–Tianjin–Hebei economic zone. Available at: www.cctv-america.com/?p=83654 (accessed February 22, 2016).

CIS (1973) Your Money and Your Life: Insurance Companies and Pension Funds. CIS Anti Report no. 7. London: CIS.

Cochrane, A. and Pain, K. (2000) A Globalizing Society? In D. Held (ed.), *A Globalizing World? Culture, Economics, Politics*, pp. 5–45. London: Routledge.

Corporation of London (2006) *The Importance of Wholesale Financial Services to the EU Economy 2006*. London: Corporation of London. Available at: www.cityoflondon.gov.uk/business/economic-research-and-information/research-publications/Documents/2007-2000/The%20importance%20of%20wholesale%20financial%20services%20to%20the%20EU%20economy_2006_executive%20summary.pdf (accessed February 22, 2016).

Fainstein, S. S. (1994) *The City Builders: Property Development in New York and London, 1980–2000*. Oxford: Blackwell.

Friedmann, J. (1986) The World City Hypothesis, *Development and Change*, 17(1), 69–83.

Giddens, A. (1999) *Runaway World: The Reith Lectures*, London: BBC. Available at: www.bbc.co.uk/radio4/reith1999/lectures.shtml (accessed April 13, 2017).

Hall, P. (1966) *The World Cities*. London: Weidenfeld and Nicolson.

Hall, P. and Pain, K. (eds) (2006) *The Polycentric Metropolis*. London: Earthscan.

Harrison, J. and Pain, K. (eds) (2012) *Global Cities, Volume IV, Planning and Governance of Cities in Globalization*. London: Routledge.

Henneberry, J. and Mouzakis, F. (2014) Familiarity and the Determination of Yields for Regional Office Property Investments in the UK, *Regional Studies*, 48(3), 530–546.

Heywood, A. (2000) *Key Concepts in Politics*. Basingstoke: Palgrave Macmillan.

Jacobs, J. (1969) *The Economy of Cities*. New York: Random House.

Knox, P. and Pain, K. (2010) International Homogeneity in Architecture and Urban Development? Special issue on International Real Estate Markets, Global Real Estate Industry, *Informationen zur Raumentwicklung (IzR)*, 34(2), 417–428,

Ling, D. C., Naranjo, A. and Scheick, B. (2015) MSA Geographic Allocations, Property Selection, and Performance Attribution in Public and Private Real Estate Markets. Available at: www.reit.com/sites/default/files/Paper%203%20Scheick.pdf (accessed February 22, 2016).

Lizieri, C. (2009) *Towers of Capital*. Oxford: Blackwell-Wiley.

Lizieri, C. and Pain, K. (2014) International Office Investment in Global Cities: The Production of Financial Space and Systemic Risk, *Regional Studies*, 48(3), 439–455.

Lizieri, C. and Pain, K. (2015) International Office Investment Networks and Capital Flows in the Financialization of City Space, Association of American Geographers (AAG) 2015 Conference. Chicago, US: Association of American Geographers.

Marriott, O. (1989, third edition) *The Property Boom*. London: Abingdon Publishing Company.

McKinsey & Co. (2011) Urban World: Cities and the Rise of the Consuming Class. Available at: www.mckinsey.com/insights/urbanization/urban_world_cities_and_the_rise_of_the_consuming_class (accessed February 22, 2016).

McKinsey & Co. (2013) Urban World: The Shifting Business Landscape. Available at: www.mckinsey.com/insights/urbanization/urban_world_the_shifting_global_business_landscape (accessed 22 February 22, 2016).

Murray, W. (2006) *Geographies of Globalisation*. London: Routledge.

Pain, K. (2007) City of London Global Village: Understanding the Square Mile in a Post-Industrial World Economy. In S. Barber (ed.), *The Geo-Politics of the City*, pp. 19–38. London: European Research Forum.

Pain, K. (2008) Spaces of Practice in Advanced Business Services: Re-thinking London–Frankfurt Relations, *Environment and Planning D Society and Space*, 26(2), 264–279.

Pain, K. (2011a) "New Worlds" for "Old"? Twenty-First-Century Gateways and Corridors: Reflections on a European Spatial Perspective, *International Journal of Urban & Regional Research*, 35(6), 1154–1174.

Pain, K. (2011b) Spatial Transformations of Cities: Global City Region? Mega City Region? In B. Derudder, M. Hoyler, P. J. Taylor and F. Witlox (eds), *International Handbook of Globalization and World Cities*, pp. 83–93. Cheltenham: Edward Elgar.

Pain, K. (2011c) Cities and Sustainability: Reflections on a Decade of World Development. In B. Derudder, M. Hoyler, P. J. Taylor and F. Witlox (eds), *International Handbook of Globalization and World Cities*, pp. 318–327. Cheltenham: Edward Elgar.

Pain. K. (2017) World City. In D. Richardson, N. Castree, M. F. Goodchild, A. L. Kobayashi, W. Liu, and R. Marston (eds), *The International Encyclopedia of Geography: People, the Earth, Environment, and Technology – Economic Geography and Regional Development*, pp. 1–9. Chichester: John Wiley & Sons.

Pain, K., Vinciguerra, S., Hoyler, M., and Taylor, P. J. (2012) Europe in the World City Network, Working Paper 3, TIGER Territorial Impact of Globalization for Europe and its Regions, Applied Research 2013/1/1. Brussels: European Spatial Observation Network. Available at: http://164.15.12.207/espon/tiger/final/ (accessed February 22, 2016).

Pain, K. and Van Hamme, G. (2014) *Changing Urban and Regional Relations in a Globalizing World: Europe as a Global Macro-Region*. Cheltenham: Edward Elgar.

Sassen, S. (1991) *The Global City: New York, London, Tokyo*. Princeton: Princeton University Press.

Scott, A. J. (2001) Globalization and the Rise of City-Regions, *European Planning Studies*, 9(7), 813–826.

Shi, S., Wall, R., and Pain, K. (forthcoming) The Conjunction of Networked Agglomeration and Location Factors in Chinese Cities: Taking FDI and Domestic Investment as an Example. In N. Pengfei (ed.), *The Urban Competitiveness Report*. Cheltenham: Edward Elgar.

Tabb, W. K. (1982) *The Long Default: New York City and the Urban Fiscal Crisis*. New York: Monthly Review Press.

Taylor, P. J. (2001) Specification of the World City Network, *Geographical Analysis*, 33, 181–194.

Taylor, P. J. (2004) *World City Network: A Global Urban Analysis*. London: Routledge.

Taylor, P. J., Beaverstock, J. V., Cook, G., Pandit, N. and Pain, K. (2003) *Financial Services Clustering and its significance for London*. London: Corporation of London. Available from: www.cityoflondon.gov.uk/business/economic-research-and-information/research-publications/Documents/2007-2000/Financial-Services-Clustering-and-its-significance-for-London-publishedreportWeb.pdf (accessed February 22, 2016).

Thompson, G. (2000) A Globalizing Economy? In D. Held (ed.), *A Globalizing World? Culture, Economics, Politics*, pp. 85–126. London: Routledge.

UN DESA (2014) *World Urbanisation Prospects: The 2014 Revision* [highlights]. New York: United Nations. Available at: http://esa.un.org/unpd/wup/Highlights/WUP2014-Highlights.pdf (accessed February 22, 2016).

UN DESA (2015) *World Urbanization Prospects: The 2014 Revision*. New York: United Nations. Available at: http://esa.un.org/unpd/wup/FinalReport/WUP2014-Report.pdf (accessed February 22, 2016).

UN DESA (2016a) *Overview: Leaving No-one Behind: Progress Towards Achieving Socially-Inclusive Development*. New York: United Nations. Available at: www.un.org/esa/socdev/documents/2016/RWSS2016ExecutiveSummary.pdf (accessed February 22, 2016).

UN DESA (2016b) *World Economic Situation Prospects*. New York: United Nations. www.un.org/en/development/desa/policy/wesp/wesp_current/2016wesp_full_en.pdf (accessed February 22, 2016).

28

Local politics, planning, land use controls, and real estate development

John McDonald

Abstract

This chapter consists of two parts; the first is a general discussion of the real estate development process, and the second contains three case studies describing how one significant actor in the development process influenced final outcomes. The general discussion begins with a description of the actors – local public officials, developers, professional planners, and citizens. The context in which development takes place is considered next, and the planning process used in the local jurisdiction follows. A discussion of specific policies used to set out the rules for development concludes the general discussion. The three case studies focus on the roles played by an individual real estate developer, a mayor, and a planning official. Donald J. Trump organized the redevelopment of an old hotel into the Grand Hyatt Hotel in New York City; Mayor Richard J. Daley originated the plan for the revitalization of downtown Chicago that began in the 1950s; and Edmond Bacon was the planning official in Philadelphia who influenced various projects to renew the central city. In each case the prime mover in the process had to convince others to agree to the projects in question and to make use of the skills of several types of professionals whose special skills contributed to the final outcomes.

Introduction

The purpose of this chapter is to provide an overview of the entire real estate development process, from setting community goals to seeing actual development on the ground. The chapter focuses on the development process as practiced in the USA. We begin by examining the actors involved and their different motivations – public officials who want to be re-elected, developers who must have vision and persuasive abilities and meet the test of the market, local residents who have their ideas about the nature of a good community, and so on. Then we consider the context in which these actors must function – from central city with declining employment and tax base to suburban boom town. Next comes the formulation of plans and the selection of particular policies to implement those plans. Do the policies selected in fact further the plan? Up to this point the chapter speaks in general terms; it does not delve into the details of development financing or construction.

The final portion of the chapter consists of three case studies to illustrate how actual real estate developments came to fruition. The case studies are drawn from three major, older central cities – New York, Chicago, and Philadelphia. The cases are chosen to illustrate how a real estate developer, a mayor, and a city planner played the leading role in the drama. The developer, Donald Trump, was the driving force behind the creation of the Grand Hyatt Hotel in midtown Manhattan. Richard J. Daley was the mayor who was most responsible for the revitalization and growth of downtown Chicago. Edmond Bacon was the city planner in Philadelphia who influenced through vision and persuasion many redevelopment projects over his 21 years as director of the Philadelphia City Planning Commission, an agency without a large budget or staff. The case studies illustrate how a leader in the development process was able to achieve results consistent with community goals such as expanding employment, enhancing the tax base, and/or improving the quality of life in the city.

The Actors

We shall concentrate on the actors at the local level, the level of a municipality. Those actors include elected officials both at the municipal level and at the level of a district within the municipality, real estate developers and their associates, the citizens of the municipality, the nearby neighbors of a proposed development, and the planners who are employed by the municipality. Each of these actors has a different motivation, and the process of reaching agreement on a development project is a process of negotiation and compromise. These actors and their motives are described in this section in a stylized manner.

Elected officials at the municipal level (e.g. the mayor) have the interests of the entire municipality in mind, and in order to pursue an agenda must have the support of a sufficient number of the city council (i.e. the officials elected at the district level). The mayor bases his or her agenda on the current and expected condition of the municipality, and seeks to build support by demonstrating how success with that agenda is of benefit to a sufficient number of city council members. In an ideal world, mayors seek to improve the municipality by enhancing employment opportunities, the tax base, the quality of public services (schools, crime prevention, and so on), and overall quality of life. The attitude of the mayor towards development projects is based on these objectives.

City council members represent their districts and operate, at least in part, with the interests of the district resident voters in mind. Different districts may have very different interests. One member may represent a high-income neighborhood consisting of home owners, while another may represent a slum where most of the residents are renters. One member in a large city may represent the downtown area, while others represent places that look like suburbs. One member may want to limit development, while another looks to attract pretty much any and all development.

Developers look to make money with projects that meet the market test and the tests imposed by the other actors. They must have the vision to imagine a project that can satisfy both the market and the other actors, and the skill to push the project along to completion. These people are the key actors in the sense that no private development will happen without them. A book by Peter H. Brown based on interviews of over 100 successful developers reveals that, in order to be successful, a developer must have a measure of "vision, tenacity, and the ability to reconcile many voices" (Brown, 2015, p. 37).

The citizens of the municipality vote for the mayor and for the members of the city council. They respond for numerous reasons, but one of those likely reasons is the agenda for development espoused by the candidates based on the goal of improving the community as noted above. For example, William Fischel (2001) is well known for the "home voter" hypothesis, in which citizen voters in suburbs act to protect their property values by policies that limit housing supply. In addition, the municipality may have public committees (official or unofficial) that take positions on development policy in general and/or particular projects.

The nearby neighbors of a proposed development often tend to have parochial interests in keeping the neighborhood as it is. They worry about something new coming in. Where I used to live, a local college wanted to install a tennis court for its students, with lighting, on its own property across the street from some homes. The home owners banded together to oppose the project, which did not happen. In other communities residents are eager for new developments that provide jobs and/or retailing.

Finally, the planners sit in the middle of the process. The old style of "top down" planning is rare these days. Instead, planners participate in and help to guide more "bottom up" planning. Kaiser, Godschalk, and Chapin, authors of a classic textbook in urban land use planning, depict the role of the planner as follows:

> Local land planners must work with many publics. To be acceptable and effective, land use plans must recognize and reconcile the pluralistic interests of these publics with those expressed by governments and markets. It is a major challenge to design a fair and orderly form for expressing and reconciling the interests of those affected by land use decisions. Managing this complex public participation process makes land planning as much an art as a science, especially when planning concerns change so often.
>
> *(Kaiser et al., 1995, p. 17)*

Another standard planning textbook by Levy (2003) generally agrees with this statement, but then states that there are different styles of planning. The planner's role is to facilitate the planning process and to aid it with expertise, rather than do a "top-down" plan. The different styles of planning include:

- planner as neutral public servant, says how to and what if, rather than should or should not;
- planner as builder of community consensus – political view of planning;
- planner as entrepreneur when planners run an agency that is task-oriented, such as an urban renewal agency;
- planner as advocate of a particular group or program;
- planner as agent of radical change, neo Marxists.

We shall meet one influential planner who was a combination of advocate and consensus builder.

Context of development and community goals

The general goals of improving employment opportunities, increasing the tax base, improving public services, and enhancing the overall quality of life in the municipality receive different emphases depending upon the situation in which the local actors find themselves. Goals may depend also upon other factors, such as the ideological bent of the actor. But surely, actors take account of the status of their relevant area. Urban municipalities vary from large, old, central cities to new, rapidly growing suburbs. They vary from poor to rich. They vary from ethnically diverse to ethnically homogeneous. They vary from centers of industry and business to bedroom communities.

Old, central cities need jobs and a tax base. They need to stop the decline. Growing suburbs need to plan the growth so that a "good" community is the result. Such places have more options than most. Poor municipalities also need jobs and a tax base, but they also likely need improvements in housing and public services in general. Rich municipalities need to balance the status quo with opportunities for development. Ethnically diverse communities need to balance the interests of their various constituencies, while ethnically homogeneous communities may wish to maintain that homogeneity while under pressure to diversify. Centers of industry and business need to preserve the economic base while seeking to enhance it while maintaining and improving the quality of life for residents. Bedroom communities may need a tax base that does not disturb the character of the community.

In simple terms, the job of the developer is to propose projects that fit the situation and meet the test of the market.

Planning to achieve goals

Most municipalities have a plan of some sort that pertains to development. The plan may be as simple as a zoning ordinance with three categories of land use – residential, commercial, and industrial. Such a simple plan at least indicates how much land is to be devoted to commercial and industrial developments and where they are to be located. Municipalities with a simple zoning ordinance also typically have a building code, a housing code, and a capital improvements plan for public facilities. These documents add up to a reasonably clear development plan.

Other municipalities may have a much more elaborate comprehensive plan for development, a detailed zoning ordinance with numerous categories for housing, commercial, and industrial use that includes height and volume restrictions, and detailed performance standards for all types of land use. They may also provide for development incentives, often in the form of property tax reductions and revenue bond programs, or for controls on growth. One study by McDonald and McMillen (2004) found that larger municipalities have more elaborate development regulations and programs, perhaps because larger municipalities can afford the expense of drafting and implementing more regulations.

The question is how these plans are created. Plans and regulations require the approval of the elected public officials. So the question is, "Who rules?" Or, perhaps, in whose interest are the rules made? This question makes the topic fall into the field of the economics of regulation. Does one particular interest group "capture" the regulatory process? If so, which one – and does the answer depend upon circumstances? Numerous scholars have answered this last question in the affirmative.

Ellickson (1977) and Fischel (1985, 2001) argue that home owners can be expected to dominate the politics of small suburbs. Most of the households in small suburbs are home owners who are not transient, engaged in local politics to some extent, and interested in their own property values and in exclusion of people or businesses seen as detrimental to the quality of the community. The home owner voters in poorer suburbs, on the other hand, can be expected to have less interest in controlling development and more interest in promoting development and local jobs. Real estate developers are outsiders who must adapt to the interests of the home owners.

Larger suburbs and central cities contain a more diverse population that includes a sizable group of renters. Ellickson (1977) argued that in such places the issues are more numerous and there tends to be vote trading and log-rolling. In this circumstance, developer interests are more likely to dominate the real estate development rules and policies. In short, the "special interests" of the developers capture the regulatory process, and local officials are more apt to "sell" favors to developer interests. In addition, voters in low-income neighborhoods may oppose development controls that are seen as restricting them. McDonald (1995) reported that a proposed zoning ordinance for the City of Houston, supported by the business community and passed by the city council, was defeated in a 1993 referendum largely by voters in lower-income African American and Latino areas of the city. Molotch (1976) is well known for calling acquiescence to developer interests as the "growth machine," but these low-income voters in Houston prevented passage of a zoning ordinance supported by the business community.

Policies

Municipalities adopt a set of policies to carry out the plan. Those policies include building and housing codes, a zoning ordinance, and a capital improvements program. Research by McDonald and McMillen (2004) on the other policies selected by suburbs in metropolitan Chicago found that these policies tend to group into three categories; quality development regulations, regulation of lower-class developments, and growth controls. Quality development regulations include architectural review, appearance regulations, and historic preservation, while lower-class uses include mobile homes, adult uses, and group homes. Growth controls come in two varieties, those that limit growth (e.g., limitations on building such as growth boundaries, development impact fees) and those that encourage growth (e.g., property tax abatements, enterprise zones, tax increment financing districts, industrial revenue bonds, etc.). Suburbs with higher incomes tend to adopt quality regulations, while suburbs with lower incomes and higher crime rates restrict lower-class developments. Suburbs with greater potential for growth tend to limit growth, while suburbs with larger minority populations and greater crime rates tend to adopt policies to encourage development.

Zoning likely is the most pervasive policy that municipalities adopt to implement a land use and development plan. It is also a policy that has been studied extensively. Zoning is a municipal ordinance that specifies the allowable use of every parcel of land in the jurisdiction. Zoning does not grow out of the common law of nuisances, but rather from the police power of local government to protect the welfare of the public. Many zoning ordinances also control the density of land use by imposing bulk regulations on structures.

What are the motives for zoning? Is zoning an effective means for reaching the desired outcomes? Zoning ordinances were adopted widely in the USA in the 1920s, and the original motivation was the promotion of the general welfare by separating land uses in order to mitigate the negative external effects of the rapid industrialization of urban areas. As stated in a model state enabling law, a zoning ordinance enables a jurisdiction to

> lessen congestion in the streets; to secure safety from fire, panic, and other dangers; to promote health and the general welfare; to provide adequate light and air; to prevent overcrowding of the land; to avoid undue concentration of population; to facilitate the adequate provision of transportation, water, sewerage, schools, parks, and other public requirements.
>
> *(Callies et al., 1999)*

In short, a zoning ordinance is an essential planning tool. Plus, an area zoned for residential use is an area into which industry cannot move; in effect, zoning is an insurance policy that costs the individual resident nothing.

However, over time municipalities realized that zoning could also be used to influence the size of the local tax base and demand for public services (i.e., fiscal zoning), and to exclude certain types of uses and people in order to maintain community homogeneity (exclusionary zoning). One can think of zoning as the appropriation of some property rights on behalf of the community. How well does zoning achieve any or all of those objectives? Given that zoning has many purposes, it is impossible to say how well it performs for all purposes.

A zoning ordinance adopted in advance of development guides that development, of course. The ordinance is a map with the street pattern and public land uses, and provides for densities of use consistent with the carrying capacity of the land. However, research by McMillen and McDonald (1990, 1991) shows that suburban zoning adopted in advance of significant development is subject to change as development unfolds. The study examined areas in north suburban Chicago in which the basic transportation infrastructure (rail lines and highways) were built or already under construction. For example, some land tracts initially zoned exclusively for single-family houses are rezoned to include commercial activity, and once some commercial activity is permitted, there is a tendency to rezone the tract exclusively for commercial use. Zoning is seen as adapting to market demand, but in a manner that attempts to anticipate potential conflicts in land use. But another study by McDonald and McMillen (2012) found that zoning is a blunt instrument when it comes to controlling external effects because it imposes simple quantity restrictions on the land market. The list of external effects of non-residential land use on residential land value is long and includes negative (pollution, noise, heavy traffic, etc.) and positive factors (e.g., proximity to employment, shopping, etc.). Suburban zoning ordinances are criticized for restricting commercial uses to the extent that residents must drive a car for simple shopping trips. One basic theoretical result is that a necessary condition for exclusive residential zoning is that the land value of the last plot of land zoned residential must exceed its value in all other uses.

A zoning ordinance in a jurisdiction that is already developed is another story. There are few detailed studies of the impact of the adoption of a zoning ordinance by a large and developed city. A study of land use in Chicago before zoning (1921) by McDonald and McMillen (1998) found that 42 percent of blocks were in mixed use (residential and non-residential), 32 percent were entirely in residential use, and 26 percent were in non-residential use. The zoning ordinance adopted in 1923 assigned virtually all the blocks devoted to one use to that use. Of the blocks in mixed use, 53 percent were assigned to residential use (but permitted the existing non-residential use to remain), and assigned the remaining 47 percent to mixed use. McMillen and McDonald (2002) shows that the value of land zoned exclusively for residential use in 1923 in Chicago increased more in value than did land zoned for other uses, which is consistent with the idea that residential zoning provides insurance against further intrusion of other uses. An empirical study by Speyer (1989) found that home owners in the Houston area are willing to pay more for property in jurisdictions with zoning than for property in the one big jurisdiction without zoning (City of Houston). The referendum on the proposed zoning ordinance for the City of Houston in 1995 was supported by the residents of middle-class neighborhoods. These same groups may still see value in zoning, but no more recent zoning ordinance has been proposed.

Is zoning an effective instrument for fiscal purposes? The basic point is that residential land use, largely because of the demand for public schools, generally imposes more costs on the jurisdiction than it produces in revenues. Commercial and industrial activity produces more revenue than costs imposed. Consequently, jurisdictions are well aware of the fiscal benefits of business land uses. Jurisdictions must balance the fiscal benefits of commercial and industrial use against community quality for residential use. As noted above, jurisdictions with higher incomes tend to restrict development, while jurisdictions with greater minority populations and higher crime rates encourage development. Providing little or no land for commercial or industrial use is effective, but setting aside a certain amount of land for commercial or industrial use does not mean those uses will exist. You can keep it out, but bringing it in is another question. Jurisdictions engage in numerous other efforts to attract the desired business development, but some of those policies, such as enterprise zones, simply tend to determine the location of developments that would have occurred in any event.

Another question that is unresolved: Does zoning follow the market? Of course, zoning generally follows the market in that downtown land is zoned for commercial activity, land adjacent to freight rail lines is zoned for industry, and much land in the suburbs is zoned for residential use. The real question is whether zoning alters the land use pattern that would have been produced by the market in a manner that achieves the objectives of the community. Not much is known about this large question. But we do know that real estate developers often request zoning changes when they meet with the planning officials to sell a proposed development, and that sometimes those requests are granted in some manner. Again, zoning is a blunt instrument that needs to be fine-tuned, and developers often are the ones who request the fine-tuning.

Case studies

This section is included to show that leadership in the development process can come from different types of actors as categorized in this chapter. Three case studies are presented to illustrate how a developer, a mayor, and a planning official each took the lead in the development process. In each case the leader had to convince the other actors of the merits of his ideas.

A developer-led project

In 1974, the 27-year-old Donald Trump was attempting to initiate development projects in Manhattan by using property owned by the Penn Central Railroad. The railroad was in the process of liquidating unproductive real estate. His contact at the railroad mentioned that there were four old Manhattan hotels that were available. The Commodore was losing money and defaulting on property taxes owed, but it was located adjacent to Grand Central Station. Trump could see almost immediately that this building had great potential. Developers must have vision to see what could be, but must also have great persistence. The Commodore reopened as a Grand Hyatt in 1980.

Trump (1987) provides the details on how the hotel was redeveloped. The first step was to obtain an option on the property, but to do so at minimal cost. He agreed to take an option to buy the property for $10 million, but subject to his obtaining a large tax abatement from the city, arranging financing for the project, and finding a hotel company partner. He agreed to pay $250,000 for the option (a large sum for him), but stalled for months over minor provisions in the contract. Since he is a developer's son and worked for his father Fred Trump, he did have some resources at his command. He hired an architect to draw up an impressive design plan to be used to sell the project to the city and the financiers. At the same time he contacted the Hyatt hotel chain out of the blue. He got the old runaround from the Hyatt executives until someone told him to contact A. N. Pritzker directly, since the Pritzker family owned the controlling interest in Hyatt. Trump worked with A. N.'s son Jay Pritzker, and by May 1975 they had put together a partnership deal contingent on obtaining the tax abatement and financing. The idea was attractive to Hyatt because it had no hotel in New York, and because of the contingencies. The final agreement had one other very important provision. Hyatt agreed not to develop any other hotel in the City of New York (except for a small luxury hotel) unless Trump gave permission. As Trump put it, "The simple fact is that Hyatt would love to build more hotels. By retaining the right to say yes or no, I own something very valuable" (1987, p. 141). The covenant basically was forced on Hyatt by one of the bankers, the people putting up a great deal of money.

With design and an agreement with Hyatt in hand, Trump approached the banks. He could not find financing without the tax abatement, and it was unlikely that a tax abatement would be granted without financing. After several unsuccessful attempts to obtain financing without the tax abatement, he went to the city to be considered for its business investment incentive program. An agreement (subject to obtaining financing) was struck that gave Trump a total abatement of property taxes for 40 years in exchange for payments based on earnings from the hotel in lieu of property taxes. The city's Board of Estimates approved a version of the deal in May 1976 under a program operated by the State of New York. There was considerable controversy over the tax abatement, but Trump argued that the renovated hotel would provide jobs and spur development in a run-down area of the city. It helped that Penn Central was in the process of closing the hotel by June 1976. Trump acknowledges that the tax abatement saved him millions, and that the project could not have been possible without it.

Trump was now able to secure financing in the amount of $35 million from Equitable Life Real Estate and $45 million from Bowery Savings Bank. Trump had still not paid the $250,000 for his option on the property, but nobody seemed to notice.

Trump paid the $10 million to Penn Central for the property, of which $6 million immediately went to the city for back taxes. As noted above, the Grand Hyatt opened in 1980, and according to Trump, "it was a hit from the first day" (1987, p. 138). Hyatt and Trump each have a 50 percent interest in the hotel, with Hyatt as manager.

What are the lessons from this case? As noted above, the developer must have vision and persistence. The developer must keep several balls in the air at the same time, be able to sell the project using a variety of methods, and have personal skills to deal with the different players. Trump had to negotiate with an architect, a hotel chain, city officials, and financiers. Each has a role to play, and speaks a particular language. The young Donald Trump no doubt learned these skills from his father.

A mayor leads downtown revitalization and growth

Richard J. Daley took office as the mayor of Chicago in 1955. The story may be apocryphal, but it is said that he looked out his office window, saw an old and rundown Loop, and resolved to do something about it. Only one major building, the Prudential Building, had been built in the downtown area since the 1920s. Over the next three years Mayor Daley put in place a series of changes to defend and revitalize a compact downtown core, attract residents to the fringes of downtown, and enhance public transit. The strategy played out successfully over the subsequent decades and helped Chicago in its transformation from an economy based on the production of goods to the provision of services. No one in the 1950s could have foreseen fully the massive changes in the economy of metropolitan areas that would take place. Indeed, the professional planners of the day in Chicago did not. The Chicago Area Transportation Study reports, undertaken in the late 1950s, projected for 1950 to 1980 small population growth of 2 percent for the city of Chicago and a 42 percent increase in manufacturing employment for the metro area. Instead the city lost 17 percent of its population and manufacturing jobs in the metro area declined by 12 percent (and 39 percent in the central city) (see McDonald (1988) for a complete retrospective assessment of the Chicago Area Transportation Study research and plans, and see McDonald (2016) for a detailed history of the economy of Chicago).

Mayor Daley operated with the support of heads of Chicago corporations and developers, which Hunt and DeVries (2013) call the growth coalition. These business leaders formed the Central Area Committee in 1956 to provide a more unified voice for their initiatives. At the same time the mayor brought the urban planning functions in the city under his command. He formed a professional Department of City Planning to work at his direction, initiated a five-year capital improvements plan, and convinced the State of Illinois to authorize cities to form public building commissions, with members appointed by the mayor. Also, he benefited from federal funding for planning work. The Department of City Planning produced the 1958 Development Plan that focused on strengthening the downtown core with office, residential, and institutional developments.

The growth coalition had already advocated a massive change in the old Chicago zoning ordinance that limited the height of downtown buildings and included so many changes as to make it a confusing mess. Mayor Daley pushed for the new ordinance that included provisions for planned unit developments and a bonus system for skyscraper developers. The new ordinance was adopted in 1957, and permitted developers to negotiate directly with the planning staff to produce the massive skyscrapers such as the First National Bank Building, the Standard Oil Building, Water Tower Place, the Sears Tower, and the residential high-rise buildings Marina City and Outer Drive East. Each of these projects required the vision, skills, and connections of a development team, of course. But the policy of the city administration was to facilitate, not to undermine. Dearborn Street in the center of the Loop was one of Mayor Daley's prime interests, and its rebirth was boosted by new public buildings – the Federal Center and the Daley Center (for local government) are downtown anchors.

One sizable failure of the overall strategy was the inability to acquire the railroad yards to the south of downtown in order to build a campus for the University of Illinois, which was intended to contain office development in central downtown (as well as to provide a first-class public university for Chicagoans). Instead the campus was built on the near west side, requiring use of the urban renewal program for the clearance of a viable ethnic neighborhood. Hunt and DeVries (2013) think that the nasty controversy that ensued took a toll on Daley and made him wary of other big fights. In short, ignoring the interests of an important actor (in this case the local residents) can be costly.

A planner shapes revitalization of a city

Edmund Bacon was appointed director of the Philadelphia City Planning Commission (PCPC) in 1949. Mr Bacon was an experienced city planner with a background in urban revitalization, but the PCPC was only advisory to the mayor and had no budget. Knowles (2009) provides several retrospective essays on the career of Edmond Bacon. Over the next 21 years he used his position to formulate and advocate revitalization plans for several major areas within Philadelphia. Bacon's method for revitalizing residential areas was to involve community members in planning, and the resulting plans were sensitive combinations of retention of the existing neighborhood, historic preservation, new developments built at a "human" scale, and pedestrian walkways. He worked with architects and designers to produce specific plans and models that he used to the sell the ideas to public officials and the public in general. He was an early and active opponent of massive urban renewal projects.

The largest of those plans (with accompanying model) was the design for the massive parcel owned by Penn Central Railroad in downtown Philadelphia. Bacon and an architect prepared a coordinated plan for the site in 1952 that included office towers, a pedestrian esplanade, retailing, and transit connections. The railroad had its own development consultant, whose plan differed from Bacon's plan. The site was developed as Penn Center (with four office towers) more along the lines of the railroad's plan, but some of Bacon's concepts were adopted. Most importantly, the development was a coordinated project and acted as a spur to future downtown development. However, Heller states that, "In the end, though, Penn Center turned out vastly different from Bacon's vision and was widely considered a disappointment" (2009, p. 39). In this case Bacon did use his powers of vision and persuasion with some (but not complete) success.

One his earliest projects as director was Society Hill, one of the oldest areas of the city that in 1950 contained tenement housing and a sizable proportion of poor residents. Bacon could see that historic preservation was a means for attracting middle-class families into the city. Not much happened on Society Hill until a new mayor, Richardson Dilworth, was elected in 1955. The new mayor favored the revitalization of Society Hill and gave Bacon a critical ally in the selling of his plan to other public officials and the business community. The mayor pushed through the creation of the Old Philadelphia Development Corporation, which would become the developer of Society Hill. This body received funding from the federal urban renewal program in 1957, and the city's redevelopment authority obtained competitive proposals for redevelopment work. Although he was not directly involved at this point, the winning proposals followed Bacon's ideas. Bacon used Mayor Dilworth's desire for creating a legacy to push his vision for Society Hill. Ultimately the redevelopment work did include some displacement of residents and demolition of some very old structures that were beyond repair, but Heller says Society Hill can be "viewed as one of the nation's most successful efforts at attracting middle-class families back from the suburbs" (2009, p. 44).

Conclusion

This chapter consisted of two parts: the general discussion of the formulation and implementation of community real estate development goals, and three case studies to illustrate the process. The general discussion was intended to provide an outline that is filled in for each development project. It has attempted to include the relevant actors and their motivations, the factors that define the goals of the community, the nature of the planning process, and the policy choices that follow. As such it perhaps is a primer for those who would consider being a player in the game.

The case studies showed how different actors can be the one of greatest influence. The real estate developer saw an opportunity that no one else perceived, and sold the idea to all of the other relevant actors. The Grand Hyatt brought jobs and helped to revitalize a portion of the central city. The mayor saw a run-down central business district and resolved to do something about it. The strategy of revitalizing downtown beginning in the 1950s was a major factor in making the Chicago downtown a leader in the newer economy based on the production of services. And the city planner formulated his own philosophy of urban redevelopment, and worked hard to persuade the other actors of the validity of his vision. He played an important role in initiating coordinated downtown development, and Society Hill attracted middle-class, tax-paying families to the central city. Leadership in the real estate development process can be, and in these cases, was consistent with important community goals.

References

Brown, Peter H. 2015, *How Real Estate Developers Think*, Philadelphia: University of Pennsylvania Press.
Callies, David, Freilich, Robert, and Roberts, Thomas, 1999, *Cases and Materials on Land Use*, 3rd edn, St. Paul, MN: West Group.

Ellickson, Robert C. 1977, Suburban growth controls: An economic and legal analysis, *Yale Law Journal*, Vol. 86, No. 3, 385–511.

Fischel, William, 1985, *The Economics of Zoning Laws*, Baltimore: Johns Hopkins University Press.

Fischel, William, 2001, *The Homevoter Hypothesis*, Cambridge, MA: Harvard University Press.

Heller, Gregory, 2009, Salesman of ideas: The life experiences that shaped Edmund Bacon, in S. Knowles, ed. *Imagining Philadelphia: Edmond Bacon and the Future of the City*, Philadelphia: University of Pennsylvania Press, pp. 19–51.

Hunt, D. Bradford, and DeVries, Jon B. 2013, *Planning Chicago*, Chicago, IL: Planners Press.

Kaiser, Edward, Godschalk, David, and Chapin, F. Stuart, 1995, *Urban Land Use Planning*, 4th edn, Chicago: University of Illinois Press.

Knowles, Scott, ed. 2009, *Imagining Philadelphia: Edmund Bacon and the Future of the City*, Pennsylvania, PA: University of Pennsylvania Press.

Levy, John M., 2003, *Contemporary Urban Planning*, 6th edn, Englewood Cliffs, NJ: Prentice-Hall.

McDonald, John, 1988, The first Chicago Area Transportation Study projections and plans for metropolitan Chicago in retrospect, *Planning Perspectives*, Vol. 3, No. 3, 245–268.

McDonald, John, 1995, Houston remains unzoned, *Land Economics*, Vol. 71, No. 1, 137–140.

McDonald, John, 2016, *Chicago: An Economic History*, New York: Routledge.

McDonald, John, and McMillen, Daniel, 1998, Land values, land use, and the first Chicago zoning ordinance, *Journal of Real Estate Finance and Economics*, Vol. 16, No. 2, 135–150.

McDonald, John, and McMillen, Daniel, 2004, Determinants of suburban development controls, *Urban Studies*, Vol. 41, No. 2, 341–361.

McDonald, John, and McMillen, Daniel, 2012, The economics of zoning, in N. Brooks, K. Donaghy, and G. Knapp (eds) *The Oxford Handbook of Urban Economics and Planning*, Oxford: Oxford University Press, pp. 438–459.

McMillen, Daniel, and McDonald, John, 1990, A two-limit model of suburban land-use zoning, *Land Economics*, Vol. 66, No. 3, 272–282.

McMillen, Daniel, and McDonald, John, 1991, A Markov chain model of zoning change, *Journal of Urban Economics*, Vol. 30, No. 2, 257–270.

McMillen, Daniel, and McDonald, John, 2002, Land values in a newly zoned city, *Review of Economics and Statistics*, Vol. 29, No. 1, 62–72.

Molotch, Harvey, 1976, The city as growth machine: Toward a political economy of place, *American Journal of Sociology*, Vol. 82, No. 2, 309–322.

Speyer, Janet, 1989, The effect of land-use restrictions on market value of single-family homes in Houston, *Journal of Real Estate Finance and Economics*, Vol. 2, No. 2, 117–130.

Trump, Donald, 1987, *The Art of the Deal*, New York: Random House.

29

Trends in land use and government policy affecting real estate development in the USA

David Hamilton and Richard Peiser

Abstract

Land use in the United States is shaped by a combination of direct regulation and financial incentives targeting development. The real estate industry is regulated at local, regional, state, and federal levels of government. While the most direct control is still exercised by local zoning authorities, policy is increasingly influenced by regional and state authorities, particularly in technical aspects of project design, but also when projects are likely to create substantial economic or fiscal impact. As regional impacts of land use have received greater scrutiny, state and federal authorities have responded with additional regulation and incentive structures, sometimes with unintended consequences. As local markets are increasingly shaped by larger forces, paradoxically much new development is established under the control of non-governmental associations, devolving operational regulation to the most local level of governance.

Finance for development has also been greatly affected by macro-economic forces, policy responses in the wake of the Great Financial Crisis of 2007–2009, and technological innovation. This chapter presents an overview of a number of recent changes affecting development finance, both on the private and public side. Programs such as the EB-5 Immigration Program, crowdfunding, and private equity funds are changing how developers raise equity. Public–private development is also evolving with increasing reliance on tax increment financing, community benefits agreements, linkages and exactions, business improvement districts and incentive zoning, and value recapture.

Introduction: Land use regulation and governance

Historically, the regulation of land use in the USA has been highly decentralized, with decision-making in the hands of local officials or boards charged with implementing county or municipal zoning and subdivision ordinances. Decisions, enforcement actions and new laws may be challenged in local, state, and federal courts, but these challenges typically involve procedural fairness or jurisdiction, rather than the underlying land use issues. Increasingly, policy preferences of regional, state, and federal governments have begun to more strongly influence the type, mix and design of development, even when these policies are enacted through local control processes. As land use policies grow more coordinated, a parallel devolution is occurring at the lower levels of governance, to extremely local non-governmental entities that have taken on functions associated with municipalities and counties.

Zoning and related controls

Local control of land use is most commonly implemented through municipal zoning, which directly enables allowable uses and densities on all parcels within a delineated zone. A typical designation, such as "R-4" indicates a residential zone allowing four units per acre, and is further defined in the ordinance to allow single-family homes, duplexes, for example, and offer conditional approval on potentially suitable uses, such as inns or small commercial use. Zoning usually includes basic dimensional requirements such as minimum setbacks from property lines and maximum building heights, and in jurisdictions with substantial development pressure the code may include exacting standards for semi-public elements such as signage and architectural features. Some rural jurisdictions do not have zoning, but where it exists, most land use decisions are simple determinations: Is the proposed use consistent with zoning? Conforming proposals may be approved administratively, by municipal or county planning staff as so-called "as-of-right" development.

Re-zoning and land use change

Land owners wishing to change use, or to increase density above the limit of their zone, may petition for re-zoning. In fact, changes to the existing zoning of land are a primary means by which land value is created in the USA. Owners may apply to change to another existing designation, or to alter the text of their zone in the law. As an example, a developer may buy a parcel in zone "R-10," allowing homes on 20,000 square foot lots, and propose that the site is more appropriately developed as multi-family mid-rise housing. Amendments like these typically cannot be reviewed administratively, and open the project to public hearings, usually before a zoning board or planning commission. This path is more formal, requires comprehensive documentation and public notice, is more susceptible to opposition, and therefore riskier.

Whether administrative or political, most zoning ordinances require a set of cross-coordinating reviews by multiple officials. These reviews may direct comment to laws outside the zoning code, such as a water protection ordinance to minimize impact on shared community groundwater. They typically involve health, fire and safety officers, and public works engineers who review and comment on more technical aspects of a proposal such as sewer capacity, fire truck access, or stormwater management. These studies range from quick checks of road and septic facilities to comprehensive review of proposed construction procedures for managing impacts, depending on the jurisdiction and project complexity. Technical review of projects continues after zoning approval in the form of building permits. Building permits are required for substantial construction in most jurisdictions, and may require detailed submission of structural, mechanical, and electrical drawings, as well as coordinated civil engineering by state-licensed professionals. Increasingly, building codes require a certification of green building standards. The building permit process for a single family home is short, often requiring minimal documentation, while complex commercial buildings may take months to review, comment, and resubmit under statutory timelines.

Following the explosive growth of many American cities in the postwar period, many larger municipalities began more closely regulating land use, both by expanding the body of zoning ordinances, and by adding new legal mandates and administrators. Large municipal planning departments may include a variety of specialized professionals addressing ordinances on affordable housing provision, fiscal, social-service, and environmental impacts of development proposals. As cities expanded outward into automobile-dependent suburbs, serious deficiencies were revealed in local control. First, ordinances intended to manage growth could be "leapfrogged" by moving beyond the jurisdiction, and second, the scale of development impacts might now extend to multiple jurisdictions, such as a city's watersheds, road traffic, or school system. As a result, many municipalities have consolidated some kinds of land use control, at least in policy, to regional entities, which may maintain mapping and planning expertise, author resource and development plans, and review proposals on behalf of member communities (Fahmy, interview). These planning entities operate in an increasingly crowded landscape of regional actors in transportation, conservation, and energy planning, and are in many cases conduits of state and federal grant funding toward specific planning priorities. It is becoming common for substantial development projects to face some level of regional-plan review.

State regulation and enforcement

In some cases, the administrative burden of development, or the scope of identified impacts, is large enough to warrant state intervention in the process. This is often the case with state public health reviews of items such as septic systems or restaurant facilities, environmental impacts addressed by state law, or other aspects of a project regulated by state or multi-state standards, such as building or energy codes. In such cases, the local jurisdiction usually remains the coordinator of permitting activities, and refers the proposal to authorities as required. Environmental review of a proposed project is commonly managed by a state Department of Environmental Quality or similar agency, which is empowered by state law to enforce regulations on land use changes in a variety of sensitive natural resources. Developments which impede or alter streams, identified wildlife habitat, agricultural or historic resources or scenic views may be required to submit documentation to receive a specific approval or permit for each. The scope of these reviews has continually increased as citizen pressure, lobbying, and legal action have led to specific state laws to address agricultural and historic preservation, regional habitat protection, and other impacts (Kassell, interview). Increasing technical review of these standards has also opened new routes of legal challenge to unpopular development, as opponents may contest approvals they feel are insufficiently rigorous under the law. Projects open to these kinds of statewide reviews can expect substantial delay and costly professional fees to clear these hurdles.

State oversight of local land use processes

State interest in land use has expanded in recent decades to include issue-specific reviews of several issues relevant to developers. The Chesapeake Bay Preservation Act is Virginia's legislative component of a larger multi-state agreement to protect coastal resources from upstream sedimentation and other pollution. To this end, state regulations can reach into local land use planning to improve development standards and comprehensive plans, and to regulate their enforcement. Other states such as Oregon and Maryland have adopted statewide smart growth laws that hold local codes and decisions to a standard for sustainable land use planning, to promote generally compact and walkable development. Some states favor carrots, offering substantial funding and expertise to local bodies who agree to update their zoning and related ordinances. These laws have had mixed success, as legislative initiative has usually run ahead of funding and implementation. Nevertheless, regional, county, and city plans promising smart or sustainable growth are proliferating in US jurisdictions, and sometimes create tension between customary ways of doing business and actual rules. Issue-specific legislation may address housing, as in Massachusetts, where statewide inclusionary zoning rules were adopted to compel provision of lower-cost housing within approvals of new projects. For more than a generation, compulsory affordable housing percentages were a feature of large urban projects, but states have shifted with planning doctrine to promote more integrated types and costs of housing, even beyond city limits. Similarly, laws attempting to address climate change have been introduced in California, with potentially broad effects in real estate, where energy intensity is linked to density and transit choice (Fahmy, interview).

Federal involvement in land use

The mechanism for federal involvement in land use is similar, but follows a generation behind states. Federal oversight over impacts to waters of the US has been in place since the 1972 expansion of the Clean Water Act. Since then, scientific understanding of non-point source pollution and sedimentation has advanced to become a major concern in the preservation of watersheds, and land disturbance is, along with the filling of wetlands, the most common reason for direct federal permit oversight. The jurisdiction of this regulation has broadened consistently, and land uses need be only tenuously related to the hydrology of navigable waters to fall under jurisdiction of the Army Corps of Engineers (USACE) permit program. Cost and delay are often sufficient reason to alter designs to achieve lower-impact development patterns when crossing even minor streams or emergent wetlands. Again, the local zoning agency most often plays a coordinating role, but developers planning a substantial project must have a professional study the parcel to determine jurisdiction and prospects for approval. Other federal reviews can include the Department of Fish and Wildlife (DFW) and the Environmental Protection Agency (EPA), and the Department of Housing and Urban Development (HUD).

Most recently, federal interest in land use has appeared in explicit smart growth initiatives of the federal government. The Sustainable Communities Initiative, a partnership of HUD, EPA, and the Department of Transportation (DOT) has, since 2010, added additional conditions to their direction of infrastructure funding. Because federal highway funding is a major driver of cities, smart growth advocates have long complained that by underweighting urban populations' per capita needs and emphasizing regional and interstate networks over mass transit, federal funding has been a driver of suburbanization and low-quality development choices, and the new initiatives attempt to direct both planning and hard construction to projects that promote a more compact, city-centric settlement pattern. Largely a product of the Obama administration, the longevity of these initiatives is uncertain, but if they persist, the effects of even a modest redirection of resources could be substantial for real estate in the USA. Many of the innovations and project types of the postwar American land use are fundamentally suburban in character, and increasing urbanization is requiring adjustment of concepts such as road standards, housing types, and retail concepts. If nascent trends toward urban living continue, and if infrastructure investments and other incentives align with trend, a substantial change in development may result.

Changes in regulatory approach

At every level, controls on land use have followed a trajectory of broadening scope and increasing scrutiny, but the implementation of these controls has followed what seems like an opposite direction. In many relevant areas of regulation, local and state actors have changed their approach from prescriptive rules to performance-based standards. Simultaneously, enforcement, and sometimes even creation and maintenance of standards may be outsourced to third parties as communities attempt to regulate issues about which municipal officials have no particular expertise. These trends reflect tension between a new quantification of land and resource use in environmental design, and the necessary pullback of public employees as governments have limited overhead in the aftermath of financial crisis. "Quantification" in this case refers to the array of data sources such as Geographic Information Systems (GIS), which allow comprehensive mapping of resources, and also to the array of modeling tools, such as ComCheck, to facilitate closer calculation of development proposals' actual impacts (Kellenberg, interview). Where laws might prohibit "deposition of fill within twenty feet of a stream bank," a similar control today might refer to a publicly available GIS layer that delineates actual hydrology of a region and limits of Total Maximum Daily Load (TMDL) in a receiving watershed, referring to Best Management Practices (BMPS), promulgated by a society of engineers, which may be appropriate to achieve compliance. Such standards offer a dual benefit, allowing design flexibility to tailor solutions to projects, while offering a measurable compliance test.

The complexity of demonstrating compliance is, however, beyond the scope of most jurisdictions' capability, so standards may be maintained as independent certifications that must be submitted by an engineer hired by the applicant. Third party certification belongs to a larger trend of government outsourcing, and may include soil and water evaluations, certification of septic or fire-protection design, hazard identification, even extending to more subjective reviews like compliance with architectural standards.

Energy codes and green buildings

The most complete example of this phenomenon is green building standards. For a generation, some level of green building has been a portfolio requirement of many government agencies and universities. Energy codes have been in place in California and some municipalities since the late 1970s, but just as interest in energy efficiency has come to public notice, prescriptive codes have given way to performance standards maintained by third parties (Kiefer, interview). Where energy codes were once fairly limited in scope, requiring, for example, minimum R-values in building envelopes and mandating that a percentage of fixtures be energy efficient, modern state or municipal codes may allow certification under GreenPoint, EnergyStar (a federal standard), or LEED (Leadership in Energy and Environmental Design) to a certain threshold and climate zone referenced by the standard. LEED is maintained and continuously updated to reflect building science and improving technology by its author, the US Green Building Council (USGBC). The result is a set of options that the designer may trade to find an affordable and achievable path to substantial reductions in resource use and other environmental criteria. Buildings and land developments can voluntarily contract with auditors to demonstrate compliance.

New zoning codes

The concept of flexibility in achieving a performance standard has reached even the basics of land use, where it seems a difficult fit with zoning's reductive legal approach. Zoning is criticized (usually fairly) for presupposing the separation of use groups. In fact, the last major innovation in zoning was the postwar idea of the Planned Unit Development (PUD), which addressed this problem. PUDs allow a mixture of uses and building types to be co-developed under a master plan, typically on a large parcel, in ways that may not meet the zoning code, but which provide a successful physical program for the site. In addition to mixed use developments, PUDs have been used for special-use facilities such as airports that simply don't fit into use-based zoning. While an improvement, the PUD approach is still parcel-by-parcel. Many planners have long pointed out that the successful creation of urbanism relies on the comingling of mixed uses, on many parcels, to create streetscapes which perform diverse functions for owner/tenants, for vehicles, and for the pedestrian public. Related to smart growth, many architects and developers have successfully promoted alternative Form Based Codes (FBC). Rather than sharply segregating use, FBC codes presuppose mixed use in many locations, and may in fact refrain from specifying particular uses, other than to prohibit nuisance. Rather, the municipality (and this is usually an urban phenomenon) designs several prototypical street types, which are specified in form and pedestrian function (Plater-Zyberk, interview). These may be more or less appropriate to different locations in the district, and suitable for a variety of uses, but if the template is followed, all development will contribute to increasing key measures of quality.

Private infrastructure and governance

Coalescing land use and performance standards around new development have coincided with a 30-year boom in the creation of non-governmental entities performing functions long associated with municipal sovereignty. Some are large, with access to capital, and may provide privately funded infrastructure, as in a public–private tollway. More common is the smaller voluntary or semi-voluntary organization, such as the Downtown Improvement District or Home Owners Associations (HOA). In both cases, property owners enter agreements (which run with the property, not the owner) to provide funding, usually in the form of a regular assessment, which is aggregated and spent in common on community priorities. In Special Districts, this usually takes the form of a levy paid on area businesses or owners that might be used to fund streetscape improvements or other infrastructure serving the area.

In Homeowner Associations (HOAs), all owners of a residential community are liable for assessments to fund maintenance and programs for the community, including capital improvements to shared resources. The concept is not new, but its proliferation is impressive: Over 60 million Americans are members of some form of HOA. These organizations have always regulated and maintained planned communities and condominiums, but over the past 20 years they have diversified into a wide array of service provision. Telecommunications and other utility services are commonly contracted through the commons, and architectural review and dispute resolution are adjudicated. New, more walkable development models, such as those promoted by Smart Growth America and the Congress for a New Urbanism (CNU) are utterly dependent upon this community "software" to successfully manage the many shared common areas that make them appealing: plazas and parks, in-community school and other public facilities (Killoren interview, 2010).

Governmental reaction to these private, hyper-local governance structures is of interest. They have risen in prominence as confidence in municipal service provision has, in many places, declined. They also relieve budget pressures. Where state and county road agencies once measured their success by miles of new road, maintenance of existing infrastructure is now the overriding concern (Kiefer, interview). A strong preference for public roads meeting common standards, sometimes financed by private developers, but always dedicated to the system after construction, has in some cases reversed. Some highway departments relish the privatization of many of these "last mile" segments under community maintenance agreements. The reviewing agency may simply confirm that the maintenance agreement is binding and complete, and count itself lucky not to have to do the job. Similarly, downtown businesses may relish the ability to complete streetscape improvements without taking the time to convince all the voters of a town to approve a bond with a narrow benefit.

Policies affecting development finance

Debt

Development in the United States has benefited from historically low interest rates since the Great Financial Crisis of 2007–2009. Traditionally, developers of income property[1] have obtained a permanent mortgage commitment before they begin development of an apartment, office, retail, industrial, or other form of income property. They would then take the permanent mortgage commitment to a bank who would provide a short-term construction loan to construct the project and carry it through lease-up. When the project achieves stabilized occupancy or reaches pre-defined levels of net operating income (NOI), then the permanent mortgage will be funded and will 'take out' (pay off) the construction loan.

In the early 1980s, interest rates soared in the United States to double-digit levels and permanent mortgage commitments became very hard to get. At that time, developers obtained "mini-permanent" loans of 3–5 years. These "mini-perm" loans were used to construct the project and to carry it for several years of operations, with the expectation that interest rates would fall and more reasonable permanent mortgages would become available. Today, some of the largest developers will construct projects without permanent loan commitments, but smaller developers still require such commitments in order to obtain construction financing, since construction lenders are rightfully concerned that there will be a permanent loan take-out of their short-term financing.

Land development, condominiums, and other for-sale projects have no permanent loans. They are constructed using only short-term development loans or construction loans and are paid off as the lots or condominiums are sold. Unlike credit lines where the loans can be re-borrowed, land and condominium loans have release provisions that stipulate the loan repayment or "release" amount as each sale is made. The releases are usually a multiple – 1.2 or 1.3 – times the pro-rata share of the loan (release price = value of lot sold/total value of land × maximum loan amount × multiple).

Equity

Equity requirements in most income properties range from 30–40 percent of the total cost. While small developers typically must raise equity from friends and family, larger developers with strong track records are able to access institutional sources. These include pension funds, private equity funds, family offices, insurance companies, endowments, and sovereign funds. Equity may also be available from publicly traded funds such as Real Estate Investment Trusts (REITs), but more often than not, REITs prefer to own the real estate rather than to joint-venture a project with a developer or sponsor. Institutional investors may be accessed through pension fund advisors and investment brokers.

Many developers will leverage their equity by borrowing mezzanine loans. Mezzanine loans typically occupy the second debt position and carry interest rates ranging from 10–12 percent or more, but they reduce the developer's equity to 10–15 percent of the total capital cost. Ever since the Global Financial Crisis started in 2008, banks have been very cautious about making construction loans (Wang and Zhang, 2014). Many developers have found that construction and development loan terms are so onerous or have such low loan-to-value ratios that they have sourced better terms borrowing from private loan funds like Mosaic Real Estate Investors who make 1–3 year first-mortgage loans to developers ranging in loan-to-value ratios from 60–75 percent with interest rates in the low teens.

Private finance

How are developers raising equity today?

Equity investment in real estate has changed dramatically since 1990. Private equity funds and REITs have been the dominant source of equity as shown in Figure 29.1. In the historically low interest rate environment that has persisted from 2009 to 2016, there has been an abundance of equity, but fewer high-quality investment opportunities. Private equity funds and foreign equity investors have poured money into US real estate, bidding prices up and cap rates down. While offering strong potential, crowdfunding remains in a state of "watchful waiting" (PwC and the Urban Land Institute, 2015).

Roy March (2012) highlights five major reasons for the explosion in new sources of equity and debt. The biggest catalyst for bringing in new capital sources was the Savings and Loan Crisis in the late 1980s and the formation of the Resolution Trust Corporation (RTC) to liquidate the assets of failed thrifts. "Seizing the opportunity to provide liquidity to an industry that had none, a handful of entrepreneurs, private equity firms, and Wall Street merchant banks formed the first so-called opportunity funds." The RTC jump-started both the real estate private equity funds and the CMBS industries. A second factor was the growth of the equity REIT market which has soared from $5.5 billion in 1990 to $846 billion in 2014 (NAREIT,

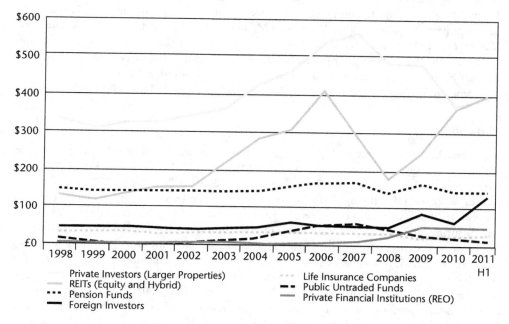

Figure 29.1 US real estate equity capital flows (1998–2011)

Source: PWC and Urban Land Institute, 2011

2016). A third factor was the very low interest rates on mortgages and other investment alternatives that have spurred institutional investors, led by university endowments, notably Yale and Harvard, to increase their allocations to real estate, moving from core investments into more value-adding and opportunistic real estate investments. A fourth factor has been the deluge of cross-border investments due in large part to the emergence of sovereign wealth funds. The US remains the "safe haven" for investments, especially after the Great Financial Crisis of 2007–2009. Finally, the fifth factor has been the evolution of technological innovation, which has helped the industry overcome its reputation for lack of transparency. Public information has provided accurate and timely data and more democratic access to it. With the growth in REITs, more REIT analysts are covering the industry (see March, 2012).

As March notes, there is a dark side to this increase in equity capital. With real estate more closely tied to global financial markets, it is more closely correlated with other financial asset classes, reducing its appeal as a counter-cyclical alternative investment. Whereas prices used to be "sticky," moving slowly relative to the stock market, the closer correlation has increased the volatility of real estate. This has made real estate investment riskier and less of a hedge for mitigating portfolio cyclicality.

EB-5 immigration program

The EB-5 program provides foreign investors with the opportunity to obtain a green card if they invest $800,000–$1.2 million in a US business venture. The program has been a particularly popular source of equity for real estate, especially for Chinese investors. To get approval, developers must demonstrate they will create 10 jobs for each visa issued. The program is limited to 3000 visas per year for qualifying immigrants who invest in Targeted Employment Area (TEA) locales, while an additional 3000 visas per year are set aside for investors in a Regional Center project. The program has been used to raise equity in a number of high-profile projects including Hudson Yards ($600M) and Atlantic Yards ($475M) in NYC and Time Warner ($272.5M) in the Los Angeles Film Regional Center. EB-5 funds can be invested both in equity and debt. Investor return requirements are very low – 2–6 percent. Real estate is popular because of the perceived safety as an investment (Jahangiri, 2014; Savills Studley, 2015). The program is controversial because it allows foreign nationals to buy US green cards.

Crowdfunding

Crowdfunding – raising money over the internet – is in its infancy as of 2017, but promises to be an increasingly important source of equity. Made possible by the JOBS Act of 2012, real estate crowdfunding has grown from $0 to $.25 billion in 2015, with 131 crowdfunding platforms in the USA (Dong, 2016). In general, real estate crowdfunding allows investors to share in any financial returns or profits of a given investment. The JOBS Act effectively removes the Securities and Exchange Commission prohibition on general solicitation and advertising for private placement offerings. Originally limited to accredited investors, Title III of the JOBS Act enables small investors to invest in real estate with certain limitations – investors with net worth or annual income below $100,000 may not be sold more than $2000 or 5 percent of their net worth or annual income in any one year (Dong, 2016, p. 21). While some crowdfunding platforms vet each deal very carefully, other platforms simply provide a marketplace for investors and those looking for money to meet with little or no screening.

The crowdfunding industry has yet to experience an economic downturn. Since regulations have been considerably relaxed for non-accredited investors, many unsophisticated investors are likely to find themselves in weak or overleveraged deals. When the downturn inevitably occurs, there will likely be a fallout in the industry and a series of lawsuits involving crowdfunding companies that have sponsored bad deals. Nevertheless, crowdfunding is here to stay and will be increasingly important over time.

Private equity for land development

Private equity funds have long concentrated on income property, but as the real estate cycle has matured in 2013–15, there has been increasing appetite for new development. Most of the investment has gone into land that is already entitled for development. However, lots for homebuilding have become harder to find in parts of the country, such as California and Texas, where the volume of homebuilding has been high. Some public homebuilders, such as Lennar and Toll Brothers, accumulated a sizeable land bank during the 2008–10 economic downturn, but others, such as D. R. Horton, the largest US homebuilder, and the Pulte Group, have a shorter inventory – 7.4 years versus 12 years. This trend to reduce inventory followed the disastrous experience of builders during the S&L crisis, where excessive land holdings caused many to face bankruptcy. Indeed, major homebuilders earned some of their highest profits on land entitlement and development during the 1980s, but when the market crashed, they could not afford to hold on to their land.

Development activity in the United Kingdom illustrates the build-up of development activity toward the end of the cycle. Figure 29.2 shows that the sales of development sites in London and the rest of the United Kingdom are well above their long-term average and in fact are back to their 2006–7 levels. Spain, by contrast, has lagged well behind the UK in its recovery following the Global Financial Crisis but the appetite for development is re-emerging.

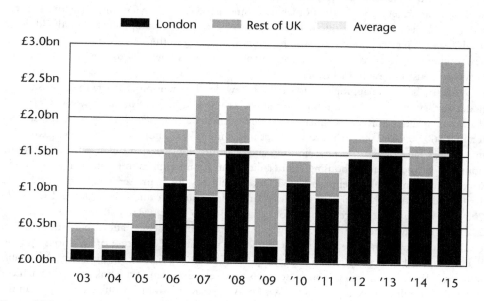

Development site sales well above average

■ London   ▨ Rest of UK   ▦ Average

Figure 29.2   United Kingdom development site sales back to 2006–7 levels

Source: Real Capital Analytics, 2015

Public–private development and public finance

This section describes several of the most popular programs for providing public financial assistance for economic development. While these programs were originally devised for urban renewal, they have become increasingly used for neighborhood stabilization and economic development undertaken by private developers.

Community Benefits Agreements

Community Benefits Agreements (CBAs) are legally enforceable contracts signed by community groups and a developer setting forth a range of community benefits that a developer agrees to provide in conjunction with a development project. They are intended to be negotiated directly between community groups and the developer. The developer agrees to provide specified community benefits in exchange for the community groups' support of the proposed project for entitlements and approvals (Gross et al., 2002).

CBAs are related to development agreements, which are contracts between a developer and a city or county outlining the subsidies that the local government will provide to a project. Development agreements are intended to lock in certain development rights for a site with respect to what the developer can build so that the developer is not subject to the whims of future changes in elected officials who may not like the project as it was conceived and approved by previous officials. In exchange, the developer agrees to provide specified benefits and investments in infrastructure as part of the development. CBAs may be incorporated into development agreements.

Linkage and exactions

Linkages and exactions are popular mechanisms that local governments use to get new development projects to pay for affordable housing, parks, fire and safety, infrastructure, schools, hospitals, libraries, and other public benefits. Historically, new developments created additional needs for such services but did not pay for them. Beginning in the 1980s with impact fees, new development projects were charged with many such fees, sometimes to the point of placing heavy burdens on new homebuyers.

While such fees are appropriate for making new development projects pay their fair share of the costs they impose on communities to provide these services, one of the downsides has been that municipalities shift an ever increasing share of these costs to new development. Impact fees totaling $60,000–$80,000 or more were not uncommon in California in the 2000s. Exorbitant impact fees raise the cost of new housing and apartments, placing an unfair burden on young homebuyers and reducing housing affordability. Many communities in California, for example, were relying on development-generated fees to help pay for many services in the 1980s and 1990s and again in the 2000s. When severe recessions came in the early 1990s and in 2007–2009, the communities were strapped for cash and had to make severe cutbacks in services.

Linkages have become increasingly popular to help generate affordable housing. For example, developers may be required to provide 15 percent or 20 percent of new housing units that are affordable to people who make 80 percent of the median income in a city. Affordable housing may also be linked to commercial development. Sometimes the linkages fail to achieve the desired result. For example, downtown Los Angeles required major office developers in the 1990s to build off-site parking garages on the edge of downtown and were limited to how much parking they could build on-site in an effort to reduce downtown congestion. The unintended consequence was that the new office buildings suffered for lack of on-site parking, and the new off-site parking garages went bankrupt when there was insufficient demand to cover their costs. Downtown Los Angeles became less competitive for tenants who preferred suburban locations where parking was accessible. The City of Los Angeles eventually did away with the requirement when the full effects became apparent – a costly lesson.

Incentive zoning and value recapture

Incentive zoning gives zoning bonuses in the form of higher FARs (Floor Area Ratios) for providing desired amenities. New York City, for example, gave developers up to 30 percent higher density for providing public open space such as plazas in front of their buildings. Mies Van Der Rohe's Seagram building on Park Avenue pioneered the concept of setting the building back from the street with a large public plaza in front – a concept that changed the face of Manhattan.

Value recapture enables cities to generate ongoing revenues from new development. The Boston Redevelopment Agency (BRA), for example, gets 2 percent of the resale price for condos built on BRA land. The funds generated through the Value Recapture program help to provide a pool of money for affordable housing in other projects.

A related mechanism that institutions use to maintain affordable housing for university faculties and employees is through deed restrictions that require faculty to sell their homes to other faculty or to restrict the resale prices of the homes. Such programs enable universities, hospitals, and other institutions to maintain affordable housing for employees in high-cost areas such as Los Angeles and Cambridge. University Hill at the University of Irvine, California, for example, restricts the resale price of housing to the original purchase price plus the cost of improvements and certain fix-up expenses incurred for resale, adjusted by certain indices as described in the Ground Sublease. Furthermore, homes must be offered to faculty and staff in the university community, and owners may have to move if they no longer work for the university. The program has been very successful in maintaining affordable housing for faculty and staff in a high-priced area (Irvine Campus Housing Authority, 2016).

Tax increment financing

Tax increment financing (TIF) has become a favored mechanism for financing public contributions associated with new development. The boundaries of a TIF district are defined by the city or redevelopment agency to support bond financing of public facilities such as parking garages for a shopping area. The incremental property value increase associated with the new development is captured by the TIF district for purposes of generating taxes to pay interest and amortization on bonds which are sold to build the new garage. For example, suppose new development adds $600 million in property value within a TIF. If the tax rate is 2 percent, and 75 percent of the tax increment is captured by the TIF district, then 2% × 75% × $600 million = $8 million per year in taxes would be available to support the bonds. Municipal bonds are tax exempt (from federal income taxes) and carry lower interest rates than corporate bonds. If interest rates are, say, 4 percent, and the bonds have a 40-year term, then the additional taxes would support $158,340,000 in bond financing.[2] Thus, the future incremental taxes can be used to support the sale of bonds which provide capital for constructing infrastructure and amenities to help improve the district.

TIFs are the most widely used local government program for financing economic development in the United States. However, the program has been the source of intergovernmental tension and conflict over the using public aid for private benefit. Its popularity stems from a number of factors: It is highly decentralized. It reinforces the fiscalization of development policy. It plays off the fragmentation of local government and the competition among municipalities for investment and economic growth. And it fits well with the entrepreneurial spirit that characterizes contemporary local economic development policy (Briffault, 2010). Squires and Hutchison (2014) points out that there are downside risks with TIFs such as potential detriment to areas outside the TIF, local control over TIF spending, and what happens to TIF support for affordable housing when Redevelopment Agencies are abolished as they were in California in 2012.

Business Improvement Districts

Business Improvement Districts (BIDs) are another popular program for improving redevelopment areas, especially retail and business centers. BIDs were started to provide security, street lighting, benches, landscaping, and other amenities to improve the safety and desirability of shopping and business districts. The businesses within a specified geographic area agree to impose a special tax – usually 1 or 2 percent of property values – to support bonds for improvements similar to TIF districts and to provide funds for ongoing operations. The improved appearance and perceived safety of the district brings new customers and helps to raise retail sales and make the area more desirable for office tenants.

Summary

There are far too many topics relating to governance and land use policy relating to development to be covered in a single chapter. The present chapter has addressed a number of current programs and policies affecting development in the United States. Because most real estate and land use policies are implemented at the state and local level, different parts of the country are confronting different issues and are promulgating different programs and policies. In general, the coasts, and especially the gateway cities such as New York City, Los Angeles, Miami, Washington, DC, Seattle, and San Francisco are confronting problems of housing affordability and rising taxes associated with rapid urban growth. Those areas are more likely to have discretionary land use approvals by local city councils that raise the time, cost, and risk associated with obtaining entitlements for development. They tend to have more restrictive urban growth policies and a more aroused public who resist new development because of congestion, pollution, and other negative externalities from rapid growth. They are also more likely to have passed laws, such as California's Proposition 13, that limit the ability of cities to raise property taxes. In a severely budget-constrained world, city officials are forever looking for new ways to raise money to pay for infrastructure and amenities to support growth. While the programs and mechanisms described in this chapter are hardly exhaustive, they provide insight into some of the recent trends and range of programs being adopted in different parts of the country.

Notes

1   Income property includes all investment property that generates lease income. It is typically held for several years. It is built using short-term construction financing, and when completed, is financed with a long-term permanent mortgage. For-sale property includes all real estate that is sold to end buyers (houses and condominiums) or to other developers (land). It is financed by short-term construction or land development loans.
2   PV (4% interest, 40 years, $8M payment) = $158,340,000.

References

Briffault, Richard. 2010. "The Most Popular Tool: Tax Increment Financing and the Political Economy of Local Government." *The University of Chicago Law Review*, Vol. 77, 65–95.

Dong, Peng. 2016. *Real Estate Crowdfunding: The New Frontier that Calls for Risk Control*, Master's in Design Studies Thesis, Harvard Graduate School of Design.

Gross, J., LeRoy, G., and Janis-Aparicio, M. 2002. "Community Benefits Agreements: Making Development Projects Accountable." Washington, DC: Good Jobs First / California Public Subsidies Project. http://community-wealth.org/sites/clone.community-wealth.org/files/downloads/report-gross.pdf, accessed December 31, 2015.

Irvine Campus Housing Authority. 2016. "UC Irving Resale Restrictions." http://icha.uci.edu/uploads/files/Resale%20Restrictions%20-%20plain.pdf, accessed February 13, 2016.

Jahangiri, Ali. 2014. "Why the EB-5 Program Works Well for Real Estate Investment." *NuWIRE Investor*, April 14. www.nuwireinvestor.com/articles/why-eb-5-program-investors-love-real-estate-61772.aspx, accessed January 20, 2016.

March, Roy H. 2012. "The Making of an Asset Class." *Wharton Real Estate Review*, Spring, 2012. http://realestate.wharton.upenn.edu/review/index.php?article=229, accessed December 29, 2015.

NAREIT. 2016. "US REIT Industry Equity Market Cap: Historical REIT Industry Market Capitalization: 1972–2014." www.reit.com/data-research/data/us-reit-industry-equity-market-cap, accessed January 19, 2016.

PwC and the Urban Land Institute. 2011. *Emerging Trends in Real Estate: United States and Canada 2012*, Washington, DC: The Urban Land Institute.

PwC and the Urban Land Institute. 2014. *Emerging Trends in Real Estate: United States and Canada 2015*, Washington, DC: The Urban Land Institute.

PwC and the Urban Land Institute. 2015. *Emerging Trends in Real Estate: United States and Canada 2016*, Washington, DC: The Urban Land Institute.

Real Capital Analytics. 2015. *European Capital Trends, Year in Review*.

Savills Studley. 2015. "EB-5 Investment and the Impact on Commercial Real Estate." Savills Studley Insights, March. www.savills-studley.com/Collateral/Documents/English-US/Research/2015/Insights/eb5_investment_and_the_impact_on_commercial_real_estate.pdf, accessed January 20, 2016.

Squires, Graham and Hutchison, Norman. 2014. "The Death and Life of Tax Increment Financing (TIF): Redevelopment Lessons in Affordable Housing and Implementation." *Property Management*, Vol. 32, No. 5, 368–377.

Wang, F. A. and Zhang, T. J. 2014. "Financial Crisis and Credit Crunch in the Housing Market." *Journal of Real Estate Finance and Economics*, Vol. 49, No. 2, 256–276.

Interviews

Fahmy, Stewart. President, Calandev, San Jose, California, interview by D. Hamilton, 2009.

Kassell, Dan. President, Granite Homes, Irvine, California, interview by D. Hamilton, 2002.

Kellenberg, Steve. Partner, AECOM (formerly EDAW), Irvine, California, interview by D. Hamilton, 2010.

Kiefer, Matthew. Partner, Goulston & Storrs, Boston, Massachusetts, interview by D. Hamilton, 2010.

Killoren, Don. Principal, Celebration Associates, Hot Springs, Virginia, interview by D. Hamilton, 2010.

Plater-Zyberk, Elizabeth. Founding Principal, Duany Plater-Zyberk & Company, Miami, Florida, interview by D. Hamilton, 2011.

Index

Printed in the United States
by Baker & Taylor Publisher Services